Giant Molecules

SPE MONOGRAPHS

Giant Molecules

Essential Materials for Everyday Living and Problem Solving

RAYMOND B. SEYMOUR
*University of Southern Mississippi
Hattiesburg, Mississippi*

CHARLES E. CARRAHER
*Florida Atlantic University
Boca Raton, Florida*

WARNER MEMORIAL LIBRARY
EASTERN COLLEGE
ST. DAVIDS, PA. 19087

A WILEY-INTERSCIENCE PUBLICATION

John Wiley & Sons, Inc.

NEW YORK / CHICHESTER / BRISBANE / TORONTO / SINGAPORE

Library of Congress Cataloging in Publication Data:

Seymour, Raymond Benedict, 1912–
    Giant molecules: essential materials for everyday living and
  problem solving/Raymond B. Seymour, Charles E. Carraher.
       p.   cm. -- (SPE monographs, ISSN 0195-4288)
    "A Wiley-Interscience publication."
    Includes bibliographical references.
    ISBN 0-471-61532-3
    1. Polymers. 2. Plastics. I. Carraher, Charles E. II. Title.
  III. Series.
  QD381.S47   1990
    668.9--dc20   89-22522
                                                                  CIP

Printed in the United States of America

10 9 8 7 6 5 4 3 2 1

The book Giant Molecules is dedicated to our deceased colleagues, Paul Flory, Daniel Fox, Carl Marvel, and Bill Bailey, and to polymer scientists, Linus Pauling and Herman Mark whose contributions to polymer science were essential for our understanding of the material in this book.

Foreword

While our ancestors were somewhat knowledgeable about polymers, such as leather, paint, rubber, celluloid, rayon, and Bakelite, they would not be well informed on the many new giant molecules without understandable references. Hence, it is timely that Drs. Seymour and Carraher have met this need in this treatise on *Giant Molecules*.

As a service to the nonscience user of plastics and other polymers, they have not assumed that the reader has any previous knowledge of chemistry or polymer science. Thus, *Giant Molecules* should serve as a source of essential knowledge for the nonscience major and the average citizen. Readers of this well-written book will be much more knowledgeable about the world's second largest industry, which will become the world's most important industry before the end of the twentieth century.

H. MARK

Polytechnic Institute of New York
Brooklyn, New York

Series Preface

The Society of Plastics Engineers is dedicated to the promotion of scientific and engineering knowledge of plastics and to the initiation and continuation of educational programs for the plastics industry. Publications, both books and periodicals, are major means of promoting this technical knowledge and of providing educational materials.

New books, such as this volume, have been sponsored by the SPE for many years. These books are commissioned by the Society's Technical Volumes Committee and, most importantly, the final manuscripts are reviewed by the Committee to ensure accuracy of technical content. Members of this Committee are selected for outstanding technical competence and include prominent engineers, scientists, and educators.

In addition, the Society publishes *Plastics Engineering, Polymer Engineering and Science (PE&S), Journal of Vinyl Technology, Polymer Composites,* proceedings of its Annual and Regional Technical Conferences (ANTEC, RETEC), and other selected publications. Additional information can be obtained from the Society of Plastics Engineers, 14 Fairfield Drive, Brookfield, Connecticut 06804.

ROBERT D. FORGER

Executive Director
Society of Plastics Engineers

Technical Volumes Committee

Robert E. Nunn, Chairman	George E. Nelson
Lee L. Blyler, Jr.	James L. Throne
Thomas W. Haas	Lewis B. Weisfeld
Louis T. Manzione	Bonnie J. Bachman

Preface

Since the production of polymers is the world's second largest industry, and the use of these products by each individual exceeds that of all other materials, it is important that a knowledge of these giant molecules be available to everyone. Many books have been written for the numerous scientists and engineers who are employed in various phases of the plastics, rubber, fiber, coatings, and adhesives industries, but prior to the writing of *Giant Molecules*, no readily understandable book was available to the nonscientist who is concerned about these materials that are essential for everyday living and problem solving.

We have attempted to satisfy the need to know about these giant molecules by using a multidisciplinary approach that includes information on biology, chemistry, physics, earth science, and engineering. It is not anticipated that readers will become experts in any of these fields, but it is hoped that they will gain a better knowledge of the world and the polymer age in which we live.

Giant Molecules includes a glossary, review questions and answers, and some history of the development of this important branch of knowledge. We have enjoyed writing this book, which should be of interest to liberal arts majors, technicians, and informed citizens. We hope that you will also enjoy reading *Giant Molecules*.

Many thanks to Debi Sones and Machell Haynes, who typed this manuscript.

RAYMOND B. SEYMOUR
CHARLES E. CARRAHER

Hattiesburg, Mississippi
Boca Raton, Florida
April 1990

Contents

Giant Molecules

CHAPTER 1

The Building Blocks of Our World

1.1   INTRODUCTION

This chapter presents a brief overview of some of the chemistry that is essential for an appreciation of the science of giant molecules, but this volume is not intended to be a science textbook.

We will be concerned with matter, that is, anything that has mass and occupies space. The term mass is used to describe a quantity of matter. However, in most cases, we will refer to weight instead of mass. Weight, unlike mass, varies with the force of gravity. For example, an astronaut in orbit may be weightless but his or her mass is the same as it was on the earth's surface.

1.2   UNITS OF MEASUREMENT

Scientists and citizens of most other nations use the meter-gram-second (mgs) or metric system for measuring distance, weight, and time. The metric system will be used occasionally in this book. However, since Americans are moving very slowly, inch by inch (25.4 mm), from the outmoded foot-pound-second (fps) system to the metric (mgs) system, we will use the English system throughout this book. A conversion table for changing fps units to mgs units is given in Table 1.1.

Table 1.1   Useful Conversions to Metric Measures

Symbol	When You Know (fps)	Multiply by	To Obtain (mgs)	Symbol
in.	Inch	2.5	Centimeter	cm
yd	Yard	0.9	Meter	m
mi	Mile	1.6	Kilometer	km
oz	Ounce	28	Gram	g
lb	Pound	0.45	Kilogram	kg
tsp	Teaspoon	5	Milliliter	mL
Tbsp	Tablespoon	15	Milliliter	mL
fl. oz	Fluid ounce	30	Milliliter	mL
c	Cup	0.24	Liter	L
qt	Quart	0.95	Liter	L
gal	Gallon	3.8	Liter	L
yd^3	Cubic yard	0.76	Cubic meter	m^3
°F	Fahrenheit	5/9 (after subtracting 32)	Celsius	°C
°C	Celsius (centigrade)	Add 273	Kelvin	K

We will use the Celsius (centigrade) temperature scale in which water freezes at 0°C and boils at 100°C, as well as the Fahrenheit temperature scale, in which water freezes and boils at 32°F and 212°F, respectively. We will also use the Kelvin (K) temperature scale (absolute temperature scale), in which water freezes and boils at 273 K and 373 K, respectively.

As shown in Table 1.2, multiples or submultiples of 10 are used as prefixes to the mgs units in the metric system. The prefixes kilo (k), mega (M), and giga (G) represent multiples of one thousand (10^3), one million (10^6), and one billion (10^9). (The exponent denotes the number of integers after the first integer, as illustrated in Table 1.2.) Other common prefixes are centi (c), milli (m), micro (μ), and nano (n) for submultiples of one hundredth (10^{-2}), one thousandth (10^{-3}), one millionth (10^{-6}), and one billionth (10^{-9}). (The negative exponent denotes the number of decimal places that precede the first integer.) It should be pointed out that 1 billion in the United States is 10^9 but is 10^{12} in the United Kingdom and many other countries.

Table 1.2   Prefixes for Multiples and Submultiples

Multiple or Submultiple		Prefix	SI Symbol
10^{12}	1 000 000 000 000	tera	T
10^9	1 000 000 000	giga	G
10^6	1 000 000	mega	M
10^3	1000	kilo	k
10^2	100	hecto	h
10^1	10	deka	da
10^0	1		
10^{-1}	0.1	deci	d
10^{-2}	0.01	centi	c
10^{-3}	0.001	milli	m
10^{-6}	0.000 001	micro	μ
10^{-9}	0.000 000 001	nano	n
10^{-12}	0.000 000 000 001	pico	p
10^{-15}	0.000 000 000 000 001	femto	f
10^{-18}	0.000 000 000 000 000 001	atto	a

1.3   SYMBOLS FOR THE ELEMENTS

The ancient Greeks represented their four elements by triangles and barred triangles, that is, fire $= \triangle$, water $= \nabla$, air $= \underline{\triangle}$, and earth $= \overline{\nabla}$. Although none of these is an element, the triangle is still used as a symbol for heat or energy in chemical equations. The ancient Babylonians and medieval alchemists represented these elements by using variations of the moon and other celestial bodies.

John Dalton used circles as symbols for elements in the eighteenth century. His symbols for some of the common elements were: oxygen $= \bigcirc$, hydrogen $= \odot$, nitrogen $= \oslash$, carbon $= \bullet$, and sulfur $= \oplus$. This cumbersome system of symbols was displaced early in the nineteenth century by Jöns J. Berzelius, who used the capitalized initial letter of the name of each element. To avoid redundancy, he used a second lowercase letter to distinguish carbon (C) from calcium (Ca), and so on. Some symbols, such as Na for sodium and Fe for iron, were derived from the Latin names, which, in these examples, are natrium and ferrum, respectively.

All the known elements are now represented internationally by specific one- or two-letter symbols. Only a few of these, such as hydrogen (H), carbon (C), oxygen (O), and nitrogen (N), are present in giant molecules.

1.4   ELEMENTS

Even in ancient times, many philosophers believed that all matter was composed of a limited number of substances or elements. According to the early Chinese philosophers, there were four elements, namely, earth, solids such as wood, yin, and yang. The ancient Greek philosophers believed that all material forms consisted of various combinations of earth, air, fire, and water. The ancient Babylonians identified seven metallic elements, and many newly discovered substances were also called elements by philosophers during the Middle Ages.

An element is now defined as a substance consisting of identical atoms. There are 108 or more known elements but we are interested in only a handul of these, namely, hydrogen, carbon, oxygen, nitrogen, and a few others.

Only a few of the over one hundred elements are common in nature. These can be remembered using the mnemonic "P. Cohn's CAFE"—that is, phosphorus, carbon, oxygen, hydrogen, nitrogen, sulfur, calcium (Ca), and iron (Fe).

1.5   ATOMS

Some ancient Greek philosophers, such as Aristotle, maintained that matter was continuous, but 2400 years ago Democritus insisted that all matter was discrete, that is, made up of indivisible particles. He named these particles *atomos*, after the Greek word meaning indivisible. Over 23 centuries later, this concept for matter was adopted by John Dalton, who coined the word atom.

According to Dalton's theory, all matter consists of small, indestructible solid particles (atoms) that are in constant motion. These atoms, which are the building units of our universe, are characteristic for each element, such as oxygen (O), hydrogen (H), carbon (C), and nitrogen (N).

The scientists of the early nineteenth century did not recognize the difference between an atom and a molecule, which is a combination of atoms. This enigma was solved by Amedeo Avogadro and his student Stanislao Cannizzaro. These Italian scientists, who coined the term molecule from the Latin name *molecula* or little mass, showed that, under similar conditions of temperature and pressure, equal volumes of all gases contained the same number of molecules. They showed that simple gases, such as oxygen, hydrogen, and nitrogen, existed as diatomic molecules, which could be written as O_2, H_2, and N_2.

The atoms of these gases are unstable and combine spontaneously to produce stable molecules, which are the smallest particles of matter that can exist in a free state. Although the oxygen (O_2), hydrogen (H_2), and nitrogen (N_2) molecules are diatomic, most compounds consist of polyatomic molecules. For example, water (HOH), which is written H_2O, is a triatomic molecule, ammonia (NH_3) is a tetraatomic molecule, and methane (CH_4) is a pentaatomic molecule. Chemical formulas show the relative number and identity of atoms in each specific molecule.

1.6   CLASSICAL ATOMIC STRUCTURE

Each atom consists of a dense, positively charged nucleus that is surrounded by a less dense cloud of negatively charged particles. The magnitude of each of these positively charged nuclear particles, called protons (after the Greek word *protos* or first), is equal to the magnitude of the negatively charged particles, called electrons (after the Greek word for amber). Thus, all atoms contain an equal number of + and − charged particles and hence are neutral. The mass of a proton is about 1840 times that of the electron and the diffuse cloud occupied by the electrons has a diameter that is about 100,000 times that of the nucleus.

The nucleus may also contain dense neutral particles called neutrons (from the Latin word *neuter*, meaning neither), which have a mass similar to that of the positively charged protons. A hydrogen atom consists of 1 proton and 1 electron, whereas the oxygen atom consists of 8 protons, 8 neutrons, and 8 electrons. These atoms have mass numbers of 1 and 16, respectively. The mass number is equal to the sum of the number of protons and neutrons in an atom. We will not be concerned with other atomic particles such as neutrinos, mesons, quarks, and gluons, and except for its contribution to mass, we can disregard the neutron.

It is generally accepted that electric current results from the flow of electrons, but the actual existence of these negatively charged atomic particles was not recognized until their presence was observed by J. J. Thomson in 1897. The neutron was discovered by James Chadwick in 1932. The proton, which was discovered by Ernest Rutherford in 1911, is simply the hydrogen atom without an electron. It is the positively charged building unit for the nuclei of all elements.

The present accepted model for the atom is based on many discoveries made by a host of scientists. Many of these investigators were recipients of Nobel prizes. Obviously, their many contributions cannot be discussed in depth in this book nor learned in an introductory science course. You may find it advantageous to scan much of the description of atomic structure and read it more carefully after you have read some of the subsequent chapters.

In the eartly part of the twentieth century, Henry Moseley showed that x rays with characteristic wavelengths were produced when metallic elements were bombarded by electrons. He assigned atomic numbers to these elements based on the wavelength of the x rays. The atomic number is equal to the number of protons, which, since the atom has a neutral charge, is also equal to the number of electrons in each atom. The atomic numbers are 1 for hydrogen, 7 for nitrogen, and 8 for oxygen. The mass numbers of these atoms are 1, 14, and 16, respectively. The difference between the mass numbers and atomic numbers is equal to the number of neutrons present in each atom.

Niels Bohr proposed an atomic model in which the electrons traveled in relatively large orbits around the compact nucleus and the energy of these electrons was restricted to specific energy levels called quantum levels. The lowest energy level was near the nucleus, but under certain conditions an electron could pass from one energy level to another and this abrupt change is called a "quantum jump."

1.7 MODERN ATOMIC STRUCTURE

The concept of principal quantum levels or shells is still accepted, and these levels are designated, in the order of increasing energy, from 1 to 7, and so on, or by the letters K, L, M, and so on. The electron exhibits some of the characteristics of a particle, like a bullet, and some of the characteristics of a wave, like a wave in the ocean.

Werner Heisenberg used the term uncertainty principle to describe the inability to locate the position of a specific electron precisely. In general, this lack of precision is related to the energy used in viewing, which causes the particle to move in accordance with the energy used by the viewer. Because of the presence of the viewer and compiler of data, sociological observations are also uncertain.

Erwin Schrödinger, working independently of Heisenberg, used wave mechanics, which can also be used for the study of waves generated in a pool of water, to describe the patterns of an electron surrounding a nucleus. His approach, which led to the description of the movement and location of electrons, has been refined and is called quantum mechanics.

The position of electrons is now described in the general terms of probability pathways called orbitals. Thus, in considering the location of an electron, it is proper to describe it in general terms of probability. This probability pathway is called an orbital, and the maximum number of electrons that can occupy a single orbital is two.

1.8 PERIODICITY

All 108 or so elements are arranged in the order of their increasing atomic numbers in a periodic table. This table is a slight modification of the one devised by Dmitry Mendeleyev in the last part of the nineteenth century. Mendeleyev arranged the elements in order of their increasing atomic weights or mass numbers and successfully used this periodic table to predict physical and chemical properties of all known and some undiscovered elements. In the modern periodic chart, the elements are arranged vertically in groups or families according to their atomic numbers instead of their mass numbers. All members of a group have the same number of electrons in the atoms of their outer or valence shells. The number of electrons in the valence shell increases as one goes from left to right in the horizontal rows or periods. We will be concerned only with the electrons in the outermost or valence shell. Valence, which is derived from the Latin word *valentia*, meaning capacity, is equal to the combining power of an element with other elements. For example, the valence of hydrogen is one and that of carbon is four.

The periodic table is shown in Table 1.3. It is called "periodic" because there is a recurring similarity in the chemical properties of certain elements. Thus, lithium, sodium, potassium, rubidium, cesium, and francium all react similarly. In the periodic table these elements are arranged in the same vertical column called a group or family. For the main group elements, those designated with the letter "A", the group also corresponds to the number of electrons in the outer or valence shell. Thus, all IA elements have a single outer, valence electron; IIA elements have two valence electrons; IIIA elements have three outer electrons, and so on.

Knowing the number of outer, valence electrons is important because these electrons are responsible for the existence of all compounds through formation of bonds. The elements designated by the letter "B" are called transition elements.

Some of the families have special names. The IA family is known as the alkali metals; the IIA family is known as the alkaline earth metals; and the Group VII elements are known as the halogens. Hydrogen has features of both Group IA and Group VIIA elements and yet has properties quite different from these elements. Thus it is often shown separately or as a member of both Groups IA and VIIA in periodic charts.

In addition to being an orderly presentation of the elements, from which all matter as we know it is composed, the periodic chart also contains a vast abundance of

Table 1.3 The Periodic Table

Legend (key box): atomic number / symbol of the element / atomic weight — shown as **1 / H / 1.01†**

Period	I A	II A	III B	IV B	V B	VI B	VII B	VIII	VIII	VIII	I B	II B	III A	IV A	V A	VI A	VII A	VIII A
1	1 H 1.01†																	2 He 4.00
2	3 Li 6.94	4 Be 9.01											5 B 10.81	6 C 12.01	7 N 14.01	8 O 16.00	9 F 19.00	10 Ne 20.18
3	11 Na 23.00	12 Mg 24.31											13 Al 26.98	14 Si 28.09	15 P 30.97	16 S 32.06	17 Cl 35.45	18 Ar 39.95
4	19 K 39.10	20 Ca 40.08	21 Sc 44.96	22 Ti 47.90	23 V 50.94	24 Cr 52.00	25 Mn 54.94	26 Fe 55.85	27 Co 58.93	28 Ni 58.71	29 Cu 63.55	30 Zn 65.37	31 Ga 69.72	32 Ge 72.59	33 As 74.92	34 Se 78.96	35 Br 79.90	36 Kr 83.80
5	37 Rb 85.47	38 Sr 87.62	39 Y 88.91	40 Zr 91.22	41 Nb 92.91	42 Mo 95.94	43 Tc* 98.91	44 Ru 101.07	45 Rh 102.91	46 Pd 106.4	47 Ag 107.87	48 Cd 112.40	49 In 114.82	50 Sn 118.69	51 Sb 121.75	52 Te 127.60	53 I 126.90	54 Xe 131.30
6	55 Cs 132.91	56 Ba 137.34	57 La 138.91	72 Hf 178.49	73 Ta 180.95	74 W 183.85	75 Re 186.2	76 Os 190.2	77 Ir 192.22	78 Pt 195.09	79 Au 196.97	80 Hg 200.59	81 Tl 204.37	82 Pb 207.2	83 Bi 208.98	84 Po* [210]	85 At* [210]	86 Rn* [222]
7	87 Fr* [223]	88 Ra* 226.02	89 Ac* [227]	104 Rf* [261]	105 Ha* [262]													

Lanthanides

58 Ce 140.12	59 Pr 140.91	60 Nd 144.24	61 Pm* [147]	62 Sm 150.4	63 Eu 151.96	64 Gd 157.25	65 Tb 158.93	66 Dy 162.50	67 Ho 164.93	68 Er 167.26	69 Tm 168.93	70 Yb 173.04	71 Lu 174.97

Actinides

90 Th* 232.03	91 Pa* 231.04	92 U* 238.03	93 Np* 237.05	94 Pu* [244]	95 Am* [243]	96 Cm* [247]	97 Bk* [247]	98 Cf* [251]	99 Es* [254]	100 Fm* [257]	101 Md* [258]	102 No* [255]	103 Lr* [256]

*All isotopes are radioactive.

† All atomic weights have been rounded to 0.1.

[ ] Indicates mass number of longest known half-life.

information. Depending on the particular periodic table, it may contain the chemical name, for example, carbon; the chemical symbol, C; the atomic, number, which is the number of protons and in a neutral atom also the number of electrons; and the atomic mass or atomic weight in atomic mass units (amu) or daltons (one dalton = one amu), which is the sum of the number of protons and the average number of neutrons that occur naturally.

atomic number = number of protons = number of electrons in a neutral atom

atomic mass = number of protons + (average) number of neutrons

For carbon, the atomic mass is not 12 but rather 12.011 since carbon exists in nature with two different numbers of neutrons. About 99% of carbon has six protons and six neutrons, and about 1% of carbon has six protons and seven neutrons. Atoms that are of the same element, that is, have the same number of protons in their nucleus, but have different numbers of neutrons are called isotopes. Thus carbon has three naturally occurring isotopes: carbon-12 (99%), carbon-13 (1%), and carbon-14 (trace).

Hydrogen's isotopes are so well known that they even have their own names. Hydrogen with one proton and no neutrons is simply called hydrogen; hydrogen with one proton and one neutron is called deuterium; and hydrogen with one proton (it would not be hydrogen if it had any number other than one proton) and two neutrons is called tritium. The beginning letters for the isotopes of hydrogen can be remembered from Hot, DoT.

The nuclei of many elements are unstable and spontaneously emit, or give off, particles, energy, or both. Such isotopes are called radioactive isotopes or radioisotopes. The three most common forms of natural radiation are shown in Table 1.4. The alpha particle is a package of two neutrons and two protons. This corresponds to the nucleus of helium. It has a positive two charge since each proton is positively charged and there are no electrons present to neutralize the positive charges. They are fast traveling (about 5 to 10% of the speed of light), but relative to the other two radioactive emissions they are slower. Alpha particles are massive, thus their destructive capability is great. Fortunately, their massiveness also allows them to be stopped by thin sheets of aluminum foil, several sheets of paper, or human skin to prevent internal damage. The beta particle travels up to about 90% of the speed of light whereas the gamma particle travels at the speed of light. Because of the small mass associated with these two emissions and their great speeds, both have penetrating powers greater than that of the alpha particle.

TABLE 1.4   Characteristics of Three Common Radioactive Emissions

Name	Identity	Charge	Relative Mass (amu)	Penetrating Power
Alpha	2 protons and 2 neutrons	+2	4.0026	Low
Beta	Electron	−1	0.0005	Low to moderate
Gamma	High-energy radiation similar to x rays	0	0	High

1.9 MOLECULAR STRUCTURE

The formation of molecules from atoms is dependent on the formation of primary chemical bonds based on a sharing or exchange of electrons. The former are called covalent bonds and the latter are called ionic bonds. Our emphasis in polymer chemistry will be on covalent bonds.

G. N. Lewis represented valence electrons as dots. Thus, hydrogen with one valence electron, oxygen with six valence electrons, and nitrogen with five valence electrons may be represented as $\cdot$H, $:\ddot{O}$, and $:\ddot{N}$.

The goal in chemistry is the formation of molecules from atoms through the attainment of stable electronic structures. We may use the Lewis representation to show the electrons in the outer shells of hydrogen, oxygen, and nitrogen molecules as H:H, $\ddot{O}::\ddot{O}$, and $:N\!\stackrel{..}{\cdot}\!N:$. The shared bonds between the atoms are usually represented by single, double, and triple bonds as follows: H—H, O=O, and N≡N.

It is important to note that nature strives to complete both the subshell levels and shell by the exchange and sharing of electrons. For H and He, this means filling the lone s orbitals. Since the maximum number of electrons in a single orbital is two, He, which already has two s electrons (i.e., $1s^2$ electronic configuration), is typically inert, whereas H, which has one s electron (i.e., $1s^1$ configuration), typically adds one electron to achieve a completed s orbital or loses one electron to become simply a proton (H^+).

The rule of eight can be particularly useful when describing the composition of many common compounds. The Lewis dot formulas for water, methane, phosgene, and ammonia are as follows, where each "dot" represents an outer or valence electron and two "dots" represent a pair of shared electrons, that is, a covalent bond.

It is not customary to show the presence of unbonded electrons but to use simple structural representations such as

In chemical formulas, one simply notes the atoms present and their relative abundance as shown by

$$H_2O \qquad CH_4 \qquad COCl_2 \qquad NH_3$$

It is important to note that both H_2O and OH_2 are correct, but it is customary to write the formula for a molecule of water as H_2O. We will not be concerned with such rules in this book, but it is critical that you remember that the water molecule contains two atoms of hydrogen bonded to one atom of oxygen by covalent bonds.

1.10 CHEMICAL EQUATIONS

In the same manner that unstable atoms, like hydrogen, oxygen, and nitrogen, combine to form stable diatomic molecules with a complete electron duet for hydrogen and a complete electron octet for oxygen and nitrogen, dissimilar atoms also enter into combinations to produce more complex molecules. Many of these reactions release energy in the form of heat and are said to be exothermic. In contrast, those in which energy must be added to the reactants to cause a chemical reaction are called endothermic.

The equation for the exothermic reaction between hydrogen and oxygen molecules for the formation of water molecules is shown as

$$H_2 + O_2 \rightarrow H_2O$$

According to the law of conservation of mass, the weight or mass of reactants (H_2 and O_2) must equal the mass of the product (H_2O). Hence, we must balance the equation by placing small integers before the symbols for the molecules, that is, we must also ascertain that the same number of atoms of each element is on each side of the arrow. In this example, we obtain a balanced equation by placing the number 2 before both H_2 and H_2O:

$$2H_2 + O_2 \rightarrow 2H_2O$$

A balanced equation is very important. For the production of water, it states that two molecules of H_2O will be produced by the combination of two molecules of H_2 and one molecule of O_2.

As a result of investigations by Avogadro, we now know that there are 6.023×10^{23} molecules in 22.4 L of oxygen or hydrogen and this number of molecules is called a mole. There are also 6.023×10^{23} particles in a mole of anything, including liquids and solids. The mass or weight of one gram mole of hydrogen, oxygen, and water is 2, 16, and 18 g, respectively. The weight of 18 g of H_2O is 2 g + 16 g or 0.04 lbs.

1.11 CHEMICAL BONDING

Ionic bonds are formed by the exchange of electrons. The tendency for ionic bond formation is greatest when the reacting elements are far apart in the rows of the periodic table. Thus, lithium (Li) tends to lose a valence electron readily and fluorine (F) tends to gain a valence electron readily to produce a stable salt, lithium fluoride (Li^+, F^-). Since the lithium lost an electron in this exchange, it became less negative or more positive by one electron and is called a lithium cation (Li^+). In contrast, the fluorine has become more negative by accepting an electron and is called a fluoride anion (F^-). (It is not essential to show the electron octet in the fluoride ion since it has a stable electron structure, i.e., an octet.)

The strongest ionic bonds are formed between ions from atoms having large differences in electronegativity, which is a measure of the tendency to attract electrons. The electronegativity values increase as one goes from left to right in the periodic table. Ionic bonds are very important in inorganic chemistry but much less important in the science of giant molecules, which are usually organic compounds containing mostly atoms bonded exclusively by sharing electrons.

The bonds between similar atoms or between atoms with similar electronegativity values are formed by sharing of electrons and are called covalent bonds. The electronegativities of hydrogen and carbon are similar and hence covalent bonds between carbon and hydrogen atoms are present in hydrocarbons, such as methane:

$$
\begin{array}{c}
H \\
| \\
H-C-H \\
| \\
H
\end{array}
\quad \text{or} \quad CH_4
$$

Polar covalent bonds are formed when the difference in the electronegativity values is greater than that in molecules, such as hydrogen and methane. Thus, methyl chloride,

$$
\begin{array}{c}
H \\
| \\
H-C-Cl \\
| \\
H
\end{array}
\quad \text{or} \quad H_3CCl
$$

is a polar molecule because of the relatively large difference between the electronegativity values of carbon and chlorine.

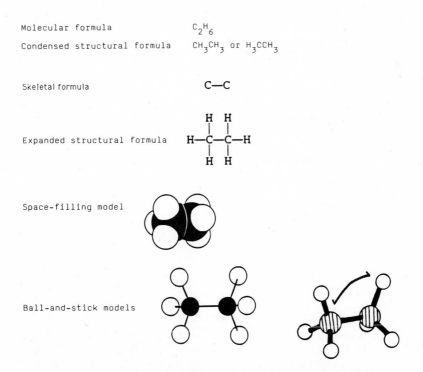

Molecular formula	C_2H_6
Condensed structural formula	CH_3CH_3 or H_3CCH_3
Skeletal formula	C—C
Expanded structural formula	$H-C-C-H$ (with H's above and below)
Space-filling model	
Ball-and-stick models	

Figure 1.1. Sample models for depicting the molecular structure of ethane.

In general, single bonds, such as those present in ethane ($H_3C—CH_3$), are called sigma bonds. Additional bonds such as those present in ethylene ($H_2C\!\!=\!\!CH_2$) are called pi bonds. The pi bonds are located above and below the bonding axis of the sigma bond. The bonds in ethylene, which are called double bonds, are not twice as strong as single (sigma) bonds. Actually, because of the presence of the pi bonds, double bonds are much more reactive than single sigma bonds.

Throughout the text, different types of formulas and models will be employed to emphasize various aspects of the chemical structures (Figure 1.1). General molecular formulas are employed for brevity, whereas skeletal formulas are used to emphasize main-chain or other desired characteristics such as branching and to show structural features related to bond angles. Generalized line drawings convey more extensive generalizations, in expanded structural formulas which emphasize the bonding among the different atoms. Ball-and-stick models (Table 1.5) are used to convey bonding, bonding angles, possible relative positions of the various atoms, and associated geometric properties of the atoms. Space-filling models are constructed from atomic models whose relative size is related to the actual volumes occupied by the particular atoms. Still other pictorial models convey further aspects of the overall geometry and shape of molecules.

Table 1.5   Geometric Models for Simple Molecules

Molecule	Geometry	Bond Angle	Structure
CO_2	Linear	180°	
O || H—C—H	Trigonal planar	120°	
CH_4	Tetrahedral	109°	
NH_3	Tetrahedral/ trigonal pyramid	107°	
H_2O	Tetrahedral/ bent or "V"	105°	

1.12   INTERMOLECULAR FORCES

Ionic bonds between atoms with large differences in electronegativity values and covalent bonds between atoms with small differences in electronegativity values are called primary covalent bonds. The length of primary covalent bonds varies from 0.09 to 0.2 nm and that of the carbon–carbon single bond is 0.15–0.16 nm. These primary bonds are strong bonds with energies usually greater than 90 kcal per mole.

There are also attractive forces between molecules, called secondary forces. These forces operate over long distances of 25–50 nm and have lower energy values (1–10 kcal/mol) than primary bonds. These secondary forces are called van der Waals forces. Intermolecular forces increase cumulatively as one goes from methane (CH_4) to ethane (C_2H_6) to propane (C_3H_6), and so on, in a homologous series. A homologous series is one in which each member differs by a methylene group (CH_2).

These secondary forces may be classified as weak London or dispersion forces (about 2 kcal/mol), dipole–dipole interactions (2–6 kcal/mol), and hydrogen bonds (about 10 kcal/mol). Since these forces are cumulative, the secondary bond energies and boiling points increase as one goes from methane (CH_4) to ethane (CH_3CH_3) to propane ($CH_3CH_2CH_3$) to butane ($CH_3CH_2CH_2CH_3$), and so on.

1.13   BASIC LAWS

All science is based on the assumption that the world about us behaves in an orderly, predictable, and consistent manner. The scientist's aim is to discover and report this behavior. It is an adventure we hope you will share with us in this course.

The scientific method involves making observations, looking for patterns in the observations, formulating theories based on the patterns, designing ways to test these theories, and, finally, developing "laws."

Observations may be qualitative (It is cool outside) or quantitative (It is 70°F outside). A *qualitative* observation is general in nature *without* attached *units*. A *quantitative* observation is more specific in *having units* attached. Gathering quantitative observations can be referred to as gathering measurements, collecting data, or performing an experiment. Patterns are often seen only after numerous measurements are made. Such patterns may be expressed by employing a mathematical relationship. Younger children like balloons but with other children about, they often resort to hiding the balloons—sometimes in the refrigerator. Later they notice that the balloons became smaller in the refrigerator. Thus the volume of the balloon, *V*, is directly related to temperature, *T*. This is expressed mathematically as

$$V \propto T$$

Our theory then is that as temperature increases, the volume of the balloon increases. This may also be called a hypothesis. We can test this hypothesis by further varying the temperature of the balloon and noting the effect on volume. We can then construct a model from which other hypotheses can be formed and other measurements performed.

Continuing with the balloon (made out of giant molecules) example, we can construct a model that says that pressure, the force per unit area, which is acting to expand the balloon, is due to gaseous particles, that is, molecules. This model can also be called a theory that resulted from interpretation, or speculation.

Eventually, a theory that has been tested in many ways over a long period is elevated to the status of a "law." We have a number of "laws" that are basic to the sciences. The following are some of these.

1. The world about us behaves in an orderly, predictable, and consistent manner. Thus, copper wire conducts an electric current yesterday, today, and tomorrow; under usual conditions water will melt at 0°C (32°F) yesterday, today, and tomorrow, and so on. We also hope that the orderly, predictable, and consistent behavior is explainable and knowable.

2. Mass/energy cannot be created or destroyed. This is called the Law of Conservation of Mass/Energy. It was originally described by Antoine Lavoisier around 1789 and referred to only as the conservation of mass. Later, Albert Einstein extended this to show that mass and energy were related by the famous equation

$$E = mv^2$$

where E is energy, m is mass, and v is velocity. Thus, while the total mass/energy is conserved, they are convertible as described by the Einstein equation.

Lavoisier was born in Paris in 1743. His father wanted him to become a lawyer but Lavoisier was fascinated by science. He wrote the first modern chemistry textbook, *Elementary Treatise on Chemistry*, in 1789. To help support his scientific work, he invested in a private tax-collecting firm and married the daughter of one of the company's executives. His connection to the tax collectors proved fatal, for eventually the French revolutionaries demanded his execution. On May 8, 1794, Lavoisier was executed on the guillotine.

3. A given compound always contains the same proportion of elements by weight and the same number of elements. Thus water molecules always contain one oxygen atom and two hydrogen atoms. Another compound that contains two oxygen atoms and two hydrogen atoms is not water, but rather is a different compound called hydrogen peroxide, often used as a disinfectant in water. This observation is a combination of two laws: first, the Law of Definite Proportions, described by the Frenchman Joseph Proust (1754–1826), and second, the Law of Multiple Proportions, initially described by the Englishman John Dalton (1766–1844). In fact, as mentioned earlier, Dalton was the first to describe what compounds, elements, and chemical reactions were. Briefly, the important aspects are:

(a) each element is composed of tiny particles called atoms;
(b) the atoms of the same element are identical; atoms of different elements differ from the atoms of the first element;
(c) chemical compounds are formed when atoms combine with each other;
(d) each specific chemical compound contains the same kind and number of atoms; and
(e) chemical reactions involve reorganization of the atoms.

John Dalton was a poor, humble man. He was born in 1766 in the village of Eaglesfield in Cumberland, England. His formal education ended at age 11 but he was clearly bright and, with help from influential patrons, began a teaching career at a Quaker school at the age of 12. In 1793 he moved to Manchester, taking up the post as tutor at New College.

He left in 1799 to pursue his scientific studies full time. On October 12, 1803, he read his now famous paper, "Chemical Atomic Theory," to the Literary and Philosophical Society of Manchester. He went on to lecture in other cities in England and Scotland. His reputation rose rapidly as his theories took hold, which laid the foundation for today's understanding of the world around us.

4. Electrons are arranged in orderly, quantized energy levels about the nucleus, which is composed of neutrons and protons. Most of us are familiar with a rainbow. The same colors can be obtained by passing light through a prism, resulting in a continuous array called a spectrum. If elements are placed between the continuous light source and the prism, certain portions of the spectrum are blank and produce a discontinuous spectrum. Different discontinuous spectra were found for different elements.

Eventually, this discovery led to an understanding that the electrons of the same elements resided in the same general energy levels and that they accepted only the specific energy (the reason for the blank spots in the spectrum) that permitted the electrons to jump from one energy level to another. These energy levels are called quantum levels. We live in a quantized universe in which movement, acceptance of energy, and emission of energy are all done in a discontinuous, quantized manner. Fortunately, the size of these allowable quantum levels decreases as the size of the matter in question increases as is the case in atomic structure.

1.14   MATTER/ENERGY

As far as we know, the universe is composed of matter/energy and space. Space, as presently understood, is contained within three dimensions. Energy may be divided according to form—magnetic, radiant, light; or its magnitude—ultraviolet, infrared, microwave; or its source—chemical energy, coal, oil, light, sugar, moving water, wind, nuclear; or its activity—kinetic or potential. Briefly, *kinetic energy* is energy in action—the lighting of a light bulb by a battery. *Potential energy* is energy at rest—a charged battery not being discharged. Potential energy can be converted to kinetic energy and conversely kinetic into potential. Thus a book on a shelf represents potential energy. If book is pushed from the bookshelf, the potential energy is converted into kinetic energy.

Matter/energy is conserved as described in the Law of Conservation of Matter/Energy. Matter can be described in terms of its physical state as solid, liquid, or gas. As shown in Figure 1.2, a *solid* has a fixed volume and a fixed shape, and does not assume the shape and volume of its container. A *liquid* has a fixed volume but not a fixed shape. It takes the shape of the portion of the container it occupies. A *gas* has neither a fixed volume nor shape. Some materials are solids, liquids, or gases depending on temperature or the time scale we use. Thus, glass acts like a solid at room temperature but begins to flow when heated to about 750°F, then acting like a

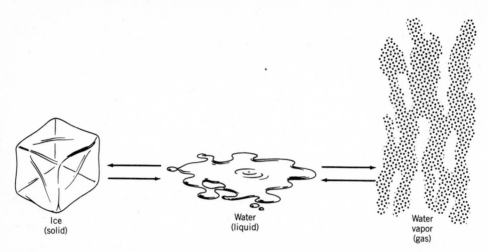

Figure 1.2. Water undergoing changes in state. *From left to right:* solid to liquid (melting) and liquid to gas (vaporization, boiling). *From right to left:* gas to liquid (condensation) and liquid to solid (freezing).

liquid. Glass acts like a solid when hit by a ball, but acts like a slow-flowing liquid when viewed over a period of a thousand years.

Most non-cross-linked matter undergoes transitions from solid to liquid to gas as temperature is increased or from gas to liquid to solid as temperature is decreased. These transitions are given names such as melting or freezing points. Thus, water below 0°C is solid, it melts (melting point) at 0°C (32°F), and boils (temperature of evaporation or boiling point) at 100°C (212°F). In turn, water above 212°F is a gas that condenses to a liquid at 212°F and freezes at 32°F.

Boiling, freezing, and melting are all physical changes. A *physical change* does not alter the chemical composition. Water can be broken into its elements of hydrogen and oxygen, however, and such a process is called a *chemical change* since the chemical composition of the matter is changed.

Physical properties are properties that can be measured without changing the chemical composition of the matter. Your height, color of hair, and weight are all physical properties. Other physical properties are density, color, boiling point, and freezing point.

Physical properties can be extensive or intensive. An *extensive property* is one that depends on the amount of matter present. Thus, mass is an extensive property. *Intensive properties* do not depend on the amount of matter present. Density, boiling point, and color are intensive properties.

Chemical properties are properties that matter exhibits when its chemical composition changes. The reaction of an iron nail with oxygen to form rust is a chemical reaction and the fact that iron reacts with oxygen is a chemical property of iron.

Matter can also be divided into components. Heterogeneous matter includes sidewalk cement, window glass, and most natural materials. Homogeneous matter or solutions include carbonated beverages, sugar in water, and brass (an alloy of zinc and copper). Examples of compounds include water, polyethylene, and table salt (NaCl). Some elements are iron (Fe), carbon (C), aluminum (Al), and copper (Cu).

GLOSSARY

Anion: A negatively charged ion.

Atomic number: A number that is equal to the number of protons in a specific atom.

Atom: The building blocks of the universe. An atom is the smallest stable part of an element.

Avogadro's number: 6.023×10^{23} particles in a mole.

Cation: A positively charged ion.

Celsius: Temperature scale: a scale in which water freezes at $0°C$ and boils at $100°C$.

Covalent bond: Bonds formed by sharing of electrons.

Dipole–dipole interaction: Moderately strong van der Waals forces.

Electron: The negatively charged building unit for all atoms.

Electronegativity: A measure of the tendency of an atom to attract electrons.

Element: A substance, such as carbon, consisting of identical atoms.

Endothermic reaction: A reaction in which energy is absorbed.

Exothermic reaction: A reaction in which energy, in the form of heat, is released.

Gram mole: Mass of 6.023×10^{23} particles in grams.

Homologous series: A series of related organic compounds, each differing by a methylene group (CH_2).

Hydrogen bond: Strong secondary forces resulting from the attraction of the hydrogen atom to an oxygen or nitrogen atom.

Ion: A charged atom.

Ionic bond: Bonds formed by an exchange of electrons.

kcal: Kilocalorie (1000 calories).

Kelvin: Temperature scale: An absolute scale in which water freezes at 273 K and boils at 373 K.

Law of Conservation of Matter: Principle stating that the total amount of mass remains unchanged in chemical reactions.

Lewis representation: Designation of outer or valence electrons as dots.

London dispersion force: Weak van der Waals dispersion forces.

Mass: A quantity of matter that is independent of gravity.

Mass number: A number that is equal to the number of protons plus the number of neutrons in a specific atom. Also called atomic weight.

Matter: Anything that has mass and occupies space.

Metric system: A decimal system of units for length in meters (m), mass in grams (g), and time in seconds (s).

Mole: 6.023×10^{23} particles.

Molecule: A combination of atoms capable of independent existence, for example, hydrogen (H_2), which is a diatomic molecule.

Neutron: An uncharged building unit for all atoms except hydrogen. The mass of the neutron is approximately 1 amu.

nm: Nanometer, 10^{-9} m.

Nucleon: Nuclear particles, that is, protons and neutrons.

Octet Rule: Rule of 8, that is, a stable compound has 8 outer electrons in the outer shell of the atom.

Orbital: The probable pathway of an electron in an atom.

Periodic Law: The arrangement of elements in order of increasing atomic numbers, which shows the periodic variation in many chemical and physical properties.

Periodic table: A systematic arrangement of atoms in the order of their increasing atomic numbers.

Periodic table families: Groups of elements arranged in vertical columns, all having the same number of valence electrons.

Periodicity: The position of an element in the periodic table.

Pi bonds: The bonds above and below the sigma bonds in double-bonded atoms, such as ethylene.

Primary chemical bond: Bonds between atoms in molecules in which the electrons are shared or exchanged.

Principal quantum number: Numbers used to describe the gross distance of electrons from the nucleus in an atom.

Proton: The positively charged building unit for all atoms. The mass of the proton is approximately 1 amu. This mass is approximately 1840 times that of the electron.

Quantum level: A specific energy level for an electron in the atomic shell.

Quantum mechanics: A description of the movement and location of an electron in an atom.

Quantum number: Numbers used to describe the average position and possible pathway of an electron.

Secondary quantum number: Numbers used to describe the shapes of the probable path of an electron. The letters s, p, d, and f are used to describe the subshells.

Sigma bond: Single covalent bonds between atoms.

Valence electron: Electrons in the outer shell of an atom.

van der Waals force: Attractive force between the nonpolar atoms. Also called London dispersion force.

Wavelength: The distance between waves.

X ray: Electromagnetic radiation of extremely short wavelengths produced by the bombardment of metals by electrons.

REVIEW QUESTIONS

1. Will the mass of an astronaut be greater or less in outer space than on the earth's surface?

2. What is 0 Kelvin (K) on the Celsius scale?

3. What is 0 Kelvin (K) on the Fahrenheit scale?

4. Which of the following are actually chemical elements: water, fire, carbon, hydrogen?

5. How many atoms are there in a molecule of methane (CH_4)?

6. How does a proton differ from a hydrogen atom?

7. Which has the greater mass: an electron or a proton?

8. Which has the longer wavelength: visible light or an x ray?

9. What is the atomic number of hydrogen, carbon, nitrogen, and oxygen?

10. What is the mass number of hydrogen, carbon, nitrogen, and oxygen?

11. What element has the same number of electrons as carbon in its outer shell? (Hint: Use the periodic table, Table 1.2.)

12. Which of the following have covalent bonds: CH_4, H_2O, C_2H_6?

13. Which of the following have ionic bonds: NaCl, LiF, HCl?

14. Show the Lewis dot representation for methane.

15. Which of the following is an exothermic reaction: boiling eggs or a burning candle?

16. If the total weight of reactants in a chemical reaction is 18 g, which is the weight of the products of this reaction?

17. How many particles in 0.1 mol?

18. What is the sign of the charge of an anion?

19. How many nanometers (nm) are there in 1 meter (m)?

20. If CH_4 and C_3H_8 are members of a homologous series, what is the formula for the homologue with two carbon atoms?

21. Which is the stronger: a dipole–dipole interaction or a London dispersion force?

22. Which is stronger: a hydrogen bond or a covalent bond in CH_4?

BIBLIOGRAPHY

Callewaert, D. M., and Genyea, J. (1989). *Fundamentals of College Chemistry*. New York: Worth.

Cherim, S. M., and Kallan, L. E. (1980). *Chemistry, An Introduction* (2nd ed.). Philadelphia: Saunders.

Ehrlich, J. (1981). *Geomicrobiology*. New York: Dekker.

Feigl, D. M., and Hill, J. W. (1983). *General, Organic, and Biological Chemistry: Foundations of Life* (1st ed.), Minneapolis: Burgess.

Herron, J. D. (1981). *Understanding Chemistry: A Preparatory Course* (1st ed.), New York: Random House.

Hill, J. W. (1980). *Chemistry for Changing Times* (3rd ed.), Minneapolis: Burgess.

Jones, M. M., et al. (1987). *Chemistry and Society*, Philadelphia: Saunders.

Jones, M. M., Netterville, J. T., Johnston, D. O., and Wood, J. L. (1983). *Chemistry, Man and Society* (4th ed.), Philadelphia: Saunders.

Kostiner, E., and Rea, J. R. (1979). *Fundamentals of Chemistry*. San Diego, CA: Harcourt Brace Jovanovich.

Makin, M. (1982). *Study Guide to Ucko's Basics for Chemistry*. San Diego, CA: Academic Press.

McQuarrie, D. A., and Rock, P. A. (1986). *General Chemistry*. San Francisco: Freeman.

Mohrig, J. R., and Child, W. C. (1987). *Chemistry in Perspective*. Rockleigh, N.J.: Allyn & Bacon.

Ouellette, R. (1987). *Understanding Chemistry*. New York: Macmillan Co.

Rayner-Canham, G., Last, A., Van Roode, M., and Perkins, R. (1983). *Foundations of Chemistry*. Reading, MA: Addison–Wesley.

Stine, W. R. (1981). *Applied Chemistry* (2nd ed.), Rockleigh, NJ: Allyn & Bacon.

Stoker, S. H. (1983). *Introduction to Chemical Principles*. New York: Macmillan Co.

Ucko, D. A. (1982). *Basics for Chemistry*. San Diego, CA: Academic Press.

Visscher, M. O. (1978). *The Ideas of Chemistry*. San Diego, CA: Harcourt Brace Jovanovich.

Williams, A. L., Embree, H. D., and DeBey, H. J. (1981). *Introduction to Chemistry*. Reading, MA: Addison–Wesley.

ANSWERS TO REVIEW QUESTIONS

1. The weight of the astronaut will be less in outer space but the mass will be unchanged.

2. $-273°C$.

3. $-459°F$.

4. Carbon and hydrogen.

5. 5.

6. The proton (H^+) has one less electron than the hydrogen atom (H).

7. The proton has a mass 1840 times that of the electron.

8. Visible light has a much longer wavelength.

9. 1, 6, 7, 8.

10. 1, 12, 14, 16.

11. Silicon, both are in the IVA family.

12. All three.

13. NaCl and LiF.

14.
$$H$$
$$H:\ddot{C}:H$$
$$H$$

15. A burning candle.

16. 18 g.

17. 6.023×10^{22}.

18. Negative.

19. 1×10^9 or 1 billion.

20. C_2H_6.

21. Dipole–dipole interaction.

22. The covalent bond.

CHAPTER II _____

Small Organic Molecules

2.1 INTRODUCTION

Because they serve as good examples, organic molecules were used to illustrate covalent bonding and intermolecular forces in Chapter 1. Some additional information on organic chemistry that should be useful in the study of giant molecules is presented in this chapter.

2.2 EARLY DEVELOPMENTS IN ORGANIC CHEMISTRY

In 1685, N. Lemery classified all matter as being animal, vegetable, or mineral. The latter class included inorganic compounds, such as salts (sodium chloride, NaCl), acids (hydrochloric acid, HCl), and alkalies (sodium hydroxide, NaOH). The former classes, that is, animal and vegetable, consisted almost entirely of organic or carbon-containing compounds.

Although Solomon referred to the reaction of vinegar with chalk in some of his proverbs several millenia ago (*Proverbs* 10:26, 25:20), the fact that all organic compounds contain carbon was not recognized until Johann Gmelin made this observation in 1848. Jöns Berzelius used the term inorganic to describe chalk and other minerals in the nineteenth century. However, both he and his contemporaries insisted that although inorganic compounds could be synthesized, it was not possible to synthesize organic compounds in the laboratory, since they believed that a "vital force" was essential for such a synthesis.

In 1828, Friedrich Wöhler demonstrated that the so-called vital force was not essential when he produced an organic compound, urea (H_2NCONH_2), by heating an inorganic compound, ammonium cyanate, as shown by the equation

$$NH_4NCO \xrightarrow{\Delta} H_2NCONH_2$$

Polymer science might have been limited to natural polymers, such as proteins, nucleic acids, starch, and cellulose, if Wöhler had not demonstrated that the vital force was not essential in the synthesis of organic chemicals. In general, organic compounds are classified as aliphatic, that is, molecules with linear chains of atoms, or as aromatic, that is, molecules with unsaturated cyclic structures. Some organic molecules occur as saturated cyclic structures, like cyclohexane (C_6H_{12}), or heterocyclic structures, like ethylene oxide,

$$H_2C\!\!-\!\!\!-\!\!\!-\!\!CH_2$$
$$\diagdown \diagup$$
$$O$$

2.3 ALKANES

Methane (CH_4), which is the major component of natural gas and one of the decomposition products of organic matter, is the simplest and one of the most abundant organic compounds. Methane is the first member of a homologous series, called the alkane or paraffin hydrocarbon series. All alkanes have the empirical formula $H\text{-}(CH_2\text{-})_nH$, where n is equal to 1 for methane and is equal to 500 or more for the giant molecule polyethylene. In spite of its name, polyethylene is a member of the alkane and not the ethylene (ethene) homologous series.

Because of different intermolecular forces, homologues have different physical properties, such as boiling points, melting points, and densities. The prefixes used for the names of the low-molecular-weight members of a homologous series are related to the number of carbon atoms present in these compounds. Thus, $H\text{-}(CH_2\text{-})_3H$ is called propane and $H\text{-}(CH_2\text{-})_4H$ is called butane after butyric acid, which is also a 4-carbon compound and is responsible for the odor of rancid butter.

The prefixes for homologues with five or more carbon atoms are similar to those used for geometrical figures. Thus, $H\text{-}(CH_2\text{-})_5H$ is the formula for pentane, $H\text{-}(CH_2\text{-})_6H$ is the formula for hexane, $H\text{-}(CH_2\text{-})_7H$ stands for heptane, and $H\text{-}(CH_2\text{-})_8H$ is the formula for octane. The residue, after the removal of a hydrogen atom ($H\cdot$) from an alkane, is called an alkyl radical. It has the general formula $H\text{-}(CH_2)_n$ and is represented by the symbol $R\cdot$. Specific radicals related to the alkane homologues are called methyl ($CH_3\cdot$), ethyl ($C_2H_5\cdot$), propyl ($C_3H_7\cdot$), and so on.

Structural formulas for alkanes are simply attempts to represent models of these molecules on paper. Experimental evidence is available to show that all the carbon–hydrogen bonds in methane are of equal length and directed toward the corners of a tetrahedron. Accordingly, the bond angles between the carbon and hydrogen atoms are 109.5° and this angle is characteristic for every carbon–hydrogen bond and every carbon–carbon bond in all alkane hydrocarbons.

We may also use the Lewis representation to show the structural formulas for methane, ethane, and propane:

Methane, Ethane, Propane

The names and structural formulas for several alkanes are shown in Table 2.1.

For reasons of simplicity, we shall represent the covalent bonds by short lines called single bonds. Thus, we may write the formula for methane as

$$H-C-H$$ with H above and below

It should be understood that each carbon atom will be surrounded by four pairs of dots representing eight electrons and these four electron pairs represent four covalent bonds joined to carbon or hydrogen atoms. Accordingly, we may simplify these structural formulas and use less detailed skeletal formulas in which the presence of the hydrogen atoms on the carbon atom is understood. Thus, methane, ethane, and

Table 2.1   Names of Unbranched Alkanes (Normal Alkanes)

Name	Number of Carbons	Geometrical Formulas	Molecular Formulas
Methane	1	CH_4	CH_4
Ethane	2	H_3CCH_3	C_2H_6
Propane	3	$H_3CCH_2CH_3$	C_3H_8
Butane	4	$H_3C(CH_2)_2CH_3$	C_4H_{10}
Pentane	5	$H_3C(CH_2)_3CH_3$	C_5H_{12}
Hexane	6	$H_3C(CH_2)_4CH_3$	C_6H_{14}
Heptane	7	$H_3C(CH_2)_5CH_3$	C_7H_{16}
Octane	8	$H_3C(CH_2)_6CH_3$	C_8H_{18}
Nonane	9	$H_3C(CH_2)_7CH_3$	C_9H_{20}
Decane	10	$H_3C(CH_2)_8CH_3$	$C_{10}H_{22}$
Undecane	11	$H_3C(CH_2)_9CH_3$	$C_{11}H_{24}$
Dodecane	12	$H_3C(CH_2)_{10}CH_3$	$C_{12}H_{26}$
Tridecane	13	$H_3C(CH_2)_{11}CH_3$	$C_{13}H_{28}$
Tetradecane	14	$H_3C(CH_2)_{12}CH_3$	$C_{14}H_{30}$
Pentadecane	15	$H_3C(CH_2)_{13}CH_3$	$C_{15}H_{32}$

propane can be represented by the skeletal formulas

$$C, \quad C—C, \quad C—C—C$$

If we proceed to write these simplified skeletal structural formulas for higher homologues, we will observe that two skeletal formulas can be written for butane and that three structures can be shown for pentane:

| *n*-Butane | Isobutane | *n*-Pentane | Isopentane | Neopentane |

These structures represent actual structural isomers that have the trivial names shown. It is important to note that the number of structural isomers increases as the number of carbon atoms in the alkane molecules increases. Physical constants for isomers of pentane and hexane are shown in Table 2.2.

Systematic nomenclature has been developed by the International Union of Pure and Applied Chemistry (IUPAC), but the trivial (common) names are used almost universally for the simple alkanes. In the IUPAC system, the compound is named as a derivative of the longest chain, and the positions of the substituents are designated by

Table 2.2   Geometrical Isomers and Physical Constants for Pentane and Hexane

Molecular Formulas	Structural Formulas	Melting Point (°C)	Density (g/mL)
C_5H_{12}	$H_3CCH_2CH_2CH_2CH_3$	-130	0.626

C_5H_{12}	$H_3C—CH—CH_2CH_3$
	        $\vert$
	      CH_3
		-160	0.620

C_5H_{12}	CH_3
	       $\vert$
	 $H_3C—C—CH_3$
	       $\vert$
	      CH_3
		-20	0.613

C_6H_{14}	$H_3CCH_2CH_2CH_2CH_2CH_3$	-95	0.660

C_6H_{14}	$H_3CCHCH_2CH_2CH_3$
	      $\vert$
	    CH_3
		-154	0.653

C_6H_{14}	$H_3CCH—CHCH_3$
	      $\vert$
	    CH_2CH_3
		-129	0.662

C_6H_{14}	CH_3
	       $\vert$
	 $H_3C—C—CH_2CH_3$
	       $\vert$
	      CH_3
		-98	0.649

appropriate numbers. Thus, the pentanes are named as follows:

$$
\begin{array}{ccc}
\text{C—C—C—C—C} & \begin{matrix} \text{C} \\ | \\ \text{C—C—C—C} \end{matrix} & \begin{matrix} \text{C} \\ | \\ \text{C—C—C} \\ | \\ \text{C} \end{matrix} \\
\text{Pentane} & \text{2-Methylbutane} & \text{2,2-Dimethylpropane}
\end{array}
$$

When a linear hydrocarbon, such as pentane, is burned, it produces carbon dioxide, water, and thermal energy, but it sputters during the combustion process. However, less sputtering (more complete combustion) is observed when branched hydrocarbons, such as 2-methylbutane and 2,2-dimethylpropane, are burned. The antiknock properties of hydrocarbons in unleaded gasoline are related to the extent of branching in the molecules.

2.4 UNSATURATED HYDROCARBONS (ALKENES)

As mentioned in Chapter 1, the carbon–carbon bond in alkanes is called a sigma (σ) bond and the bond angle is 109.5°. Thus, ethane ($H(CH_2)_2H$) contains seven bonds, six of which are carbon–hydrogen bonds. When ethane is heated in the presence of an appropriate catalyst, it loses two atoms of hydrogen, and the product is called by the trivial name of ethylene. The Greek symbol for fire ($\triangle$) is used to show that heat is added to the reactants and that this is an endothermic reaction. The equation for this dehydrogenation is

$$
\begin{matrix}
\text{H} & \text{H} \\
| & | \\
\text{H—C—C—H} \\
| & | \\
\text{H} & \text{H}
\end{matrix}
\xrightarrow{\triangle} \text{H}_2 +
\begin{matrix}
\text{H} & \quad & \text{H} \\
\diagdown & & \diagup \\
& \text{C=C} & \\
\diagup & & \diagdown \\
\text{H} & \quad & \text{H}
\end{matrix}
$$

The reverse reaction is called catalytic hydrogenation.

Ethylene is the principal starting material for the petrochemical industry. This hydrocarbon, which is called ethylene in IUPAC systematic nomenclature, is the first member of the alkene homologous series and may be represented by the empirical formulas of $H(CH_2)_n CH{=}CH_2$ or C_2H_{2n} or $R—CH{=}CH_2$, where R is an alkyl radical. Since these homologues have fewer hydrogen atoms than the alkanes, they are sometimes called unsaturated hydrocarbons and, accordingly, the alkanes are called saturated hydrocarbons. The alkenes are also called by the trivial name of olefins. Ball-and-stick models for some aliphatic hydrocarbons are shown in Figure 2.1.

The names of the alkene homologues containing three, four, and five carbon atoms are propene or propylene, butene or butylene, and pentene. As shown by the following skeletal formulas, there are three butenes that have systematic IUPAC names:

$$
\begin{array}{ccc}
\underset{1 \quad 2 \quad 3 \quad 4}{\text{C=C—C—C}} & \underset{1 \quad 2 \quad 3 \quad 4}{\text{C—C=C—C}} & \underset{1 \quad 2 \quad 3}{\begin{matrix} \text{C} \\ | \\ \text{C=C—C} \end{matrix}} \\
\text{1-Butene (butylene)} & \text{2-Butene} & \text{2-Methylpropene (isobutylene)}
\end{array}
$$

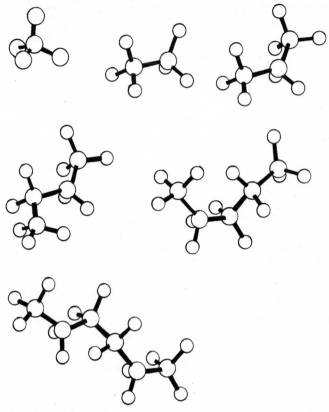

Figure 2.1. Ball-and-stick models of the first six linear members of the alkane hydrocarbon family (from left to right, top to bottom): methane, ethane, propane, butane, pentane, and hexane.

We could make simple models of 2-butene using two toothpicks to hold two gumdrops together for carbons 2 and 3. If single toothpicks were used to bond gumdrops representing carbons 1 and 4 at an angle of 109.5°, we would find that these two carbon atoms could be either on one side of the plane of the double-bonded carbon atoms or on opposite sides of this plane. These models represent actual geometrical isomers that are called cis ("on this side") and trans ("across") isomers, as shown in the skeletal structures

$$
\begin{array}{cc}
\text{C} & \text{C} \\
| & | \\
\text{C} & = \text{C}
\end{array}
\qquad\qquad
\begin{array}{c}
\text{C} \\
| \\
\text{C} = \text{C} \\
| \\
\text{C}
\end{array}
$$

cis-2-Butene   *trans*-2-Butene

The word isomer is derived from the Greek word *iso*, meaning equal, and *mer*, meaning parts. Isomers have the same chemical composition and empirical formulas (in this case C_4H_8) but have different structural formulas.

As shown in Table 2.3, cis and trans isomers may be distinguished from each other by physical properties, such as melting and boiling points. It is of interest to note that naturally occurring rubber is an olefinic hydrocarbon made up of repetitive cis linkages. In contrast, another naturally occurring hydrocarbon with a similar empirical (simplest) formula, $(C_5H_8)_n$, is a nonelastic, rigid material. The latter, which is called gutta-percha, consists of repetitive trans linkages.

Table 2.3 Properties of Selected Cis–Trans Isomers

	Melting Point (°C)	Boiling Point (°C)	Density (g/mL)
cis-2-Butene	−139	3.7	0.621
trans-2-Butene	−106	0.9	0.604
cis-1-Chloro-1-butene	—	63	0.915
trans-1-Chloro-1-butene	—	68	0.921
cis-2-Chloro-2-butene	−117	71	0.924
trans-2-Chloro-2-butene	−106	63	0.914

The empirical formula $(C_5H_8)_n$ is the general formula for many naturally occurring materials called terpenes. The value for n may range from 1 to over 1000. When there are many repeating units, such as in rubber and gutta-percha, the product is called a giant molecule, macromolecule, or polymer. The prefix macro is from the Greek word *makrus*, meaning big one.

The U.S. production for these polymers of ethylene and propylene in 1988 was 9 million and 3.6 million tons, respectively. The nonpolymeric olefins, such as ethylene, which are called monomers, are also used for the production of many other organic compounds (e.g., petrochemicals).

The cycloalkanes, for example, cyclopropane $(CH_2)_3$, cyclobutane $(CH_2)_4$, and cyclopentane $(CH_2)_5$, represent another homologous hydrocarbon series. Asphalt, which is used on roof and for road surfaces, contains high-molecular-weight cycloalkanes. As shown in Table 2.4, in contrast to the linear structures of the alkenes, the skeletal structures of the lower cycloalkene homologues are simple geometrical figures, such as triangles, squares, pentagons, and hexagons.

The alkynes constitute another homologous series. Acetylene $(HC{\equiv}CH)$, which is the first and most important member of this series, may be produced by the catalytic cracking or decomposition of saturated hydrocarbons. However, as shown by the following equation, acetylene may also be readily produced by the addition of water to calcium carbide (CaC_2):

$$CaC_2 \; + \; 2H_2O \; \rightarrow \; HC{\equiv}CH \; + \; Ca(OH)_2$$

Calcium carbide   Water   Acetylene   Calcium hydroxide

Acetylene is burned in oxyacetylene welding torches. It is also the starting material for many other organic chemicals. Some typical organic chemical compounds are discussed in Section 2.5.

Table 2.4   Structures of Some Common Cycloalkanes

Name	Molecular Formula	Structural Formula
Cyclopropane	C_3H_6	$\begin{array}{c} H_2 \\ C \\ H_2C - CH_2 \end{array}$
Cyclobutane	C_4H_8	$\begin{array}{c} H_2C - CH_2 \\ \mid \quad \mid \\ H_2C - CH_2 \end{array}$
Cyclopentane	C_5H_{10}	$\begin{array}{c} H_2 \\ C \\ H_2C \quad CH_2 \\ H_2C - CH_2 \end{array}$
Cyclohexane	C_6H_{12}	$\begin{array}{c} H_2 \\ C \\ H_2C \quad CH_2 \\ H_2C \quad CH_2 \\ C \\ H_2 \end{array}$

2.5   ALIPHATIC COMPOUNDS

Alkanes, which are saturated hydrocarbons, form compounds through substitution of a hydrogen atom by another moiety. For simple substituted alkanes, the particular radical formed is named by replacing the -ane suffix by -yl. Thus we have

Hydrocarbon	Radical
Methane (CH_4)	Methyl (CH_3-)
Ethane (CH_3CH_3)	Ethyl (CH_3CH_2-)
Propane ($CH_3CH_2CH_3$)	Propyl ($CH_3CH_2CH_2$-)

A functional group, such as a hydroxyl group (OH), may replace a hydrogen atom in an alkane in the same manner that the hydrogen atom in propane is replaced by a methyl radical in isobutane (C—C(C)—C). Organic compounds containing hydroxyl groups are called by the trivial name of alcohols or by the systematic name of alkanols.

The alcohols and organic compounds, with other functional groups, belong to characteristic homologous series having nomenclature related to that used for the alkanes. Thus, the names of the lower alkanol homologues are methanol (H_3COH), ethanol ($H(CH_2)_2OH$), and propanol ($H(CH_2)_3OH$). As shown by the following skeletal formulas, there are two propanol isomers:

$$C—C—C—OH$$

1-Propanol

$$\overset{\overset{\displaystyle OH}{|}}{C—C—C}$$

2-Propanol
(isopropyl alcohol)

Since most nonscientists are not aware of the many homologues in the alkanol series, they may erroneously assume that any alcohol may be used as a beverage. Hence, to avoid mistaken identity, one should use the systematic names and not call an alkanol, such as methanol, by the name methyl alcohol. The latter and all other alkanols, except ethanol, are extremely toxic. For example, the poisonous fusel oils present in improperly distilled liquor are pentanols, which are also called by the trivial name of amyl alcohols. Ethanol, of course, is toxic and actually lethal in large quantities. Alcohols are designated as primary (RCH_2OH), secondary (R_2CHOH), and tertiary (R_3COH) based on the decreasing number of hydrogen atoms present on the carbon atom in addition to the hydroxyl group.

When two alkyl radicals are joined by an oxygen atom, such as in $H_5C_2OC_2H_5$, these compounds are called alkyl ethers. The following formula is for ethyl ether or diethyl ether, which was used for over a century as an anesthetic:

$$CH_3—CH_2—O—CH_2—CH_3$$

When a hydrogen atom or an alkyl radical is joined to a carbonyl group (C=O), the resulting compound, such as $H_2C{=}O$, is called an aldehyde. The systematic name for this molecule is methanal, but the trivial name formaldehyde is used universally. When two alkyl groups are joined to a carbonyl group, the compound, such as $H_3C(CO)CH_3$, is called a ketone. The preceding formula is for acetone.

Acetic acid (H_3CCOOH), which is the major constituent of vinegar, consists of the methyl radical (CH_3) and the carboxyl group (COOH). The first member of this homologous series is called formic acid (HCOOH) after the Latin word *formica*, meaning ant. The blister that forms as a result of an ant bite is caused by formic acid.

When the hydrogen atom of the carbonyl group is replaced by an alkyl radical, a neutral ester (RCOOR′) is obtained. Esters, such as amyl acetate ($H_3CCOO(CH_2)_5H$), which is also called banana oil, are responsible for many characteristic fruit odors.

Alcohols and ethers may be considered to be derivatives of water (HOH) in which either one or both hydrogen atoms are replaced by alkyl groups. Likewise, amines, which are organic bases, may be considered as derivatives of ammonia (NH_3) in which one or more of the hydrogen atoms are replaced by alkyl radicals, as shown by the following formulas for methylamines. These compounds are classified as primary, secondary, and tertiary amines in accordance with the number of hydrogen atoms displaced by alkyl groups:

$$H_3C—NH_2$$

Methylamine
(primary)

$$\overset{\overset{\displaystyle H}{|}}{H_3C—N—CH_3}$$

Dimethylamine
(secondary)

$$\overset{\overset{\displaystyle CH_3}{|}}{H_3C—N—CH_3}$$

Trimethylamine
(tertiary)

Alkyl groups may also replace the hydrogen atom in hydrogen halides (such as hydrogen chloride, HCl), in hydrocyanic acid (HCN), and in hydrogen sulfide (H_2S),

Table 2.5 Typical Functional Groupings

Acid	$\underset{\text{R—COH}}{\overset{\displaystyle O}{\overset{\|}{}}}$	Ether	R—O—R
Acid chloride	$\underset{\text{R—CCl}}{\overset{\displaystyle O}{\overset{\|}{}}}$	Isocyanate	R—NCO
Alcohol	R—OH	Ketone	$\underset{\text{R—C—R}}{\overset{\displaystyle O}{\overset{\|}{}}}$
Aldehyde	$\underset{\text{R—CH}}{\overset{\displaystyle O}{\overset{\|}{}}}$	Nitrile	R—CN
Amide	$\underset{\text{R—C N—R}}{\overset{\displaystyle O\ H}{\overset{\|\ \|}{}}}$	Sulfide	R—S—R
Amine	R—NH$_2$	Thiol	R—SH
Anhydride	$\underset{\text{R—COC—R}}{\overset{\displaystyle O\ \ O}{\overset{\|\ \ \|}{}}}$	Urea	$\underset{\text{R—N—C—N—R}}{\overset{\displaystyle \ \ \ O}{\overset{\text{H }\|\text{ H}}{}}}$
Carbonate	$\underset{\text{R—OCO—R}}{\overset{\displaystyle O}{\overset{\|}{}}}$	Urethane	$\underset{\text{R—N CO—R}}{\overset{\displaystyle \ \ \text{H O}}{\overset{\text{| |}}{}}}$
Ester	$\underset{\text{R—CO—R}}{\overset{\displaystyle O}{\overset{\|}{}}}$		

Polymer Type	Interunit Linkage	Polymer Type	Interunit Linkage
Polyester	$\underset{\text{—C—O—}}{\overset{\displaystyle O}{\overset{\|}{}}}$	Polyamide	$\underset{\text{—C—NH—}}{\overset{\displaystyle O}{\overset{\|}{}}}$
Polyanhydride	$\underset{\text{—C—O—C—}}{\overset{\displaystyle O\ \ \ \ O}{\overset{\|\ \ \ \ \|}{}}}$	Polyurethane	$\underset{\text{—O—C—NH—}}{\overset{\displaystyle O}{\overset{\|}{}}}$
Polyether	$\underset{\text{R}}{\overset{\text{H}}{\overset{\|}{\text{—C—O—}}}}$	Polyurea	$\underset{\text{—NH—C—NH—}}{\overset{\displaystyle O}{\overset{\|}{}}}$

and the hydroxyl radical in nitric acid (HONO$_2$) and sulfuric acid (HOSO$_2$OH). These replacements would produce compounds such as ethyl chloride (C$_2$H$_5$Cl), methyl cyanide or acetonitrile (CH$_3$CN), ethyl mercaptan (C$_2$H$_5$SH), nitromethane (CH$_3$NO$_2$), and ethyl sulfonic acid (C$_2$H$_5$SO$_2$OH). A list of typical functional groups is shown in Table 2.5.

It is important to note that more than one substituent may be present in an organic compound and that different substituents may be present in the same molecule. Ethylene glycol, which is used as an antifreeze, has the formula HO(CH$_2$)$_2$OH, and D-glucose, which is a simple sugar or carbohydrate, contains five alcohol groups and one ether group:

β-D-Glucose

2.6 BENZENE AND ITS DERIVATIVES (AROMATIC COMPOUNDS)

Benzene, which has the formula C_6H_6, and its derivatives are called aromatic compounds. The symbol for benzene is

The circle within the hexagon indicates that this is a resonance hybrid in which all bonds have equal angles and length and, as in ethylene, there is a region of high electron density above and below the flat hexagonal ring. The two contributing forms for this hybrid are

The true structure is actually a combination of these two forms.

The six hydrogen atoms that are not shown in the skeletal formula for benzene may be replaced by alkyl groups (R), aromatic aryl (Ar) groups, and any of the functional groups cited in the preceding discussion of aliphatic chemistry. Some typical aromatic compounds are shown in Figure 2.2.

2.7 HETEROCYCLIC COMPOUNDS

In addition to linear and cyclic aliphatic compounds, such as hexane and cyclohexane, and aromatic compounds, such as benzene, there are also heterocompounds that have other atoms besides carbon in their molecules. The heterocompounds may be linear, such as ethyl ether, or heterocyclic, such as ethylene oxide (oxirane). Structures of these typical compounds are

$$H(CH_2)_2O(CH_2)_2H \qquad H_2C\!\!-\!\!\!-\!\!\!-\!\!CH_2 \qquad or$$

Ethyl ether   Ethylene
                       oxide
                       (oxirane)

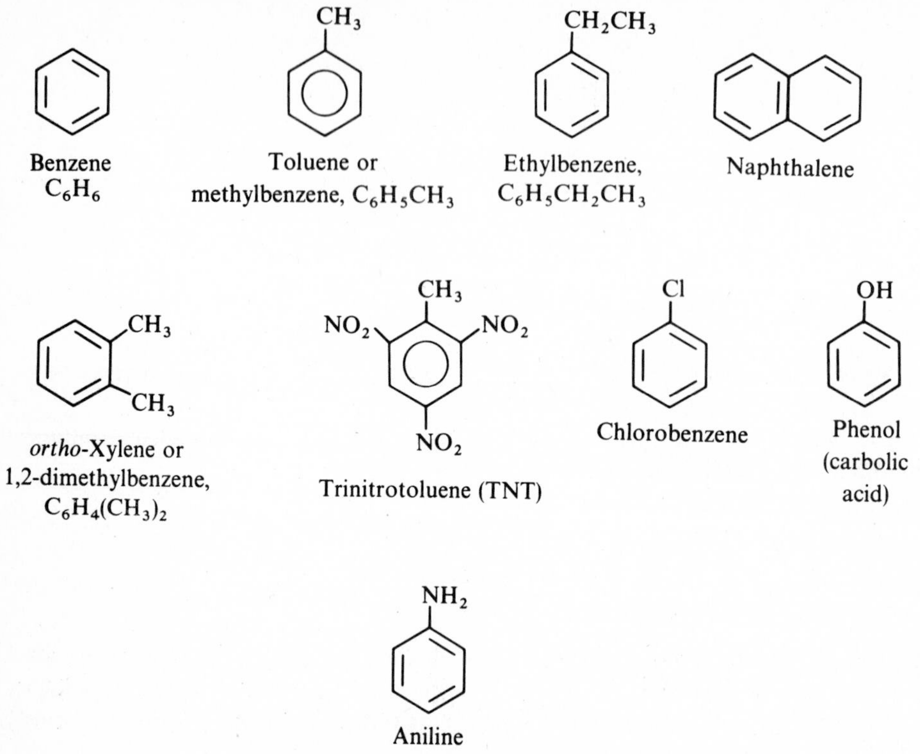

Benzene
C_6H_6

Toluene or
methylbenzene, $C_6H_5CH_3$

Ethylbenzene,
$C_6H_5CH_2CH_3$

Naphthalene

ortho-Xylene or
1,2-dimethylbenzene,
$C_6H_4(CH_3)_2$

Trinitrotoluene (TNT)

Chlorobenzene

Phenol
(carbolic
acid)

Aniline

Figure 2.2. Selected aromatic compounds.

Some of the more important cyclic and heterocyclic compounds are shown in Table 2.6.

2.8 POLYMERIC STRUCTURE

As will be described in Chapter 3, the structure of organic polymers is similar to that of small organic compounds. The principal difference is that polymers are made up of long sequences of the smaller molecules, which are called repeating units. Thus, as shown by the following structural formulas, the principal difference between the small molecule decane and a selected polyethylene molecule is the number of repeating units:

$$H-(CH_2)_{10}H \qquad H-(CH_2CH_2)_{500}H$$

Decane   Polyethylene

Obviously, there is much more to organic chemistry than we have discussed in this chapter. However, the brief discussions in these first two chapters should provide sufficient background for an appreciation of giant molecules, which are discussed in subsequent chapters in this book.

You may also gain some appreciation of the relative size of giant molecules, as

Table 2.6   Structures of Selected Simple Cyclic and Heterocyclic Compounds

	Saturated	Unsaturated

Cyclic

5-Membered

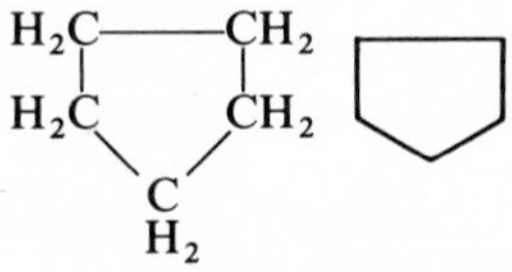

Cyclopentane

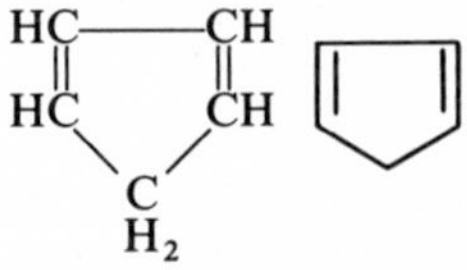

Cyclopentadiene

6-Membered

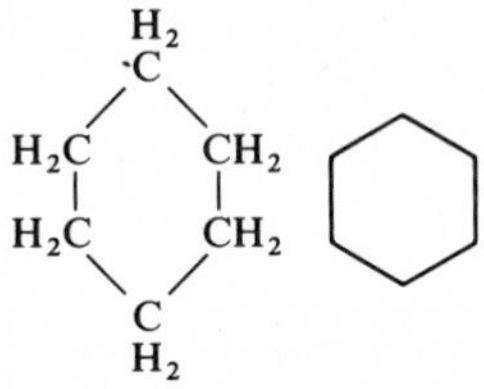

Cyclohexane

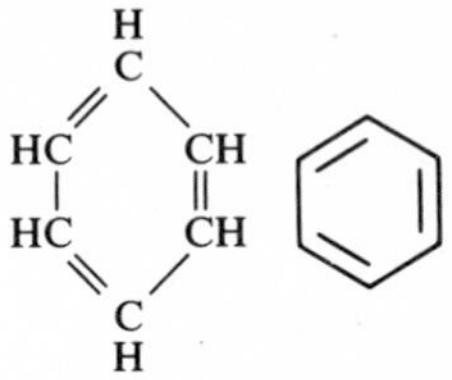

Benzene

Heterocyclic

5-Membered

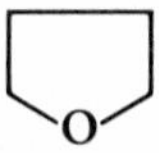

Tetrahydrofuran

Furan

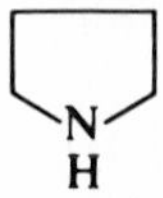

Pyrrolidine

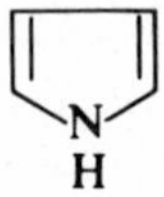

Pyrrole   Imidazole

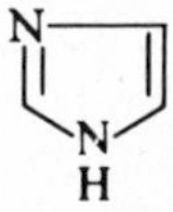

6-Membered

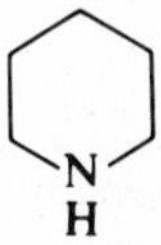

Piperidine

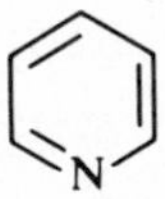

Pyridine   Pyrimidine

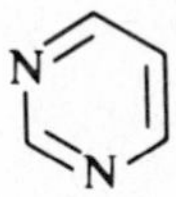

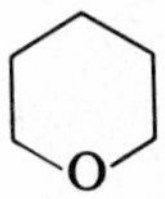

Tetrahydropyran

Pyran

Fused Rings

6-Membered
plus 5-membered

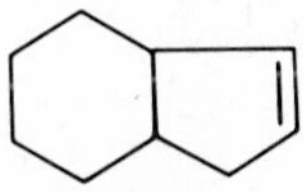

Hydrindane

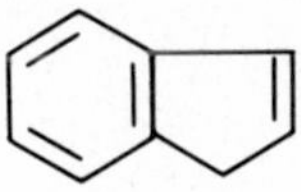

Indene

Table 2.6 *(Cont.)*

	Saturated	Unsaturated

Fused Rings

Three 6-membered
plus one 5-membered

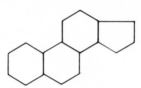

Sterane

Fused Heterocyclic

6-Membered
plus 5-membered

Indole

Purine

6-Membered
plus 6-membered

Quinoline

Pteridine

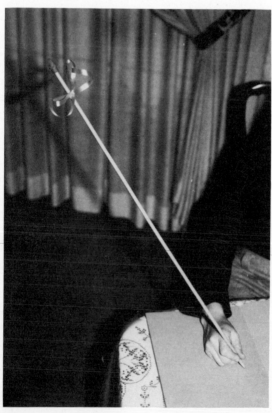

Figure 2.3. Giant polymer pencil. Courtesy of Kenrich Petrochemicals, Inc.

compared to ordinary molecules, by observing the giant polymer pencil shown in Figure 2.3. This pencil was made by forcing a mixture of wood flour and polystyrene through a hot circular die in a process called extrusion. For the convenience of the photographer, a 3-ft section was cut from the continuous extrudate. Had a 200-ft section been photographed, it would be about 300 times the length of an ordinary pencil and in the range of the ratio of the length of a giant molecule to the length of an ordinary molecule.

GLOSSARY

Acetaldehyde: H_3CCHO.

Acetic acid: H_3CCOOH.

Acetone: H_3CCOCH_3.

Acetylene: $HC{\equiv}CH$.

Alcohol: Compounds with hydroxyl (OH) substituents.

Aldehyde: $H(CH_2)_n CHO$ or

Aliphatic: Open chains of atoms like C_2H_6 and C_4H_{10}.

Alkane: Belonging to the series having the empirical formula $H(CH_2)_nH$.

Alkene homologous series: Unsaturated hydrocarbons having the formula $H(CH_2)_n CH{=}CH_2$.

Alkyl: $H(CH_2)_{\overline{n}}$.

Alkyne: Compounds having the empirical formula $H(CH_2)_n C{\equiv}CH$.

Amine: RNH_2, for example, CH_3NH_2, R_2NH, or R_3N.

Amine, primary: An amine with two hydrogen atoms on the nitrogen atom, that is, RNH_2.

Amine, secondary: An amine with one hydrogen atom on the nitrogen atom, that is, R_2NH.

Amine, tertiary: An amine with no hydrogen atom on the nitrogen atom, that is, R_3N.

Aniline: Aminobenzene ($C_6H_5NH_2$).

Aromatic: Cyclic unsaturated molecules, like benzene (C_6H_6).

Bond, single: Bond formed by sharing two electrons, represented by a single bar.

Branch: Substituents or chain extensions on the main chain of an organic compound.

Butane: C_4H_{10}.

Carboxyl: —COOH.

Catalyst: A substance that accelerates the attainment of equilibrium in a chemical reaction. Only a small amount of catalyst is required and this substance can be recovered unchanged.

Cellulose: A naturally occurring carbohydrate made up of repeating units of D-glucose ($C_6H_{12}O_6$).

Chalk: Calcium carbonate ($CaCO_3$).

cis: An unsaturated organic compound with substituents on the same side of the plane of the double bond.

Ester: RCOOR', for example, amyl acetate ($H(CH_2)_5OOCCH_3$).

Ethanol: C_2H_5OH.

Ether: ROR; also used as a trivial name for ethyl ether (($C_2H_5)_2O$).

Ethylene glycol: $CH_2(OH)CH_2(OH)$.

Formic acid: HCOOH.

Formula, skeletal: A structural formula in which the hydrogen atoms have been omitted, for example, ethane (C—C).

Formula, structural: Two-dimensional representation of molecules on paper, for example,

$$
\begin{array}{c}
H \\
| \\
H-C-H \\
| \\
H
\end{array}
$$

Functional group: A group capable of reacting further.

Fusel oil: Pentanols.

Gutta-percha: A rigid, naturally occurring hydrocarbon polymer with trans arrangement around the double bonds.

H·: Hydrogen atom.

H(CH$_2$)$_n$·: Alkyl radical, for example, CH_3.

Heterocyclic compound: Cyclic compounds with other atoms in addition to carbon atoms in the ring.

Homologous series: A series of related compounds with formulas differing by a constant unit like CH_2.

Hydroxyl group: —OH.

Inorganic chemistry: The chemistry of minerals and related compounds.

iso: An organic compound with a substituent on carbon number 2.

Isomer, geometrical: Unsaturated compounds with substituents on each of the double-bonded carbon atoms. Because of lack of free rotation, part of these substituents may be on one side of the plane of the double bond or on alternate sides.

IUPAC: International Union of Pure and Applied Chemistry.

IUPAC System: A preferred systematic nomenclature for organic compounds.

Ketone: $H[(CH_2)_n]_2CO$ or

$$
\begin{array}{c}
O \\
\| \\
R-C-R
\end{array}
$$

Monomer: An organic compound capable of forming a giant molecule.

neo: An organic compound with two substituents on the same carbon atom, for example, neopentane ($CH_3C(CH_3)_2CH_3$).

Nitric acid: $HONO_2$ of HNO_3.

Normal: A straight or continuous (linear) chain structure.

Olefin: Alkenes or unsaturated hydrocarbons,

Organic chemistry: The chemistry of carbon-containing compounds.

Pentane: C_5H_{12}.

Petrochemical: Compounds derived from petroleum.

Phenol: Hydroxybenzene (C_6H_5OH).

Polyethylene: A giant molecule belonging to the alkane homologous series.

Polymer: A giant molecule or macromolecule.

Propane: C_3H_8.

Propene: Propylene ($H_3CCH{=}CH_2$).

R·: Alkyl radical, that is, $H(CH_2)_n\cdot$.

Resonance hybrid: A molecule that can be represented by two or more structures that differ only in the disposition of electrons. The true formula (hybrid) is one that is in between the two contributing forms and is unusually stable.

Rubber, natural: An elastic, naturally occuring hydrocarbon with cis arrangement around the double bonds.

Starch: A naturally occurring carbohydrate made up of repeat units of D-glucose ($C_6H_{12}O_6$).

Sulfuric acid: $HOSO_2OH$.

Terpene: Compounds with the empirical formula $(C_5H_8)_n$.

Toluene: Methylbenzene ($C_6H_5CH_3$).

trans: An unsaturated organic compound with substituents on opposite sides of the plane of the double bond.

Vital force: An essential force formerly believed to be associated with living organisms.

REVIEW QUESTIONS

1. Which of the following are organic chemicals: $CaCO_3$, CH_4, C_6H_6 (benzene), $C_6H_{12}O_6$ (glucose)?

2. Which of the following are polymers: ethylene, protein, cellulose, polyethylene?

3. What is the empirical (simplest) formula for hexane?

4. How many electrons are present in the hydrogen atom?

5. What is the formula for the propyl radical?

6. What is the structural formula for ethylene?

7. What is the IUPAC name for isobutane?

8. What is the structural formula for propylene (propene)?

9. What is the difference between an alkene, an olefin, and an unsaturated hydrocarbon?

10. What is the structural formula for *trans*-2-butene?

11. What is the difference in the structure of elastic natural rubber and rigid gutta-percha?

12. What is the structural formula for acetylene?

13. What functional group is always present in an alcohol?

14. What is the formula for ethyl ether?

15. What is the general formula for an aliphatic aldehyde?

16. What is the formula for propionic acid?

17. What is the formula for ethyl acetate?

18. Is diethylamine $((C_2H_5)_2NH)$ a primary or secondary amine?

19. How many hydroxyl groups are there in ethylene glycol?

20. The hybrid benzene can be represented by two different structures:

     Which is correct?

21. What is the formula for ethylbenzene?

22. Is aniline a primary or secondary amine?

BIBLIOGRAPHY

Allinger, N. L., *et al.* (1976). *Organic Chemistry* (2nd ed.), New York: Worth.

Bailey, P., and Bailey, C., (1989). *Organic Chemistry: A Brief Survey of Concepts and Applications* (2nd ed.). Rockleigh, NJ: Allyn & Bacon.

Brown, W. H. (1982). *Introduction to Organic Chemistry* (3rd ed.). Pacific Grove, CA: Brooks/Cole.

Fessenden, R. J., and Fessenden, J. S. (1986). *Organic Chemistry* (3rd ed.). Pacific Grove CA: Brooks/Cole.

Kemp, D. S., and Vellacio, F. (1980). *Organic Chemistry*. New York: Worth.

Loudon, G. M. (1988). *Organic Chemistry*. Reading, MA: Benjamin/Cummings.

Marmot, S. (1987). *Organic Chemistry: A Brief Course*. San Diego, CA: Harcourt Brace Jovanovich.

Morrison, R. T., and Boyd, R. N. (1987). *Organic Chemistry* (4th ed.). Rockleigh, NJ: Allyn & Bacon.

Scott, R. M. (1980). *Introduction to Organic and Biological Chemistry*. New York: Harper & Row.

Seymour, R. B. (1971). *General Organic Chemistry*. New York: Barnes & Noble.

Solomons, T. W. S. (1988). *Fundamentals of Organic Chemistry* (4th ed.). New York: Wiley.

Streitwieser, A., and Heathcock, C. H. (1985). *Introduction to Organic Chemistry* (2nd ed.). New York: Macmillan Co.

Ternay, A. L. (1979). *Contemporary Organic Chemistry* (2nd ed.). Philadelphia: Saunders.

Wingrove, A. S., and Caret, R. L. (1981). *Organic Chemistry*. New York: Harper & Row.

ANSWERS TO REVIEW QUESTIONS

1. CH_4, C_6H_6, $C_6H_{12}O_6$.

2. Protein, cellulose, polyethylene.

3. C_6H_{14} $(H(CH_2)_6H)$.

4. One.

5. C_3H_7.

6.
$$
\begin{array}{ccc}
H & & H \\
\diagdown & & \diagup \\
& C{=}C & \\
\diagup & & \diagdown \\
H & & H
\end{array}
$$

7. 2-Methylpropane.

8.
$$
\begin{array}{ccc}
H & H & H \\
| & | & | \\
H{-}C{-}C{=}CH \\
| \\
H
\end{array}
$$

9. They are identical.

10.
$$
\begin{array}{c}
CH_3 \\
| \\
H{-}C{=}C{-}H \\
| \\
CH_3
\end{array}
$$

11. Gutta-percha is a *trans*-polyisoprene; rubber is a *cis*-polyisoprene.

12. $HC{\equiv}CH$.

13. The hydroxyl group (OH).

14. $C_2H_5OC_2H_5$.

15. RCHO.

16. H_3CCH_2COOH.

17. $H_3CCOOC_2H_5$.

18. Secondary.

19. Two.

20. Both.

21. ⬡—C_2H_5.

22. Primary.

CHAPTER III

Introduction to the Science
of Giant Molecules

3.1   A BRIEF HISTORY OF CHEMICAL SCIENCE AND TECHNOLOGY

The science of giant molecules is relatively new, and many living polymer scientists have spent their entire lifetimes in the development of our present knowledge. Many of the developments in polymer science have taken place in the twentieth century, and most of these have occurred during the last half of this century.

Of course, humans have always been dependent on giant molecules, that is, starch, protein, and cellulose, for food, shelter, and clothing, but little was known about these essential products until recently. Organic chemistry was poorly understood until

1828, when Friedrich Wöhler demonstrated that it was possible to synthesize organic molecules.

Progress in organic chemistry was extremely slow until the 1850s and 1860s, when Friedrich August Kekulé discovered a new way to write the structural formulas for organic compounds. Many breakthroughs in organic chemistry occurred in the last years of the nineteenth century, when chemists recognized the practicability of synthesis and were able to write meaningful structural formulas for organic compounds.

Most giant molecules are organic polymers, but little progress was made in polymer science until the 1930s since few organic chemists accepted the concepts of polymer molecules giant molecules as formulated by Hermann Staudinger, he did not receive the Nobel prize for his elucidation of the molecular structure of polymers until 1953. Many of his contemporaries maintained that polymers were simply aggregates of smaller molecules held together by physical rather than chemical forces.

Nevertheless, in spite of the delays in the development of polymer science, there were several important empirical discoveries in the technology of giant molecules in the nineteenth century. Charles Goodyear and his brother Nelson separately transformed natural rubber (*Hevea braziliensis ulei*) from a sticky thermoplastic to a useful elastomer (vulcanized rubber, Vulcanite) and a hard thermoset plastic (Ebonite or Vulcacite), respectively, by heating natural rubber with controlled amounts of sulfur in the late 1830s. Thomas Hancock, who discovered the process of curing natural rubber via reverse research, that is, by an examination of the Goodyears' product, coined the term vulcanization after the Roman god Vulcanos (Vulcan).

Likewise, Christian F. Schönbein produced cellulose nitrate by the reaction of cellulose with nitric acid, and J. P. Maynard made collodion by dissolving the cellulose nitrate in a mixture of ethanol and ethyl ether in 1847. Collodion, which was used as a liquid court plaster (Nuskin), also served in the 1860s as Parkes and Hyatt's reactant for making celluloid (the first man-made thermoplastic) and Chardonnet's reactant in 1884 for making artificial silk (the first man-made fiber). This "Chardonnet silk" was featured at the World Exposition in Paris in 1889.

Although most of these early discoveries were empirical, they may be used to explain some terminology and theory in modern polymer science. It is important to note that, like the ancient artisans, all of these inventors converted naturally occurring polymers to more useful products. Thus, in the transformation of heat-softenable thermoplastic castilla rubber to a less heat-sensitive product, Charles Goodyear introduced a relatively small number of sulfur cross-links between the long individual chainlike molecules of natural rubber (polyisoprene).

Nelson Goodyear used sulfur to introduce many cross-links between the polyisoprene chains so that the product was no longer a heat-softenable thermoplastic but rather a heat-resistant thermoset plastic. Thermoplastics are two-dimensional (linear) molecules that may be softened by heat and returned to their original states by cooling, whereas thermoset plastics are three-dimensional network polymers that cannot be softened and reshaped by heating. The prefix thermo is derived from the Greek word *thermos*, meaning warm, and *plasticos* means to shape or form. Since these pioneers did not know what a polymer was, they had no idea of the complex changes that had taken place in the pioneer production of these useful man-made rubber, plastic, and fibrous products.

It was generally recognized by the leading organic chemists of the nineteenth century that phenol would condense with formaldehyde. Since they did not recognize the essential concept of functionality, that is, the number of available reactive sites in a molecule, Baeyer, Michael, Kleeburg, and other eminent organic chemists produced worthless cross-linked goos, gunks, and messes and then returned to their classical research on reactions of monofunctional reactants. However, by the use of a large excess of phenol, Smith, Luft, and Blumer were able to obtain useful thermoplastic condensation products.

Although there is no evidence that Leo Baekeland recognized the existence of macromolecules, he did understand functionality, and by the use of controlled amounts of trifunctional phenol and difunctional formaldehyde he produced thermoplastic resins that could be converted to thermoset plastics (Bakelite). Other polymers had been produced in the laboratory before 1910, but Bakelite was the first truly synthetic plastic. The fact that the processes used today are essentially the same as those described in the original Baekeland patents demonstrates this inventor's ingenuity and knowledge of the chemistry of the condensation of trifunctional phenol with difunctional formaldehyde.

Prior to World War I, celluloid, shellac, Galalith (casein), Bakelite, cellulose acetate, natural rubber, wool, silk, cotton, rayon, and glyptal polyester coatings, as well as bitumen/asphalt, coumarone/indene, and petroleum resins, were all commercially available. However, as shown chronologically in Table 3.1, because of the lack of knowledge of polymer science, there were few additional significant developments in polymer technology prior to World War II.

The following advice was given to Dr. Staudinger by his colleagues in the 1920s: "Dear Colleague: Leave the concept of large molecules well alone. . . . There can be no such thing as a macromolecule." Fortunately, this future Nobel laureate disregarded their unsolicited advice and laid the groundwork for modern polymer science in the 1920s when he demonstrated that natural and synthetic polymers were not aggregates, like colloids, or cyclic compounds, like cyclohexane, but instead were long, chainlike molecules with characteristic end groups. In 1928, Kurt H. Meyer and Herman F. Mark reinforced Staudinger's concepts by using x-ray techniques to determine the dimensions of the crystalline areas of macromolecules in cellulose and natural rubber.

While Staudinger was arguing the case for his concepts of macromolecules in Germany, a Harvard professor working for DuPont was actually producing giant molecules in accord with Staudinger's concepts. In the mid-1930s Wallace Carothers, along with Julian Hill, synthesized a polyamide that they called nylon 6,6. In contrast to Chardonnet's fiber, which was made by the regeneration of naturally occurring cellulose, nylon fiber was a completely synthetic polymer.

Nylon was produced by the condensation of two difunctional reactants, namely, a dicarboxylic acid and a diamine. As shown by the following empirical equation, each product produced in the stepwise reactions was capable of further reaction to produce a linear giant molecule:

$$A—A + B—B \rightarrow A—A—B—B—$$

where $A = —COOH$ and $B = —NH_2$

$$A—A—B—B + A—A—B—B \rightarrow A—A—B—B—A—A—B—B.$$

Table 3.1   Chronological Development of Commercial Polymers

Date	Material (Brand/Trade Name and/or Inventor)	Typical Application
Before 1800	Cotton, flax, wool and silk fibers; bitumen caulking materials; glass and hydraulic cements, leather, cellulose sheet (paper); balata, shellac, gutta-percha, *Hevea braziliensis*.	
1839	Vulcanization of rubber (Charles Goodyear)	Tires
1846	Nitration of cellulose (Schönbein)	Coatings
1851	Ebonite (hard rubber; Nelson Goodyear)	Electrical insulation
1860	Molding of shellac and gutta-percha	Electrical insulation
1868	Celluloid (CN: Hyatt)	Combs, mirror, frames
1889	Regenerated cellulosic fibers (Chardonnet)	Fabric
	Cellulose nitrate photographic films (Reichenbach)	Pictures
1890	Cuprammonia rayon fibers (Despeisses)	Fabric
1892	Viscose rayon fibers (Cross, Bevan, and Beadle)	Fabric
1893	Cellulose recognized as a polymer (E. Fischer)	
1907	Phenol–formaldehyde resins (PF: Bakelite; Baekeland)	Electrical
1908	Cellulose acetate photographic films (CA)	
1912	Regenerated cellulose sheet (cellophane)	Sheets, wrappings
1923	Cellulose nitrate automobile lacquers (Duco)	Coatings
1924	Cellulose acetate fibers	
	Concept of macromolecules (H. Staudinger)	
1926	Alkyd polyesters (Kienle)	Electrical insulators
1927	Polyvinyl chloride (PVC; Semon; Koroseal)	Wall covering
1927	Cellulose acetate sheet and rods	Packaging Films
1929	Polysulfide synthetic elastomer (Thiokol; Patrick)	Solvent-resistant rubber
1929	Urea–formaldehyde resins (UF)	Electrical switches and parts
1931	Polymethyl methacrylate plastics (PMMA; Plexiglas; Rohm)	Display signs
1931	Polychloroprene elastomer (Neoprene; Carothers)	Wire coatings
1933	Polyethylene (LDPE; Fawcett and Gibson)	Cable coating, packaging, squeeze bottles
1935	Ethylcellulose	Moldings
1936	Polyvinyl acetate (PVAc)	Adhesives
1936	Polyvinyl butyral (PVB)	Safety glass
1937	Polystyrene (PS)	Kitchenware, toys, foam
1937	Styrene–butadiene (Buna-S; SBR), acrylonitrile (Buna-N), copolymer elastomers (NBR)	Tire treads
1938	Nylon 6,6 fibers (Carothers)	Fibers
1938	Fluorocarbon polymers (Teflon; Plunkett)	Gaskets, grease-repellent coatings
1939	Melamine–formaldehyde resins (MF)	Tableware
1938	Copolymers of vinyl chloride and vinylidene chloride (Pliovic)	Films, coatings
1939	Polyvinylidene chloride (PVDC; Saran)	Films, coatings

Table 3.1 *(Continued)*

Date	Material (Brand/Trade Name and/or Investor)	Typical Application
1940	Isobutylene–isoprene elastomer (butyl rubber; Thomas and Sparks)	Adhesives, coatings, caulkings
1941	Polyester fibers (PET; Whinfield and Dickson)	Fabric
1942	Unsaturated polyesters (Foster and Ellis)	Boat hulls
1942	Acrylic fibers (Orlon; Acrylan)	Fabrics
1943	Silicones (Rochow)	Gaskets, caulkings
1943	Polyurethanes (Baeyer)	Foams, elastomers
1944	Styrene–acrylonitrile–maleic anhydride, engineering plastic (Cadon)	Moldings, extrusions
1947	Epoxy resins (Schlack)	Coatings
1948	Copolymers of acrylonitrile butadiene and styrene (ABS)	Luggage, electrical devices
1955	Polyethylene (HDPE; Hogan, Banks, and Ziegler)	Bottles, film
1956	Polyoxymethylenes (acetals)	Moldings
1956	Polypropylene Oxide (Hay; Noryl)	Moldings
1957	Polypropylene (Hogan, Banks, and Natta)	Moldings, carpet fiber
1957	Polycarbonate (Schnell and Fox)	Appliance parts
1959	cis-Polybutadiene and cis-polyisoprene elastomers	Rubber
1960	Ethylene–propylene copolymer elastomers (EPDM)	Sheets, gaskets
1962	Polyimide resins	High-temperature films and coatings
1965	Polybutene	Films, pipe
1965	Polyarylsulfones	High-temperature thermoplastics
1965	Poly-4-methyl-1-pentene (TPX)	Clear, low-density (0.83 g/l) moldings
1965	Styrene–butadiene block copolymers (Kraton)	Shoe soles
1970	Polybutylene terephthalate (PBT)	Engineering plastic
1970	Ethylene–tetrafluoroethylene copolymers	Wire insulation
1971	Polyphenylene sulfide (Ryton; Hill and Edmonds)	Engineering plastic
1971	Hydrogels, hydroxyacrylates	Contact lenses
1972	Acrylonitrile barrier copolymers (BAREX)	Packaging
1974	Aromatic nylons (Aramids; Kwolek and Morgan)	Tire cord
1980	Polyether ether ketone (PEEK; Rose)	High-temperature service
1982	Polyether imide (Ultem)	High-temperature service

As a result of Carothers' contributions and subsequent discoveries, polymerization, that is, the production of giant molecules from small molecules, has been recognized as one of the greatest discoveries of all time. As was true in the nineteenth century, the art usually preceded the science, but many developments in the mid-twentieth century were based on macromolecular concepts championed by Staudinger, Mark, and Carothers.

Many discoveries in polymer technology were serendipitous or by chance, but in many cases scientists applied polymer science concepts to these accidental discoveries to produce useful commercial products. Among these accidental discoveries are the following: J. C. Patrick obtained a rubberlike product (Thiokol) when he was attempting to synthesize an antifreeze in 1929. Fawcett and Gibson heated ethylene under very high pressure, in the presence of traces of oxygen, and obtained polyethylene (LDPE) in 1933. When the gaseous tetrafluoroethylene did not escape through the open valve in a pressure cylinder, Roy J. Plunkett cut open the cylinder and found a solid product that was polytetrafluoroethylene (Teflon) in 1938.

The leading polymer scientists of the 1930s agreed that all polymers were chainlike molecules and that the viscosities of solutions of these macromolecules were dependent on the size and shape of the molecules in these solutions. Although the large-scale production of many synthetic polymers was accelerated by World War II, it must be recognized that the production of these essential products was also dependent on the concepts developed by Staudinger, Carothers, Mark, and other polymer scientists prior to World War II.

3.2 POLYMERIZATION

In addition to the step reaction polymerization described in Section 3.1, synthetic polymers may also be prepared by chain reactions, that is, addition polymerization reactions. In step reaction polymerization, difunctional reactants, such as ethylene glycol and phthalic acid, react to produce products with reactive end groups that are capable of further reaction:

$$HO{-}(CH_2)_2{-}OH + HO{-}\overset{O}{\overset{\|}{C}}{-}\bigcirc{-}\overset{O}{\overset{\|}{C}}{-}OH \rightarrow$$

$$H_2O + HO{-}(CH_2)_2{-}O{-}\overset{O}{\overset{\|}{C}}{-}\bigcirc{-}\overset{O}{\overset{\|}{C}}{-}OH$$

Polyesters, nylons (polyamides), polyurethanes, epoxy resins, phenolic resins, and melamine resins are produced by step reaction polymerization.

Most elastomers (rubbers), some fibers (polyacrylonitrile), and many plastics are produced by chain reaction polymerization. These reactions include three steps: initiation, propagation, and termination. Polymerization chain reactions may be initiated by anions, such as butyl anions $(C_4H_9{:}^-)$, by cations, such as protons (H^+), or by free radicals, such as the benzoyl free radicals $(C_6H_5COO\cdot)$. As shown in the following equations, the initiator, such as a free radical $(R\cdot)$, adds to a vinyl monomer, such as vinyl chloride, to produce a new free radical.

$$\text{R}\cdot\ +\ \overset{\text{H}\ \text{H}}{\underset{\text{H}\ \text{Cl}}{\text{C}{\cdots}\text{C}}}\ \rightarrow\ \text{RC}{-}\overset{\text{H}\ \text{H}}{\underset{\text{H}\ \text{Cl}}{\text{C}}}\cdot$$

Free radical	Vinyl chloride monomer	Vinyl chloride radical

Then, as shown by the following equation, the new free radical adds to another vinyl chloride monomer molecule to produce a dimer radical, and this reaction continues rapidly and sequentially to produce larger and larger macroradicals (n = number of repeating units).

$$
\begin{array}{ccc}
\underset{\substack{\text{Vinyl}\\\text{chloride}\\\text{radical}}}{RC-C\cdot} & + & \underset{\substack{\text{Vinyl}\\\text{chloride}\\\text{monomer}}}{C\cdot\cdot C} \rightarrow \underset{\substack{\text{Dimer}\\\text{radical}}}{-R-C-C-C-C\cdot}
\end{array}
$$

or

$$
\underset{\substack{\text{Vinyl}\\\text{chloride}\\\text{radical}}}{RC-C\cdot} + \underset{\substack{\text{Vinyl}\\\text{chloride}\\\text{monomer}}}{nC\cdot\cdot C} \rightarrow \underset{\text{Macroradical}}{R-\left[C-C\right]_n C-C\cdot}
$$

The reaction may be terminated by the collision of two macroradicals to produce a dead polymer (inactive polymer) in a coupling reaction or the macroradical may abstract a hydrogen atom from another molecule, called a telogen, to produce a dead polymer and a new radical.

$$
\underset{\substack{\text{Vinyl chloride}\\\text{macroradical}}}{R\left[C-C\right]_n C-C\cdot} + \underset{\substack{\text{Vinyl chloride}\\\text{macroradical}}}{\cdot C-C\left[C-C\right]_n R} \rightarrow \underset{\text{Dead polymer}}{R\left[C-C\right]_{n+1} \left[C-C\right]_{n+1} R}
$$

$$
\underset{\substack{\text{Vinyl chloride}\\\text{macroradical}}}{R\left[C-C\right]_n C-C\cdot} + \underset{\substack{\text{Dodecyl}\\\text{mercaptan}}}{HSC_{12}H_{25}} \rightarrow \underset{\text{Dead polymer}}{R\left[C-C\right]_{n+1} H} + \underset{\substack{\text{New free}\\\text{radical}}}{\cdot SC_{12}H_{25}}
$$

It is important to note that although the step reactions are relatively slow, the chain reactions are extremely rapid. Thus, a free radical may react with as many as 1000 or more monomer molecules in a sequential reaction in a fraction of a second. These reactions are exothermic (give off heat) and hence must be controlled by removing the heat that is produced.

3.3 IMPORTANCE OF GIANT MOLECULES

There are numerous ways to measure the importance of a specific discipline. One way is to consider its pervasiveness. Polymer science and technology are essential for our housing, clothing, and food and health needs, for polymeric materials are common and integral in our everyday lives. We are concerned with natural polymers, such as proteins, in meats and dairy products, and starches in our vegetables, and we use them as building blocks and agents of life. Synthetic polymers serve as floor coverings, laminated plastics, clothing, gasoline hoses, tires, upholstery, records, dinnerware, and many other uses.

Another way to measure the importance of a specific discipline is to consider the associated work force. The U.S. polymer industry employs more than 1 million people indirectly and directly. This corresponds favorably to the employment in the entire metal-based industry. Further, about one-half of all professional chemists and chemical engineers are engaged in polymer science and technology, including monomer and polymer synthesis and polymer characterization, and this need will increase as the industry is predicted to double in size before the beginning of the twenty-first century.

Still another way to measure the importance of an industry is to study its growth. The number of new opportunities in polymer science and technology is on a par with those in the fastest growth areas. A fourth possible consideration is the marketplace influence. After food-related materials, synthetic polymers comprise the largest American export market, both bulkwise and moneywise. Synthetic polymers are produced at an annual rate of over 150 lb for every man, woman, and child in the United States, and this $120 billion industry represents an outlay of about $1000 for each American family.

A fifth consideration is the influence of this science with respect to other disciplines. The basic concepts and applications of polymer science apply equally to natural and synthetic polymers, and thus are important in medical, health, nutrition, engineering, biology, physics, mathematics, computer, space, and ecological sciences and technology.

3.4 POLYMER PROPERTIES

There is a basic question that needs to be answered. Why has polymer science and technology grown into such a large industry, and why has nature chosen the macromolecule to be the very fabric of life and material construction? The obvious answer, and only the tip of the iceberg, is molecular size. Other answers relate to physical and chemical properties exhibited by polymers. We will briefly describe two of these properties.

A. Memory

Some polymers, such as rubber, return to their original shape and dimensions after being distorted. This "memory" is related to physical and/or chemical bonds (cross-links) between polymer chains for large distortions and to the high cumulative secondary bonding forces present between chains (intermolecular forces) for small distortions. The degree of cross-linking affects many physical properties of polymers. Thus, many elastomers, including natural rubber, change from soft to hard as the amount of cross-linking increases from 1 to 1000 units in the polymer chain.

In nature, this "memory" is utilized to restrict flow of materials and to transmit information. Memory is also exhibited by the ability of certain macromolecules to pass on impulses (nerve transmissions and electrical conductivity).

B. Solubility

The large size of polymer molecules contributes to their relatively poor solubility compared to smaller molecules. This fact is dependent on both kinetic (motion) and thermodynamic (energy) factors. Solvent molecules are not able to readily penetrate to the interior of a polymer molecule, and undissolved segments of a polymer prevent the continuous "moving away" of the dissolved segments.

The number of geometric arrangements of connected polymer segments in a chain is considerably less than if the segments were free to act as individual species. Thus, there is a decrease in the tendency to achieve random orientation, thereby decreasing the tendency for a polymer to dissolve. Nature tends to go from ordered to disordered, that is, toward random arrangements or entropy, if left uncontrolled. A good example is the tendency of our rooms to get messy if we do not expend a great deal of effort (energy) to prevent or correct this situation.

The resistance of a polymer to dissolve permits pseudosolutions or semisolubility to occur. In animals, the proteins retain flexibility through entrapment of water. Thus, moist skin is flexible, and organs can stretch and bend. In plants, water permits leaves and grass to "flow in the breeze." Synthetic polymers often utilize added plasticizing agents to achieve flexibility.

An additional reason why both nature and industry have chosen to "major in polymers" is the abundance of the building blocks of polymers readily found in nature, making polymers inexpensive and readily constructible. It is interesting to note that carbon is one of the few elements that readily undergoes catenation (forming long chains) and that both natural and synthetic polymers have high carbon content. Further, this catenation of carbon atoms can be both controlled and varied, permitting both synthesis of materials with reproducible properties and polymers with quite divergent properties.

3.5   A FEW DEFINITIONS OF POLYMERS (MACROMOLECULES)

Briefly, polymer science is the science that deals with large molecules consisting of atoms connected by covalent chemical bonds. Polymer technology is the practical application of polymer science. The word polymer is derived from the Greek *poly* (many) and *meros* (parts). The word macromolecule, that is, giant molecule, is often utilized synonymously for polymer and vice versa.

Some scientists differentiate between the two terms by using the word macromolecule to describe large molecules such as DNA and proteins, which cannot be derived from a single, simple unit, and using the term polymer to describe a large molecule such as polystyrene, which is composed of repetitive styrene units. This differentiation is not always observed and will not be used in this text. The process of forming a polymer is called polymerization.

The degree of polymerization (DP) or average degree of polymerization $(\overline{DP})$ is the number of repeating units (mers) in a polymer chain. The term chain length is used as a synonym for DP. The DP of a dimer is 2, that of a trimer is 3, etcetera. Chains with DPs below 10 to 20 are referred to as oligomers (small units) or telomers. Many polymer properties are dependent on chain length, but the change in polymer properties with changes in DP, for most commercial polymers, is small when the DP is greater than 100.

3.6   POLYMER STRUCTURE

The terms configuration and conformation are often confused. Configuration refers to arrangements fixed by chemical bonding, which cannot be altered except through primary bond breakage. Terms such as head to tail, D and L isomers, and cis and trans isomers refer to configurations of isomers in a chemical species. Conformation, on the other hand, refers to arrangements around single primary bonds. Polymers in solutions or in melts continuously undergo conformational changes, that is, changes in shape. The principal difference between a hard-boiled egg and a raw egg is an irreversible conformational change.

Monomer units in a growing chain usually form what is referred to as a head to tail arrangement in which the repeating polymer unit ($-CH_2-CHX-$) in the polymer chain can be shown simply as

$$-CH_2-\underset{\underset{X}{|}}{CH}-CH_2-\underset{\underset{X}{|}}{CH}-CH_2-\underset{\underset{X}{|}}{CH}-$$

Even with head to tail configuration, a variety of structures are possible. For illustrative purposes, we will consider possible combinations derived from the homopolymerization of monomer A and the copolymerization of A with another monomer B. Homopolymerization involves one repetitive monomeric unit in the chain.

$$nA \rightarrow +A-A-A-A-A)_n$$
Linear polymer

$$n\Lambda-(A-A-\underset{\underset{\underset{\underset{A)_x}{|}}{A}}{\overset{|}{A}}-A-A-A-\underset{\underset{\underset{\underset{A)_x}{|}}{A}}{\overset{|}{A}}-A-A)_{n-2x}$$
Branched polymer

$$2nA + \text{cross-linking agent} \rightarrow \begin{array}{c} | \quad\quad | \\ +A-A-A-A)_n \\ | \\ +A-A-A-A)_n \\ | \quad\quad | \end{array}$$
Cross-linked polymer

Copolymerization involves more than one monomeric unit in the chain, and the copolymer structure may differ:

$$—A—A—B—A—B—B—A—A—B—A—B—$$

Linear random copolymer

$$—A—B—A—B—A—B—A—B—A—B—$$

Linear alternating copolymer

$$—A—A—A—A—B—B—B—B—B—B—A—A—A—A—B—B—$$

Linear block copolymer

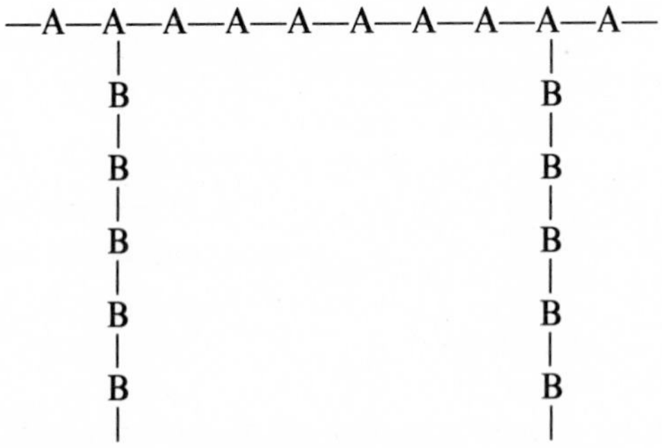

$$—A—A—A—A—A—A—A—A—A—A—$$

Graft copolymer

It is currently possible to tailor-make polymers of these structures to obtain almost any desired property by utilizing combinations of many of the common monomers.

The term configuration refers to structural regularity with respect to the substituted carbon atoms in the polymer chains. For linear homopolymers derived from monomers of the form $H_2C{=}CHX$, configurations from monomeric unit to monomeric unit can vary randomly (atactic) with respect to the geometry (configurations) about the carbon atom to which the pendant group X is attached or can vary alternately (syndiotactic), or be alike in having all the pendant X groups placed on the same side of a backbone plane (isotactic). These configurations are shown by

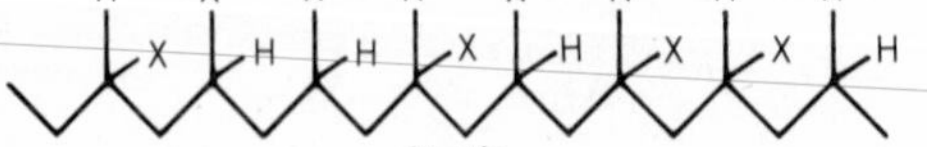

Atactic 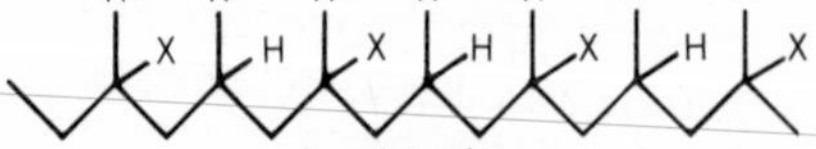   Syndiotactic

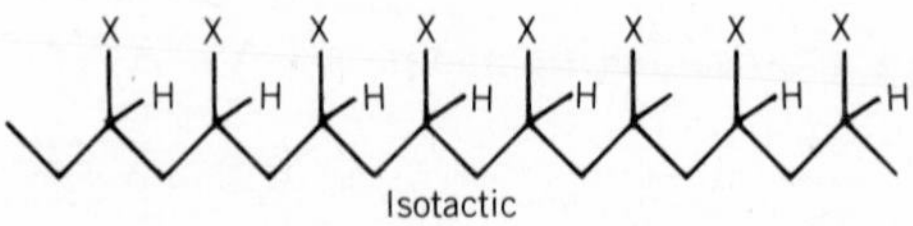

Isotactic

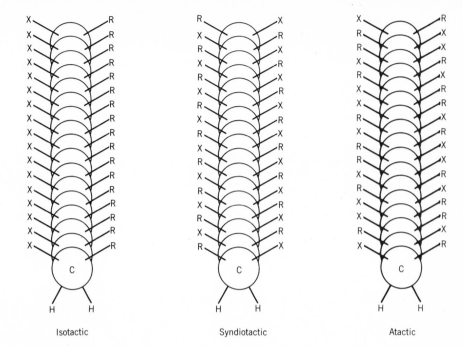

Isotactic   Syndiotactic   Atactic

Another type of stereogeometry is illustrated by polymers of 1,4-dienes, such as 1,4-butadiene, in which rotation in the polymer is restricted by the presence of the double bond. Polymerization can occur through a single static double bond to produce 1,2 molecules that can exist in the stereoregular forms of isotactic and syndiotactic and irregular, atactic forms. The stereoregular forms are rigid, crystalline materials, whereas the atactic forms are soft, amorphous elastomers.

cis-1,4   *trans*-1,4

$$
\begin{array}{cc}
\underset{H}{\overset{H}{\Big|}}\quad \underset{}{\overset{R}{\Big|}} & \underset{H}{\overset{H}{\Big|}}\quad \underset{}{\overset{H}{\Big|}} \\
-\text{C}-\text{C}- \quad H & -\text{C}-\text{C}- \quad H \\
\end{array}
$$

-1,2-   -3,4-

$$
\begin{array}{c}
\text{CH}_3 \\
|\\
\text{CH}_2\!=\!\text{C}\!-\!\text{CH}\!=\!\text{CH}_2 \rightarrow
\end{array}
$$

2-Methyl-1,3-butadiene
(isoprene)

$$
\left[\begin{array}{c}\text{CH}_3\\|\\ \text{CH}_2\text{C}\\|\\ \text{CH}\\ \|\\ \text{CH}_2\end{array}\right] + \left[\begin{array}{c}\text{CH}_2\text{CH}\\|\\ \text{CH}_3\!-\!\text{C}\\ \|\\ \text{CH}_2\end{array}\right]
$$

1,2   3,4

$$
+ \left[\begin{array}{c}\text{CH}_2 \quad\quad \text{CH}_2\\ \diagdown\;\;C\!=\!C\;\;\diagup \\ \text{CH}_3 \quad\quad H\end{array}\right] + \left[\begin{array}{c}\text{CH}_2 \quad\quad H\\ \diagdown\;\;C\!=\!C\;\;\diagup \\ \text{CH}_3 \quad\quad \text{CH}_2\end{array}\right]
$$

cis-1,4   *trans*-1,4

Polyisoprenes

Polymerization of dienes can also produce polymers in which carbon moieties are on the same side of the newly formed double bond (cis) or on the opposite side (trans). The cis isomer of poly-1,4-butadiene is a soft elastomer with a glass transition temperature (T_g) of $-108°C$. The glass transition temperature of the isomer of poly-1,4-butadiene is $-83°C$. The glass transition temperature is the temperature at which a glassy polymer becomes flexible when heated. T_g is a characteristic value for amorphous (noncrystalline) polymers.

3.7 MOLECULAR WEIGHTS OF POLYMERS

Polymerization reactions may produce polymer chains with different numbers of repeating units or degrees of polymerization (DP). Most synthetic polymers and many naturally occurring polymers consist of molecules with different molecular weights and are said to be polydisperse. In contrast, specific proteins and nucleic acids consist of molecules with a specific molecular weight and are said to be monodisperse.

Since typical molecules with DPs less than the critical value required for chain entanglement are weak, it is apparent that certain properties are related to molecular weight. The melt viscosity of amorphous polymers is dependent on the molecular weight distribution. In contrast, density, specific heat capacity, and refractive index are essentially independent of the molecular weight at molecular weight values above the critical molecular weight, which is typically a DP of about 100.

Viscosity is the resistance of a substance to flow when subjected to a shear stress. When applied to solutions of polymers and melts, viscosity is measured by a device called a viscometer. Shear or tangential stress is a force that is applied parallel to the surface, like spreading butter on a piece of toast.

The melt viscosity (η) is usually proportional to the 3.4 power of the average molecular weight at values above the critical molecular weight required for chain entanglement, that is, $\eta = \bar{M}^{3.4}$. ($\bar{M}$ or $\overline{DP}$ represents an average value for poly-disperse macromolecules.) The melt viscosity increases rapidly as the molecular weight increases, and hence more energy is required for the processing and fabrication of these large molecules. However, as shown in Figure 3.1, the strength of a polymer increases as its molecular weight increases, then tends to level off.

Thus, although a value above the threshold molecular weight value (TMWV) is essential for most practical applications, the additional cost for energy required for processing higher-molecular-weight polymers is seldom justified. Accordingly, it is customary to establish a commercial polymer range above the TMWV but below the extremely high molecular weight range. However, it should be noted that since toughness increases with molecular weight, polymers such as ultrahigh-molecular-weight polyethylene (UHMWPE) are used for the production of strong items such as trash barrels.

The value of TMWV is dependent on the glass transition temperature, the intermolecular forces, expressed as cohesive energy density (CED) of amorphous polymers, the extent of crystallinity in crystalline polymers, and the extent of reinforcement present in polymer composites. Although a low-molecular-weight amorphous polymer may be satisfactory for use as a coating or adhesive, a much

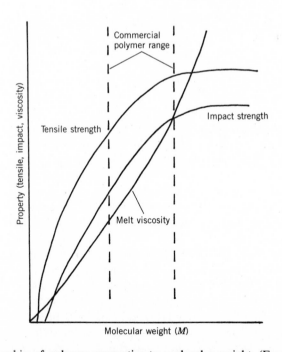

Figure 3.1. Relationship of polymer properties to molecular weight. (From *Introduction to Polymer Chemistry* by R. B. Seymour, McGraw-Hill, New York, 1971. Used with permission of McGraw-Hill Book Company.)

higher $\overline{DP}$ value may be required if the polymer is used as an elastomer or plastic. With the exception of polymers with highly regular structures, such as isotactic polypropylene, strong hydrogen intermolecular bonds are required for fibers. Because of the higher CED values resulting from stronger intermolecular forces, lower $\overline{DP}$ values are usually satisfactory for polar polymers used as fibers.

3.8   POLYMERIC TRANSITIONS

Polymers can exhibit a number of different conformational changes, each change accompanied by differences in polymer properties. Two major transitions are the glass transition temperature (T_g), which is dependent on local, segmental chain mobility in the amorphous regions of a polymer, and the melting point (T_m), which is dependent on large-scale chain mobility. The T_m is called a first-order transition temperature, whereas T_g is often referred to as a second-order transition temperature. The values for T_m are usually 33 to 60% greater than those for T_g, with T_g values being low for typical elastomers and flexible polymers and higher for hard amorphous plastics. The T_g for silicones is $-190°F$ and that for E-glass is $1544°F$. The T_g values for most other polymers are in between these extremes.

3.9   TESTING OF POLYMERS

Public acceptance of polymers is usually associated with an assurance of quality based on a knowledge of successful, long-term, and reliable tests. In contrast, much of the dissatisfaction with synthetic polymers is related to failures that possibly could have been prevented by proper testing, design, and quality control. The American Society for Testing and Materials (ASTM), through its committees D-1 on paint and D-20 on plastics, for example, has developed many standard tests that are available to all producers and large-scale consumers of finished polymeric materials. There are also testing and standards groups in many other technical societies throughout the world.

Much of the testing performed by the industry is done to satisfy product specifications using standardized tests for stress–strain relationships, flex life, tensile strength, abrasion resistance, moisture retention, dielectric constant, hardness, thermal conductivity, and so on. New tests are continually being developed, submitted to ASTM, and, after adequate verification through "round robin" testing, finally accepted as standard tests.

Each standardized ASTM test is specified by a unique combination of letters and numbers, along with exacting specifications regarding data gathering, instrument design, and test conditions, thus making it possible for laboratories throughout the world to compare data with confidence. The Izod test, a popular impact test, has the ASTM number D256-56 (1961), the latter number being the year it was first accepted. The ASTM instructions for the Izod test specify test material shape and size, exact specifications for the test equipment, detailed description of the test procedure, and how results should be reported. More complete information on testing and characteristics of polymers is provided in Chapter 8.

3.10 CHEMICAL NAMES OF POLYMERS

The International Union of Pure and Applied Chemistry (IUPAC) formed a subcommission on Nomenclature of Macromolecules in early 1952 and has continued to periodically study the various topics related to polymer nomenclature. Many of the names that scientists employed for giant molecules are *source-based*, that is, they are named according to the common name of the repeating units in the giant molecule, preceded by the prefix poly. Thus, the name polystyrene (PS) is derived from the common name of its repeating unit, and the name poly(methyl methacrylate) (PMMA) is derived from the name of its repeating unit:

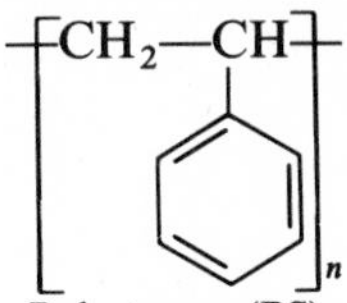

Polystyrene (PS)

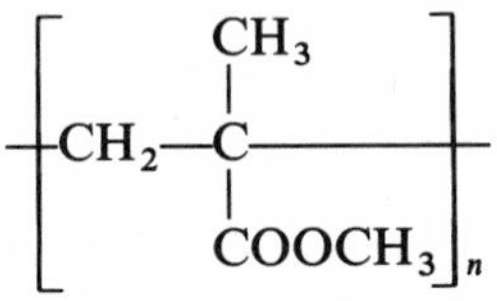

Polymethyl methacrylate (PMMA)

Little rhyme or reason is associated with *common-based* names. Some common names are derived from the "discoverer," for example, Bakelite was commercialized by Leo Baekeland in 1905. Others are based on the place of origin, such as *Hevea braziliensis*, literally "rubber from Brazil," the name given for natural rubber (NR).

For some important groups of polymers, special names and systems of nomenclature were invented. For example, the nylons were named according to the number of carbons in the diamine and carboxylic acid reactants (monomers) used in their synthesis. The nylon produced by the condensation of 1,6-hexamethylenediamine (6 carbons) and sebacic acid (10 carbons) is called nylon 6,10. Industrially, nylon 6,10 has been designated nylon 6,10, nylon 6 10, or 6-10 nylon.

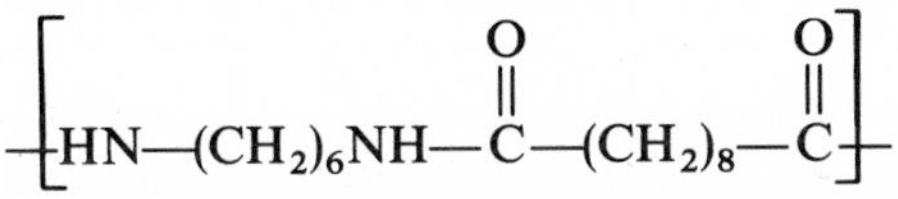

Polyhexamethylenesebacamide (nylon 6,10)

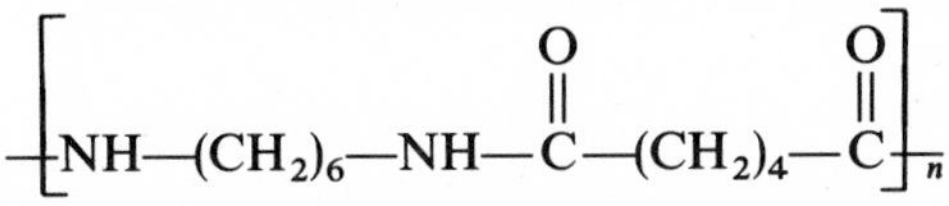

Polyhexamethyleneadipamide (nylon 6,6)

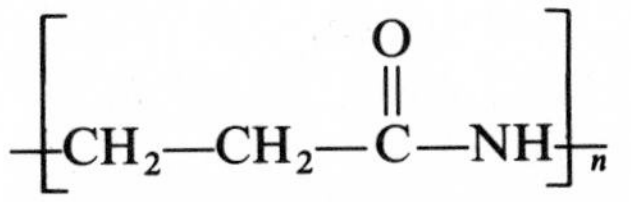

Polyalanine (nylon-3)

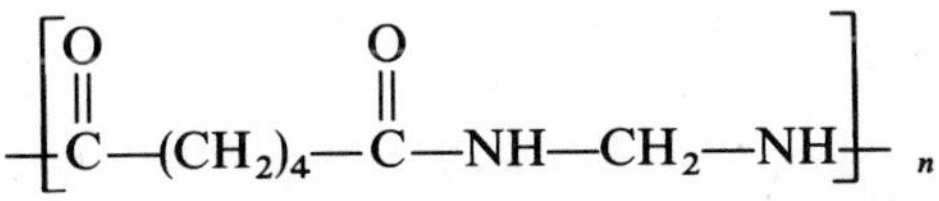

Polymethyleneadipamide (nylon 1,6)

Similarly, the polymer produced from the single reactant caprolactam (6 carbons) is called nylon-6. The structure-based name for nylons is polyamide because of the presence of the amide grouping. Thus, scientists are talking about the same family of polymers if they are talking about nylons or polyamides.

Abbreviations are also widely employed. Thus PS represents polystyrene and PVC represents polyvinyl chloride. The media have given abbreviations to some common monomers such as vinyl chloride (VCM) and styrene (SM).

$$\left[\begin{array}{c} CH_2CH \\ | \\ Cl \end{array} \right]_n$$

Polyvinyl chloride (PVC)

3.11   TRADE NAMES OF POLYMERS

Many firms use trade names to identify specific polymeric products of their manufacture. However, generic names, such as rayon, cellophane, polyesters, and polyurethane, are used more universally. For example, Fortrel polyester is a polyethylene terephthalate (PET) fiber produced by Fiber Industries, Inc. The generic term polyester indicates that the composition of this fiber is based on a condensation product of a dihydric alcohol (glycol, $R(OH)_2$) and terephthalic acid (an aromatic dicarboxylic acid, $Ar(COOH)_2$). Many generic names for fibers, such as polyester, are defined by the Textile Fiber Products Identification Act. This act also controls the composition of fibers such as rayon and polyurethane. An extensive list of trade names is found at the end of the book.

3.12   IMPORTANCE OF DESCRIPTIVE NOMENCLATURE

Unfortunately, there are also many trivial names that tend to cause some confusion. For example, when a nonscientist says alcohol, he or she means ethanol, which is just one of hundreds of alcohols. Likewise, the nonscientist uses the term sugar to indicate a specific sugar (sucrose), salt to indicate a specific salt (sodium chloride), and vinyl to indicate PVC.

The uninformed consumer may not recognize that there are numerous alcohols, sugars, salts, vinyl polymers, synthetic fibers, and plastics. After reading subsequent chapters, you will be aware of the many different polymers whose properties cover the entire spectrum, from insulators to conductors, from liquids to solids, from water-soluble to water-insoluble, and from those that soften at room temperature to those that can be used in combustion engines. Additional information on plastics, fibers, and elastomers is given in Table 3.2.

3.13   MARKETPLACE

Giant molecules account for most of what we are (proteins, nucleic acids, polysaccharides, enzymes, DNA, RNA), what we eat (food), and the society in which we

Table 3.2 Structures of Industrially Important Addition Polymers

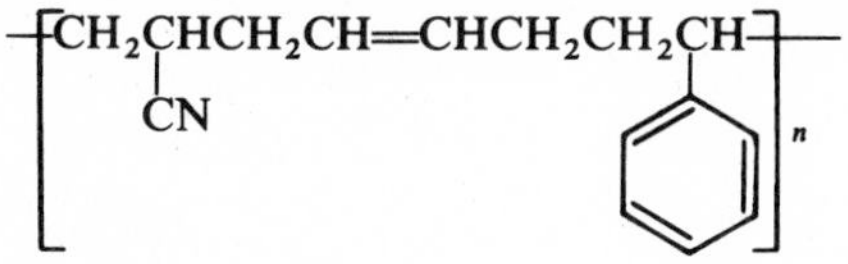

$$\left[CH_2CHCH_2CH{=}CHCH_2CH_2CH\right]_n$$
|
CN

Acrylonitrile–butadiene–styrene terpolymer (ABS)

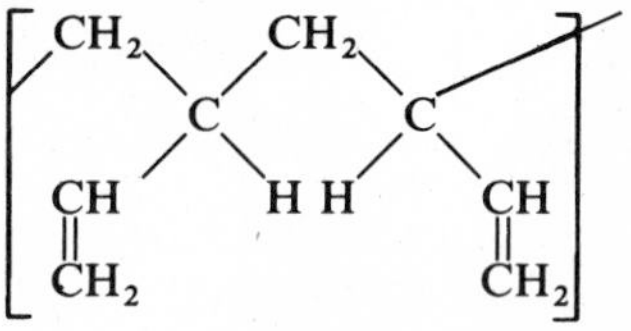

1,2-Polybutadiene

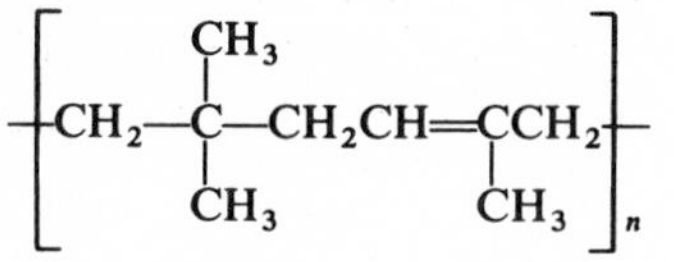

Butyl rubber

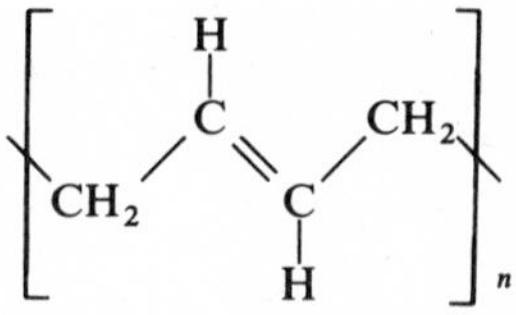

$trans$-1,4-Polybutadiene

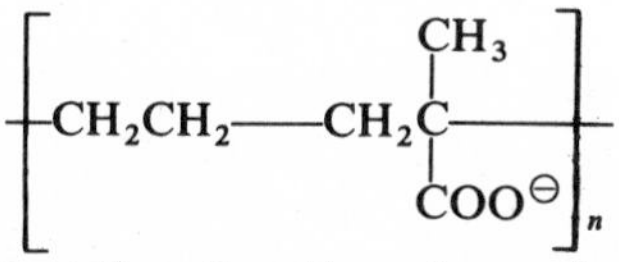

Ethylene–methacrylic acid copolymers (ionomers)

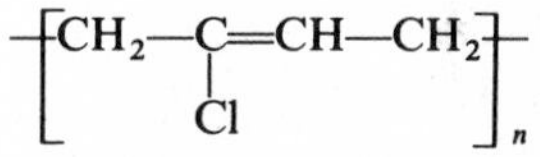

Polychloroprene

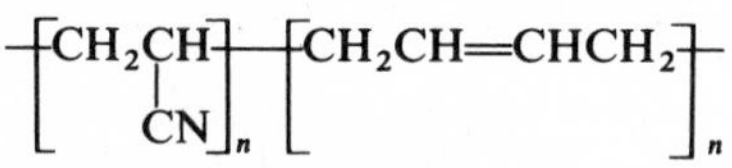

Nitrile rubber (NBR)

$\{CH_2CH_2\}_n$

Polyethylene (PE)

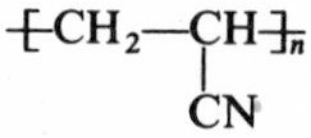

Polyacrylonitrile

$\{OCH_2CH_2\}_n$

Polyethylene glycol (PEG)

$\{(CH_2)_6{-}S\}_n$

Polyhexamethylene thioether

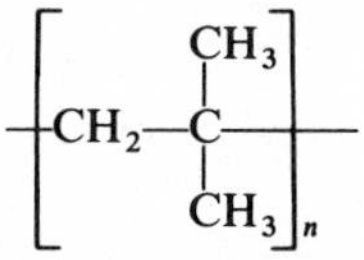

Polyisobutylene (PIB)

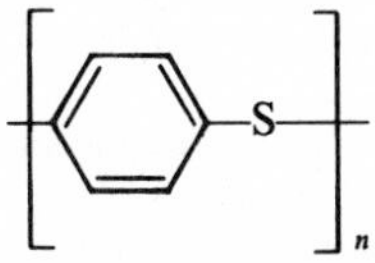

Polyphenylene sulfide (PPS)

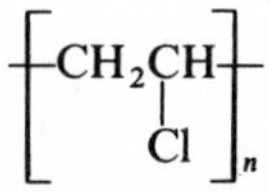

Polyvinyl chloride (PVC)

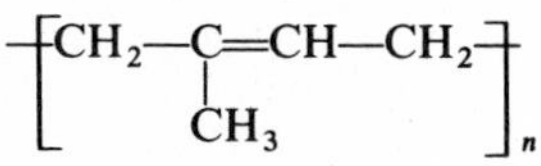

Polyisoprene

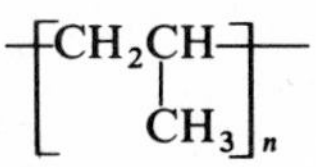

Polypropylene (PP)

$\{CH_2CCl_2\}_n$

Polyvinylidene chloride

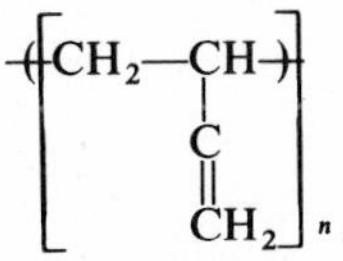

3,4-Polyisoprene

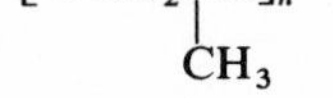

$\{OCH_2CH\}_n$
|
CH$_3$

Polypropylene glycol (PPG)

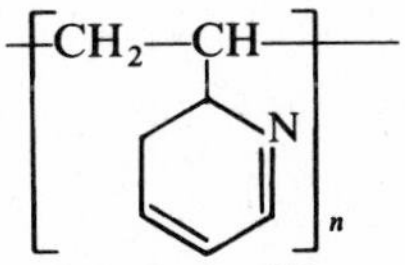

Polyvinylpyridine

Table 3.2 (*Continued*)

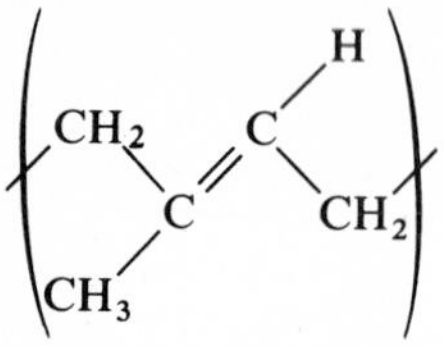

trans-1,4-Polyisoprene

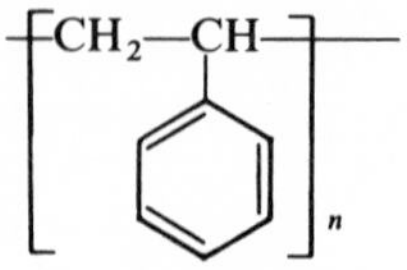

Polystyrene (PS)

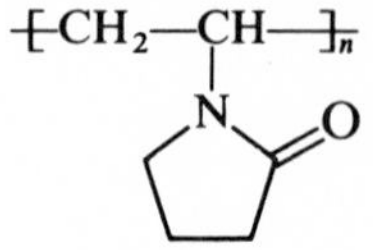

Polyvinylpyrrolidone

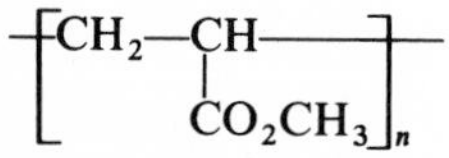

Polymethyl acrylate

$+CF_2CF_2+_n$
Polytetrafluoroethylene (PTFE)

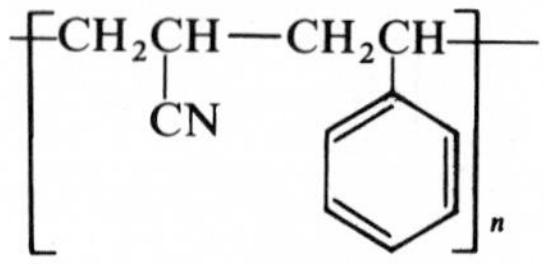

Styrene–acrylonitrile copolymer (SAN)

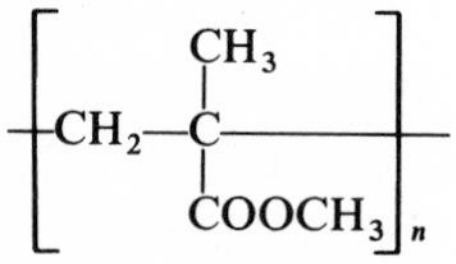

Polymethyl methacrylate (PMMA)

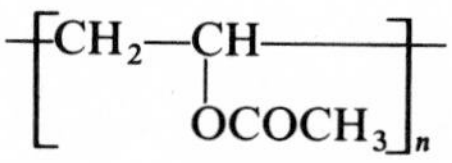

Polyvinyl acetate (PVAc)

$+OCH_2+_n$
Polyoxymethylene polyacetal

Polyvinyl alcohol (PVA)

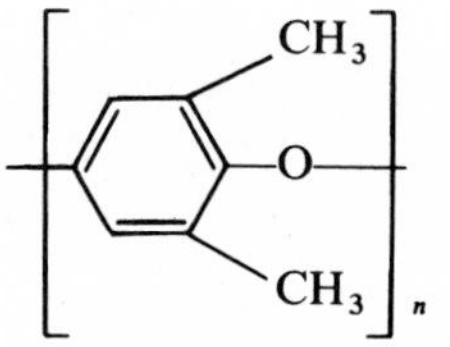

Polyphenylene oxide (PPO)

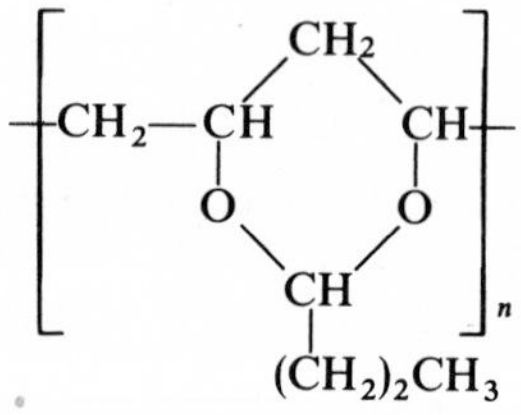

Polyvinyl butyral (PVB)

Table 3.3 U.S. Annual Production of Man-Made Fibers (millions of pounds)

Noncellulosic	
Acrylics	700
Nylons	3000
Olefins	1100
Polyesters	5600
Cellulosic	
Acetate	200
Rayon	420
Glass	1100

live (plants, buildings, roads, animals, clothing, tires, coatings, rugs, newspaper, etc.). Wood products pervade our society as construction materials and paper products, the latter being produced at an annual rate of 65 million tons at a value of $80 billion ($10^9$) in the United States. Portland cement is utilized at the rate of 70 million tons per year. In the United States synthetic polymers are produced at an annual rate of 35 million tons. These three areas alone translate to over 500 lb of paper products, over 500 lb of Portland cement, and about 150 lb of synthetic polymers annually for every person in the United States (see Tables 3.3, 3.4, and 3.5).

On the manufacturing level, the number of persons employed in the *synthetic polymer industry* alone is on par with those employed in all the metal-based industries combined. More than 60% of all chemical industrial employment in the United States involves synthetic polymers (see Table 3.6).

Table 3.4   U.S. Annual Production of Plastics (millions of pounds)

Thermosetting resins	
Epoxies	470
Polyesters	1375
Ureas	1300
Melamines	200
Phenolics	3000
Thermoplastics	
Polyethylene, low density	9900
Polyethylene, high density	8250
Polypropylene	7300
Styrene–acrylonitrile	135
Polystyrene	5100
Acrylonitrile–butadiene–styrene	1240
Polyamides	560
Polyvinyl chloride and copolymers	8300

Table 3.5   U.S. Annual Production of Synthetic Rubber (millions of pounds)

Synthetic rubber (74.4%)	
Styrene–butadiene	2100
Polybutadiene	850
Nitrile	140
Ethylene–propylene	410
Neoprene	168
Other	900
Natural rubber (25.6%)	1920

Table 3.6 U.S. Chemical Industrial Employment (in thousands)

	1975	1985
Industrial inorganic	149	143
Drugs	167	205
Soaps, cleaners, etc.	142	148
Industrial organics	150	164
Agriculture	65	60
Synthetic polymers	888	1026

Source: U.S. Department of Labor.

GLOSSARY

ABS: A terpolymer of acrylonitrile, butadiene, and styrene.

Alkyd: Polyesters produced by the condensation of a dicarboxylic acid (phthalic acid), a dihydric alcohol (ethylene glycol), and an unsaturated oil, such as linseed oil.

Amorphous: Shapeless.

Anion: (A:$^-$) A negatively charged atom or molecule.

ASTM: American Society for Testing and Materials.

Atactic: A random arrangement of pendant groups in a polymer chain.

Baekeland, Leo: Inventor of phenol–formaldehyde plastics (Bakelite), the first truly synthetic plastic (1910).

Balata: A rigid, naturally occurring *trans*-polyisoprene.

Block copolymer: A polymer made up of a sequence of one repeating unit followed by a sequence of another repeating unit.

Branched copolymer: One with branches on the main chain.

Butadiene:

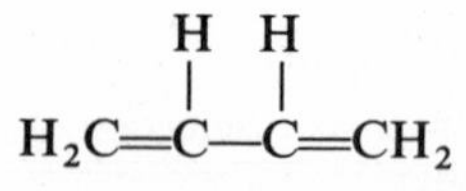

$$H_2C{=}C{-}C{=}CH_2$$ with H, H

Carothers, W. H.: Inventor of nylon 6,6.

Catenation: Chain formation.

Cation (C$^+$): A positively charged atom or molecule.

Cellophane: Regenerated cellulose film.

Celluloid: Plasticized cellulose nitrate.

Chloroprene:

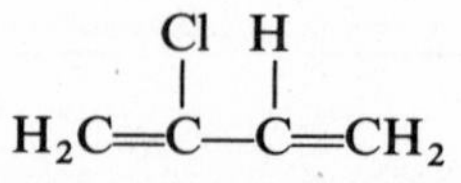

$$H_2C{=}C{-}C{=}CH_2$$ with Cl, H

cis: A geometrical isomer with both constituents on the same side of the plane of the double bond.

Cohesive energy density (CED): Internal pressure of a molecule, which is related to the strength of the intermolecular forces of the molecules.

Configuration: Arrangement of bonds in a molecule. Changes in configurations require breaking and making of covalent bonds.

Conformation: Arrangement of groups about a single bond, that is, shape that changes rapidly without bond breakage as a result of the mobility of the molecule.

Copolymer: A polymer made up of more than one repeating unit.

Coupling: The joining of two macromolecules to produce a dead polymer.

Critical molecular weight: Minimum molecular weight required for chain entanglement.

Cross-links: Chemical bonds between polymer chains, for example, bonds between *Hevea* rubber molecules produced by heating natural rubber with sulfur.

Degree of polymerization (DP): Number of repeating units (mers) in a polymer chain.

Dicarboxylic acid: An organic compound with two carboxylic acid groups.

Dimer: A combination of two smaller molecules.

DNA: Deoxynucleic acid.

Dope: Solution of cellulose acetate.

Elastomer: A rubbery polymer.

Functionality: The number of reactive groups in a molecule.

Glass transition temperature (T_g): Temperature at which segmental motion occurs when a polymer is heated, for example, glassy polymers become flexible.

Glyptal: Polyester protective coating.

Goodyear, Charles: Vulcanized *Hevea* rubber by heating it with small amounts of sulfur (1839).

Graft copolymer: A copolymer in which polymeric branches have been grafted onto the main polymer chain.

HDPE: Linear polyethylene, of higher density than LDPE.

Hevea braziliensis: Natural rubber.

Homopolymer: A polymer made up of similar repeating units.

Impact strength: Resistance to breakage, degree of lack of brittleness.

Initiation: The first step in a chain reaction.

Isoprene:

$$\underset{H_2C=C}{\overset{\displaystyle CH_3}{|}}\!\!-\!\!-\!\!-\!\!\underset{C=CH_2}{\overset{\displaystyle H}{|}}$$

Isotactic: An arrangement in which the pendant groups are all on one side of the polymer chain.

IUPAC: International Union of Pure and Applied Chemistry.

Kekulé, Friedrich A.: Developed methods for writing structural formulas of organic compounds (1850s and 1860s).

Kinetic: Related to motion of molecules.

LDPE: Low-density polyethylene, a highly branched polymer.

Linear low-density polyethylene (LLDPE): Low-density polyethylene consisting of copolymers of ethylene and 1-butene or 1-hexene.

Linear polymer: A polymer consisting of a continuous straight chain.

$\bar{M}$: Average molecular weight.

Macro: Large.

Macroradical: An electron-deficient macromolecule.

mer: Repeating unit.

Monodisperse: A macromolecule in which all molecules have identical molecular weights.

η **(eta):** Viscosity.

Neoprene: Polychloroprene.

Nylon 6,6: A polymer produced by heating the salt from the reaction of hexamethylenediamine ($H_2N(CH_2)_6NH$) and adipic acid ($HOOC(CH_2)_4COOH$).

Oligomer: Polymer consisting of 10 to 20 repeating units.

Patrick, J. C.: Inventor of America's first synthetic elastomer (rubber).

Phenol: Hydroxybenzene (C_6H_5OH).

Plasticizer: An additive that enhances the flexibility of plastics.

Polyacetals (POM): Polymers of formaldehyde with the repeating unit $+O—CH_2+$.

Polyamide (PA): A polymer with repeating amide units, such as nylon 6,6.

Polyamide–imide (PAI): A high-temperature resistant polymer with alternating amide and imide groups.

Polyarylate: A high-temperature resistant polymer produced by the condensation of bisphenol A and an equimolar mixture of iso and terephthalic acids.

Polybutylene terephthalate (PBT): High-performance polymer produced by the condensation of terephthalic acid and 1,4-dihydroxybutane.

Polycarbonate (PC): Tough, high-performance polymer produced by the condensation of bisphenol A and phosgene.

Polychloroprene: An elastomer with the repeating units

$$+\overset{\displaystyle H}{\underset{\displaystyle H}{C}}—\overset{\displaystyle Cl}{C}=\overset{\displaystyle H}{C}—\overset{\displaystyle H}{\underset{\displaystyle H}{C}}+ \quad \text{(neoprene)}$$

Polydisperse: A mixture of macromolecules with different molecular weights.

Polyether imide (PEI): A high-performance polymer with alternating ether and imide groups.

Polyether ketone (PEEK): A high-performance polymer containing the carbonyl ($C=O$) stiffening group in the polymer chain.

Polyethylene: A polymer with the repeating unit $+CH_2—CH_2+$.

Polyethylene terephthalate (PET): High-performance polymer produced by the condensation of terephthalic acid and ethylene glycol.

Polyimide (PI): High-temperature resistant polymer produced by the condensation of an aliphatic diamine and an aromatic dianhydride.

Polymerization: A process in which large molecules (giant molecules or macromolecules) are produced by a combination of smaller molecules.

Polymethyl methacrylate (PMMA): A polymer with the repeating unit

$$
\begin{array}{c}
\quad H \quad CH_3 \\
\quad | \quad\quad | \\
+C-C-\!\!-\!\!-\!\!-\!\!\!\longrightarrow\!\!) \\
\quad | \quad\quad | \\
\quad H \quad C-OCH_3 \\
\quad\quad\quad \| \\
\quad\quad\quad O
\end{array}
$$

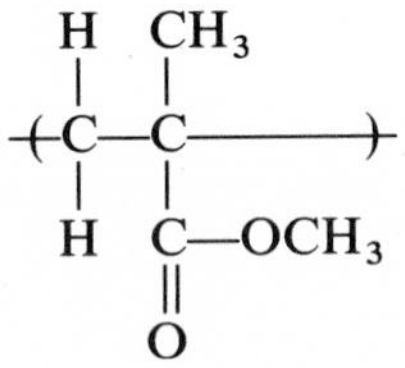

Polymethylpentene (TPX): A polyolefin with the repeating unit

$$
\begin{array}{c}
H_3C \qquad CH_3 \\
\quad\searrow\quad\swarrow \\
\quad CH \\
\quad | \\
\quad H \quad CH_2 \\
\quad | \quad\quad | \\
+C-C-\!\!-\!\!\!\longrightarrow\!) \\
\quad | \quad\quad | \\
\quad H \quad H
\end{array}
$$

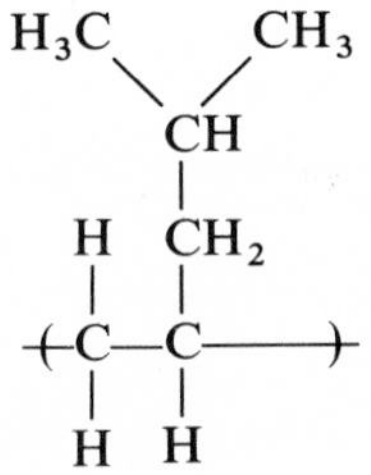

Polyphenylene oxide (PPO): A high-temperature resistant polymer with phenylene and oxygen units in the chain.

Polyphenylene sulfide (PPS): A high-temperature resistant polymer with the repeating unit $+C_6H_4-S+$.

Polyphosphazene: Inorganic polymer with the repeating unit $+N{=}P(OR)_2+$.

Polypropylene (PP): A polymer with the repeating unit

$$
\begin{array}{c}
CH_3 \quad H \\
\quad | \quad\quad | \\
+C-\!\!-\!\!-C+ \\
\quad | \quad\quad | \\
\quad H \quad\quad H
\end{array}
$$

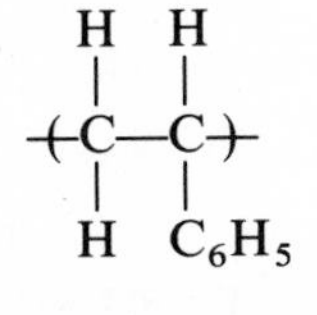

Polystyrene (PS): A polymer with the repeating unit

$$
\begin{array}{c}
\quad H \quad H \\
\quad | \quad\quad | \\
+C-C+ \\
\quad | \quad\quad | \\
\quad H \quad C_6H_5
\end{array}
$$

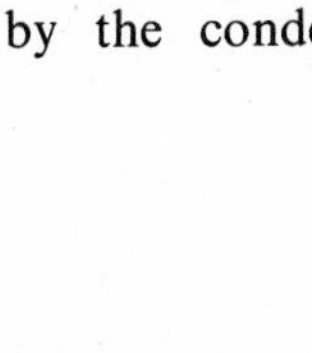

Polysulfone (PES): A high-performance polymer with the repeating unit $+C_6H_4SO_2C_6H_4+$ produced by the condensation of bisphenol A and a dichlorodiphenyl sulfone.

Polyurethane (PUR): A polymer produced by the reaction of a diisocyanate ($Ar(CNO)_2$) and a dihydric alcohol ($R(OH)_2$).

Polyvinyl chloride (PVC): A polymer with the repeating unit $+CH_2CHCl+$.

Propagation: The growth steps in a chain reaction.

Radical (R·): An electron-deficient molecule.

Rayon, cupraammonia: Cellulose fibers regenerated from a solution of cellulose in cupraammonium hydroxide.

Rayon, viscose: Cellulose fibers regenerated from cellulose xanthate.

Round-robin testing: Independent testing by different individuals.

Saran: Trade name for polymers of vinylidene chloride (PVDC).

Schönbein, Christian F.: Produced cellulose nitrate by the nitration of cellulose (1846).

Silicones: Inorganic polymers produced by the hydrolysis of dialkyldimethoxysilanes ($R_2Si(OCH_3)_2$), the repeating unit of which is

$$+O-\underset{\underset{R}{|}}{\overset{\overset{R}{|}}{Si}}+$$

SMA: Copolymers of styrene and maleic anhydride.

Staudinger, Hermann: Developed modern concepts of polymer macromolecular science (1920s).

Step reaction polymerization: Polymerization that occurs by a step-wise condensation of reactants.

Syndiotactic: An alternate arrangement of pendant groups on a polymer chain.

T_g: Glass transition temperature.

T_m: Melting point.

Technology: Applied science.

Teflon: Trade name for polytetrafluoroethylene (PTFE).

Telomer: A low-molecular-weight polymer.

Termination: The final step in a chain reaction.

Thermoplastic: A linear polymer that can be softened by heat and cooled to reform the solid.

Thermoset plastic: A cross-linked (three-dimensional) polymer that does not soften when heated.

Thiokol: Trade name for polyethylene sulfide rubber.

trans: A geometrical isomer with substituents on alternate sides of the double bond.

Transition: Change.

UHMWPE: Ultrahigh-molecular-weight polyethylene.

Wöhler, Friedrich: First chemist to synthesize an organic molecule from an inorganic compound (1828).

REVIEW QUESTIONS

1. How many functional groups are present in glycerol?

$$
\begin{array}{ccc}
H & H & H \\
| & | & | \\
(HC & \!\!\!-C- & \!\!\!CH) \\
| & | & | \\
OH & OH & OH
\end{array}
$$

2. What is *Hevea braziliensis*?

3. Which has more cross-links: flexible vulcanized rubber or hard rubber?

4. Which of the following are thermoplastics: hard rubber, Bakelite, PVC, polystyrene, polyethylene?

5. Which of the following are thermoset plastics: Melamine dishware, Bakelite, hard rubber?

6. What is the functionality of phenol?

7. How does rayon differ from cotton from a chemical viewpoint?

8. What is the principal structural difference between LDPE and HDPE?

9. Why was former President Reagan called the Teflon President?

10. Which is the faster reaction: step reaction or chain reaction polymerization?

11. What is the propagating species in cationic polymerization?

12. What is the molecular weight of polyethylene with a DP of 1000?

13. Which has the higher value for a specific polymer with both amorphous and crystalline regions: T_g or T_m?

14. Is a protein a polydisperse or monodisperse polymer?

15. Why should the molecular weight of structural polymers be greater than the critical molecular weight required for chain entanglement?

BIBLIOGRAPHY

Books

Billmeyer, F. W. (1971). *Textbook for Polymer Science*. New York: Wiley-Interscience.

Cowie, J. W. (1974). *Polymer Chemistry and Physics of Modern Materials*. New York: Intext.

Kaufman, H. W., and Falcetta, J. J. (1977). *Introduction to Polymer Science and Technology: An SPE Textbook*. New York: Wiley.

Kirshenbaum, G. S. (1973). *Polymer Science Study Guide*. New York: Gordon & Breach.

Odian, G. (1970). *Principles of Polymerization*. New York: McGraw-Hill.

Rodriguez, F. (1970). *Principles of Polymer Systems*. New York: McGraw-Hill.

Seymour, R. B. (1970). *Modern Plastics Technology*. Reston, VA: Reston.

Seymour, R. B., and Carraher, C. E. (1987). *Polymer Chemistry: An Introduction* (2nd ed.). New York: Dekker.

Stevens, M. S. (1975). *Polymer Chemistry*. New York: Addison-Wesley.

Audio Courses

Hawkins, W. L. (1975). *Polymer Synthesis*. Washington, DC: American Chemical Society.

Seymour, R. B. (1981). *Introduction to Polymer Science and Technology*. Washington, DC: American Chemical Society.

Volpe, A. (1980). *Introduction to Polymer Chemistry*. Hoboken, NJ: Plastics Institute of America.

Wallace, T. (1981). *Molecular Characterization of Polymers*. Washington, DC: American Chemical Society.

Video-Computer Courses

Smith, S. S., Gibson, H. W., and Pochan, J. M. (1980). *Introduction to Polymer Chemistry*. Washington, DC: American Chemical Society.

ANSWERS TO REVIEW QUESTIONS

1. Three.

2. Natural rubber.

3. Hard rubber.

4. PVC, polystyrene, polyethylene.

5. All are thermosets.

6. Three.

7. No difference; rayon is regenerated cellulose.

8. LDPE is highly branched and therefore has a lower density (higher volume) than linear HDPE.

9. Teflon (polytetrafluoroethylene) is slippery because of the four fluorine pendant groups on each repeating unit. Hence, few things will stick to PTFE.

10. Chain reaction polymerization.

11. A macrocation.

12. 28,000 (1000 × 28).

13. T_m.

14. Monodisperse.

15. In order to achieve strength through entanglement.

CHAPTER IV_____

Relationships between the Properties and Structure of Giant Molecules

4.1 GENERAL

Many of the properties of giant molecules (polymers) are unique and not characteristic of other materials, such as metals and salts. Polymer properties are related not only to the chemical nature of the polymer, but also to such factors as extent and distribution of crystallinity, distribution of polymer chain lengths, and nature and amount of additives. These factors influence polymeric properties, such as hardness,

67

biological response, comfort, chemical resistance, flammability, weatherability, tear strength, dyeability, stiffness, flex life, and electrical properties.

In this chapter we briefly describe the chemical and physical nature of polymeric materials that permits their classification into broad "use" divisions, such as elastomers or rubbers, fibers, plastics, adhesives, and coatings. Descriptions relating chemical and physical parameters to general polymer properties and structure are included.

4.2 ELASTOMERS

Elastomers are giant molecules possessing chemical and/or physical cross-linking. For industrial applications, the "use" temperature of an elastomer must be above the T_g (to allow for segmental "chain" mobility), and the polymer must be amorphous in its normal (unextended) state. The restoring force, after elongation, is largely due to entropy effects. As the elastomer is elongated, the random chains are forced to occupy more ordered positions, but on release of the applied force, the chains tend to return to a more random state. Entropy is a measure of the degree of randomness or lack of order in a material.

Elastomers possess what is referred to as memory—that is, they can be deformed, misshaped, and stretched, and after the stressing (applied) force is removed, they return to their original, prestressed shape.

The actual mobility of polymer chains in elastomers must be low. The cohesive energies density forces (CED) between chains should be low enough to permit rapid and easy extension of the random-oriented chain. In its extended (stretched) state, an elastomeric polymer chain should have a high tensile strength, whereas at low extensions it should have a low tensile strength. Polymers with low cross-linked density usually meet the desired property requirements. After deformation, the material should return to its original shape because of the presence of the cross-links, which limit chain slippage to the chain sections between the cross-links (principal sections).

South American Indians use the names "hhevo" and "Cauchuc," which mean "weeping wood," to describe the native rubber tree. The French continue to use the word "caoutchouc," but when it was found to be more effective than bread crumbs in removing pencil marks, E. Nairne and J. Priestley called it rubber. The term elastomer is now used to describe both natural and synthetic rubbers.

4.3 FIBERS

Characteristic fiber properties include high tensile strength and high modulus (high stress for small strains, i.e., stiffness). These properties are related to high molecular symmetry and high cohesive energy density forces between chains. Both of these properties are related to a relatively high degree of crystallinity present in fiber molecules.

Fibers are normally linear and drawn (oriented) in one direction to enhance mechanical properties in the direction of the draw. Typical condensation polymers, such as polyesters and nylons, often exhibit these properties.

If the fiber is to be ironed, its T_g should be above 350°F, and if it is to be drawn from the melt, its T_g should be below 570°F. Branching and cross-linking in fibers are undesirable since they disrupt crystalline formation, but a small amount of cross-linking may increase some physical properties if introduced after the material has been drawn and processed.

Cotton, linen, wool, and silk were used for over 2000 years before cellulose nitrate filaments were spun by H. Chardonnet. Regenerated cellulose produced by spinning cellulose xanthate was introduced in 1892 by C. Cross, E. Bevan, and C. Beadle. Cellulose xanthate is produced by the reaction of cellulose and carbon disulfide (CS_2) in the presence of alkali. The term rayon is now used to describe all regenerated cellulose, including derivatives such as acetate rayon. Nylon, which was the first synthetic fiber, was produced by W. Carothers and J. Hill in the 1930s.

4.4 PLASTICS

Materials with properties that are intermediate between those of elastomers and fibers are grouped together under the general term "plastics." Thus, plastics exhibit some flexibility and hardness and varying degrees of crystallinity. The molecular requirements for a thermoplastic are that it have little or no cross-linking and that it be used below its glass transition temperature, if amorphous, and/or below its melting point, if crystalline. Thermoset plastics must be sufficiently cross-linked to severely restrict molecular motion. The term cross-linked density is used to describe the extent of cross-linking in a material.

4.5 ADHESIVES

Adhesives can be considered to be coatings sandwiched between two surfaces. Early adhesives were water-susceptible and biodegradable animal and vegetable glues obtained from hides, blood, and starch. Adhesion may be defined as the process that occurs when a solid and a movable material (usually in a liquid or solid form) are brought together to form an interface and the surface energies of the two substances are transformed into the energy of the interface.

Starch was used to glue sheets of papyrus by the Egyptians 6000 years ago, and hydrolyzed collagen from bones, hides, and hooves (carpenter's glue) was used as an adhesive in 1500 B.C. Starch, which was partially degraded by vinegar, was used as an adhesive for paper in 120 B.C. These early adhesives continue to be used but have been largely displaced by solutions and hot melts of synthetic polar polymers.

A unified science of adhesion has yet to be developed. Adhesion can result from mechanical bonding and chemical and/or physical forces between the adhesive and adherend. Contributions through chemical and physical bonding are often more important and illustrate why nonpolar polymeric materials, such as polyethylene, are difficult to bond, whereas polar polycyanoacrylates, such as butyl-2-cyanoacrylate, are excellent adhesives. There are numerous types of adhesives, including solvent-based, latex, pressure-sensitive, reactive, and hot-melt adhesives.

$$\begin{array}{c} H \qquad\quad CN \\ \diagdown \qquad | \\ \quad C=C \\ \diagup \qquad \diagdown \\ H \qquad\qquad C-O-C_4H_9 \\ \qquad\qquad\quad || \\ \qquad\qquad\quad O \end{array}$$

Butyl-α-cyanoacrylate

The combination of an adhesive and adherend is a laminate. Commercial laminates are produced on a large scale with wood as the adherend and phenolic, urea, epoxy, resorcinol, or polyester resins as the adhesives. Some wood laminates are called plywood. Laminates of paper or textile include items with the trade names Formica and Micarta. Laminates of phenolic, nylon, or silicone resins with cotton, asbestos, paper, or glass textiles are used as mechanical, electrical, and general-purpose structural materials. Plastic composites of mat or sheet fibrous glass and epoxy or polyester resins are widely employed as fiber-reinforced plastic (FRP) structures.

4.6 COATINGS

The annual cost of corrosion is over \$75 billion (\$$10^9$) in the United States. With the exception of metal and ceramic types, nearly all surface coatings are based on polymeric films. The surface-coating industry originated in prehistoric times. By 1000 B.C., naturally occurring resins and beeswax were used as constituents of paints. The coatings industry used drying oils, such as linseed oil, and natural resins, such as rosin, shellac, and copals, prior to the early 1900s.

Linseed oil, which is obtained from the seeds of flax (*Lininum usitatissium*), was the first vegetable oil binder used for coatings. This unsaturated (drying) oil hardens (polymerizes) in air when a heavy metal salt (drier, siccative) is present. Presumably, some free oleic acid reacts with the white lead pigment to produce a drier (catalyst) when the linseed oil and pigment are heated. Subsequently, ethanolic solutions of shellac were displaced by collodion (a solution of cellulose nitrate), but oleoresinous paints continue to be used.

Phenolic, alkyd, and urea resins were used as coatings in the 1920s. Interior paints based on lattices of poly(vinyl acetate), poly(methyl methacrylate), and styrene–butadiene copolymers were introduced after World War II. Latex paints for exterior use were marketed in the late 1950s.

The fundamental purposes of coatings as being decorative and protective are giving way to more complex uses in energy collection devices and burglar alarm systems. Even so, the problems of the coating's adhesion, weatherability, permeability, corrosion inhibition, flexural strength, endurance, application, preparation, and application procedures continue to be the major issues. Effective coatings generally yield tough, flexible films with moderate to good adhesion to metal or wood surfaces.

$$CH_2—O—\overset{\overset{\textstyle O}{\|}}{C}—R$$

$$CH—O—\overset{\overset{\textstyle O}{\|}}{C}—R'$$

$$CH_2—O—\overset{\overset{\textstyle O}{\|}}{C}—(CH_2)_7—CH=CH—CH—CH=CH—CH_2—CH=CH—CH_2CH_3$$

Linolenic acid

allylic carbon atoms

$$\left.\begin{array}{c} O \\ | \\ | \\ O \end{array}\right\}$$ peroxide linkage

$$CH_3(CH_2)_4—CH=CH—CH—CH=CH—(CH_2)_7—\overset{\overset{\textstyle O}{\|}}{C}—O—CH_2$$

Linoleic acid

A drying oil

$$CH—O—\overset{\overset{\textstyle O}{\|}}{C}—R'$$

$$CH_2—O—\overset{\overset{\textstyle O}{\|}}{C}—R$$

4.7   POLYBLENDS AND COMPOSITES

Polyblends are made by mixing components together in extruders or intensive mixers or on mill rolls. Most heterogeneous systems consist of a polymeric matrix in which another polymer is imbedded. Whereas the repeating units of copolymers are connected through primary bonds, the components of polyblends are connected through secondary bonding forces. In contrast to polyblends, which are blends of polymers, composites consist of a polymeric matrix in which a nonpolymeric material is dispersed. Composites typically contain fillers, such as carbon black, wood flour, and talc, or reinforcing materials, such as glass fibers, hollow spheres, and glass mats.

4.8   CRYSTALLINE–AMORPHOUS STRUCTURES

There are numerous theories associated with crystallization tendencies and the form(s) and mix(es) of crystalline–amorphous regions within polymers. We will only briefly consider several and describe a few contributing factors.

A three-dimensional crystalline polymer may be described using a fringed micelle concept (chains packed like a sheaf of grain) or a form of folded chains. Regions where the polymer chains exist in an ordered array are called crystalline domains. Imperfections in these polymer crystalline domains are frequent and one polymer chain may reside both within a crystalline domain and within an amorphous region.

Connective chains between these domains are responsible for the toughness of a polymer. Sharp boundaries between the ordered (crystalline) and disordered (amorphous) portions are the exception but do occur in some instances, such as with certain

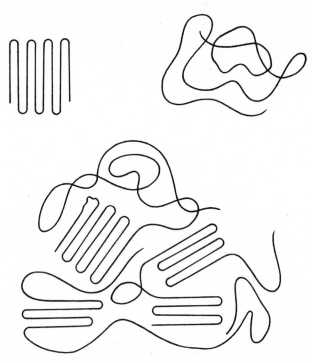

Figure 4.1. Schematic two-dimensional representations of a crystalline polymer region (top left), an amorphous region (top right), and the crystalline–amorphous regions (bottom) of a polymeric material.

proteins, poly(vinyl alcohol), and some cellulosic materials. Highly crystalline polymers exhibit high melting points and high densities, resist dissolution and swelling in solvents, and have high moduli or rigidity relative to polymers with less crystallinity.

The amount and kind of crystallinity depend on both the polymer structure and its treatment. The latter is illustrated by noting that the proportion of crystallinity can be effectively regulated for many common polymers by controlling the rate of crystallization. Thus, polypropylene can be heated above its melting point and cooled quickly (quenched) to produce a product with only a moderate amount of crystalline domains. However, if it is cooled at a slower rate (such as 1°F/5 min), the resulting polypropylene will be largely crystalline. Two-dimensional representations of crystalline and amorphous regions of polymers are shown in Figures 4.1 and 4.2. Factors that contribute to the inherent crystalline–amorphous-forming tendencies of polymers are discussed next.

A. Chain Flexibility

The tendency toward crystallinity in some polymers increases as flexibility is increased. Polymers containing regularly spaced single C—C, C—N, and C—O bonds allow rapid conformational changes that contribute to the flexibility of a polymer chain and the tendency toward crystal formation. This is also true in the case

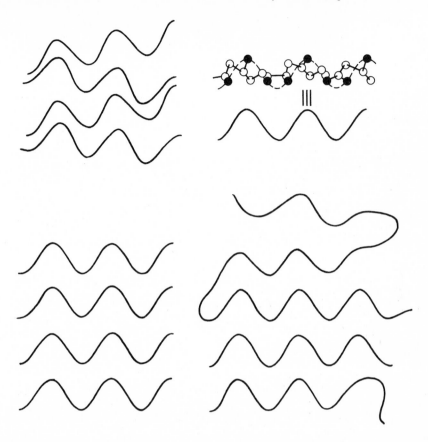

Figure 4.2. Schematic two-dimensional representations of helical conformations of crystalline (top left), amorphous (top right), and crystalline–amorphous (bottom left) regions of polypropylene.

of linear polyethylene, polypropylene, and poly(vinyl chloride), whose structures are shown in Figure 4.3.

Chain stiffness may also enhance crystalline formation by permitting only certain "well-ordered" conformations to occur within the polymer chains. Thus, poly-*p*-phenylene is a linear chain that cannot "fold over" at high temperatures. Hence, such species are crystalline, high-melting, rigid, and insoluble.

B. Intermolecular Forces

Crystallization is favored by the presence of regularly spaced units that permit strong intermolecular interchain associations. The presence of moieties that carry dipoles or are highly polarizable promotes strong interchain exchanges. This is particularly true for interchain hydrogen bond formation. Thus, the presence of regularly spaced carbonyl ($C=O$), amine (NH_2), amide ($CONH_2$), sulfoxide (SO_2), and alcohol (OH) moieties promotes crystallization.

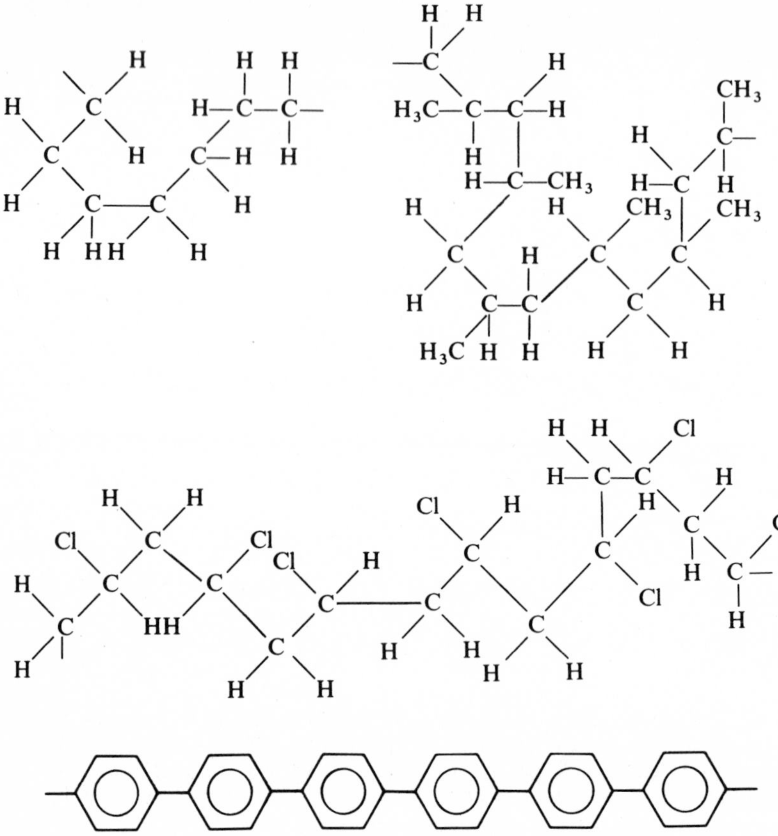

Figure 4.3. Segmental portions of linear polyethylene (top left), polypropylene (top right), and polyvinyl chloride (middle), illustrating chain flexibility, and poly-*p*-phenylene (bottom), illustrating chain stiffness.

C. Structural Regularity

Structural regularity also enhances the tendency for crystallization. Thus, it is difficult to obtain linear polyethylene (HDPE) in any form other than a highly crystalline one. Low-density, branched polyethylene (LDPE) is typically largely amorphous. The linear polyethylene chains are nonpolar, and the crystallization tendency is mainly based on its flexibility, which permits it to achieve a regular, tightly packed conformation, which takes advantage of the special restrictions inherent in the dispersion forces. Simulated structures of HDPE and LDPE are shown in Figure 4.4.

Monosubstituted vinyl monomers ($CH_2=CHX$) can produce polymers with different configurations, that is, two regular structures (isotactic and syndiotactic) and a random, atactic form. Polymers with regular structures exhibit greater rigidity and are higher melting and less soluble than the atactic form.

Extensive work with condensation polymers and copolymers confirms the importance of structural regularity on crystallization tendency and associated properties. Thus, copolymers containing regular alternation of each copolymer unit,

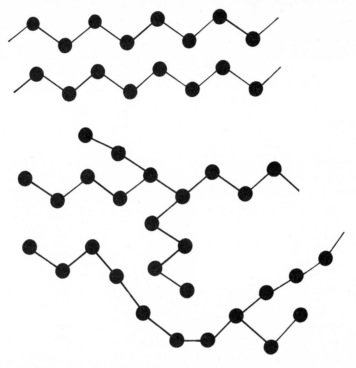

Figure 4.4. Simulated structure of high-density, linear polyethylene, emphasizing the tendency toward intrachain regularity (top), and low-density, branched polyethylene, illustrating the inability for intrachain regularity (bottom).

either ABABAB type or block type, show a distinct tendency to crystallize, whereas corresponding copolymers with random distributions of the two monomers are intrinsically amorphous, less rigid, and lower melting, and have greater solubility. The concept of micelles in natural fibers was first expressed by C. Nagele in 1858, but since most nineteenth-century scientists preferred to consider polymers as aggregates of molecules (colloids) rather than as individual giant molecules, the existence of polymer crystals was deemphasized until 1917 when H. Ambronn suggested that cellulose nitrate fiber had crystalline characteristics. In 1920, R. Herzog and M. Polanyi used x-ray diffraction techniques to show the presence of crystallites in flax fiber.

D. Steric Effects

The effect of substituents on polymer properties depends on the location, size, shape, and mutual interactions of the substituents. Methyl and phenyl substituents (pendant groups) tend to lower chain mobility but prevent good packing of chains. These substituents produce unit dipoles, which contribute to the crystallization tendency.

Aromatic substituents contribute to intrachain and interchain attraction tendency through the mutual interactions of the aromatic substituents. Their bulky size retards crystallization and promotes rigidity because of increased interchain distances. Thus, polymers containing bulky aromatic substituents tend to be rigid, high melting, less soluble, and amorphous.

Substituents from ethyl to hexyl tend to lower the tendency for crystallization, since they increase the average distance between chains and decrease the contributions of secondary bonding forces. Thus, LLDPE is an amorphous polymer. If these linear substituents are larger (12 to 18 carbon atoms), these side chains form crystalline domains on their own (side chain crystallization).

4.9 SUMMARY

Polymer properties are directly dependent on both the inherent shape of the polymer and its treatment. Contributions of polymer shape to polymer properties are often complex and interrelated but can be broadly divided into terms related to chain regularity, interchain forces, and steric effects.

GLOSSARY

Additive: Substances added to polymers to improve properties, such as strength, ductility, stability, and resistance to flame.

Adherend: A substance whose surface is adhered by an adhesive.

Adhesive: A substance that bonds two surfaces together.

Amorphous: Shapeless, noncrystalline.

Cohesive energy density (CED): A measure of intermolecular forces between molecules.

Composite: A mixture of a polymer and an additive, usually a reinforcing fiber or filler.

Density, cross-linked: A measure of the extent of cross-linking in a polymer network.

Elastomer: Amorphous, flexible polymers that are usually cross-linked to a small extent.

Entropy: A measure of the degree of disorder or randomness in a polymer.

Fiber: A threadlike substance in which the ratio of the length to diameter is at least 100:1. Fibers are characterized by strong intermolecular forces.

Fringed micelle concept: A diagrammatic representation of aligned polymer chains (crystalline) separated by regions of nonaligned or amorphous areas.

FRP: Fiberglass-reinforced plastic.

HDPE: High-density (linear) polyethylene.

Laminate: A composite resulting from adhering two surfaces together.

Latex: A stable dispersion of a polymer in water.

LDPE: Low-density (branched) polyethylene.

LLDPE: A low-density linear polyethylene, usually a copolymer of ethylene and 1-butene or 1-hexene.

Modulus: The ratio of strength to elongation, a measure of stiffness.

Paint: A mixture of a pigment, unsaturated oil, resin, and drier (catalyst).

Plastic: Substances with properties in between those of elastomers and fibers.

Plywood: A laminate of thin sheets of wood and adhesives.
Principal section: Portion of a polymer chain between cross-links.
Steric: Arrangement in space.

REVIEW QUESTIONS

1. What is the function of additives in polymers?

2. What are the characteristics of an elastomer?

3. Which has the higher entropy: stretched or unstretched rubber?

4. Which has the higher cohesive energy density (CED): an elastomer or a fiber?

5. Which has the higher cross-linked density: soft vulcanized rubber or hard rubber?

6. Which has the longer principal sections: soft vulcanized rubber or hard rubber?

7. Which has the higher modulus: soft vulcanized rubber or hard rubber?

8. Which has a higher degree of crystallinity: HDPE or LDPE?

9. What are the adhesive and adherend widely used in reinforced plastics?

10. What is the trade name of a laminate used for kitchen countertops?

11. What is the difference between a paint and a protective coating?

12. Why are latex-based coatings popular?

BIBLIOGRAPHY

Agranoff, J. (Ed.). (1980). *Modern Plastics Encyclopedia.* New York: McGraw–Hill.

Allcock, H. R., and Lampe, F. W. (1981). *Contemporary Polymer Chemistry, Part III.* Englewood Cliffs, NJ: Prentice–Hall.

Cowie, J. M. G. (1973). *Polymers: Chemistry and Physics of Modern Materials* (Chap. 14). New York: Intext.

Deanin, R. (1975). *Structure vs. Property Relationships.* Boston: Cahners Books.

Dostal, C. A. (1987). *Composites.* Metals Park, OH: ASM International.

Dubois, J. H., and John, F. W. (1982). *Plastics* (6th ed.). New York: Van Nostrand–Reinhold.

Epel, J. M., Margolis, J. M., Newman, S., and Seymour, R. B. (1988). *Engineered Plastics.* Metals Park, OH: ASM International.

Flory, P. J. (1953). *Principles of Polymer Chemistry.* Ithaca, NY: Cornell University Press.

Frados, J. (1976). *SPT Plastics Engineering Handbook* (4th ed.). New York: Van Nostrand–Reinhold.

Harper, C. A. (Ed.). (1975). *Handbook of Plastics and Elastomers.* New York: McGraw–Hill.

Milby, R. V. (1973). *Plastics Technology.* New York: McGraw–Hill.

Miles, D. C., and Briston, J. H. (1979). *Polymer Technology.* New York: Chemical Publishing Co.

Rosen, S. L. (1982). *Fundamental Principles of Polymer Materials* (Chap. 8). New York: Wiley–Interscience.

Sauer, J. A., and Pae, K. D. (1977). In H. S. Kaufman and J. J. Falcetta (Eds.), *Introduction to Polymer Science and Technology* (Chap. 7). New York: Wiley–Interscience.

Seymour, R. B. (1975). *Modern Plastics Technology.* Reston, VA: Reston.

Seymour, R. B. (1987). *Polymers for Engineering Applications.* Metals Park, OH: ASM International.

Seymour, R. B., and Carraher, C. E. (1984). *Structure–Property Relationships in Polymers.* New York: Plenum.

Seymour, R. B., and Carraher, C. E. (1987). *Polymer Chemistry: An Introduction* (2nd ed.). New York: Dekker.

Seymour, R. B., and Mark, H. F. (1988). *Applications of Polymers.* New York: Plenum.

ANSWERS TO REVIEW QUESTIONS

1. They improve properties.
2. It is amorphous when unstretched, has weak intermolecular forces, and usually has a low cross-linked density.
3. Unstretched rubbers have a greater degree of randomness or disorder.
4. A fiber usually contains intermolecular hydrogen bonds.
5. Hard rubber.
6. Soft vulcanized rubber.
7. Hard rubber.
8. HDPE has a more ordered structure.
9. The resin (polyester, epoxy) is the adhesive and the fiberglass or graphite is the adherend.
10. Micarta or Formica.
11. Paint is a protective coating, but there are many other types of protective coatings.
12. They are easy to produce and do not affect the environment adversely as do solvent-based coatings. They have a low volatile organic concentration (VOC).

CHAPTER V _____

Nature's Giant Molecules: The Plant Kingdom

5.1 INTRODUCTION

Natural polymers (giant molecules) are well known and have been essential for human life for thousands of years. Starch, cellulose, lignin, and rubber are polymers of the plant kingdom and are essential for food and shelter. These polymers are produced at an annual rate of millions of tons. Natural rubber will be discussed in Chapter 12.

Polymers of the animal kingdom, namely, proteins, nucleic acids, chitin, and glycogen, which are also essential products, are produced at annual rates of over a million tons. Some of these naturally occurring giant molecules are classified as biopolymers, but their behavior follows the same laws as those followed by synthetic polymers. These naturally occurring giant molecules of the animal kingdom are discussed in Chapter 6.

5.2   SIMPLE CARBOHYDRATES (SMALL MOLECULES)

In 1812, Gay-Lussac showed that carbohydrates, such as starch and cellulose, contained 45% carbon, 49% oxygen, and 6% hydrogen by weight. After dividing these percentages by the appropriate mass numbers, one obtains the empirical or simplest formula of CH_2O. Hence, Gay-Lussac called starch and cellulose "watered carbon" or carbohydrates. Both starch and cellulose are polymers of D-glucose, but they differ in the manner in which these building blocks or repeating units are joined together in these giant molecules.

As shown by the chemical and skeletal formulas for glyceraldehyde (2,3-dihydroxy-propanal), this simple compound and all higher-molecular-weight aldo-carbohydrates contain hydroxyl (OH) and carbonyl (C=O) groups. The OH is represented by — in the skeletal formulas.

```
         H
         |
1   H—C=O                    C=O
         |                    |
2   H—C—OH                   C—
         |                    |
3    H₂COH                   C—

    Glyceraldehyde          Skeletal formula
```

If one were to make a model of glyceraldehyde using toothpicks and gumdrops, one would discover that there are two possible arrangements for the hydroxyl group and hydrogen atom around carbon atom No. 2. The two arrangements are not as obvious as those in other isomers unless one looks at three-dimensional models, which differ like right- and left-hand gloves, that is, they are mirror images of each other.

Since they rotate the plane of polarized light in equal but opposite directions, these so-called stereoisomers are optical isomers. Ordinary light waves vibrate in all directions, but polarized light, such as that passing through a Polaroid lens vibrates in a single plane that is perpendicular to the ray of light.

The original optical isomers were labeled dextro after the Latin word *dexter*, meaning right, and *levo* after the Latin word *laevus*, meaning left. Accordingly, the trivial names dextrose and levulose have been used for the principal D-hexoses, namely, D-glucose and D-fructose.

Nature synthesizes dextro-carbohydrates almost exclusively. When we discuss proteins in Chapter 6, we will note that nature also synthesizes L-amino acids almost exclusively. The skeletal formulas that are mirror images for the simplest hydroxy aldehyde (glyceraldehyde) are

```
   C=O                      C=O
    |                        |
   C—                       —C
    |                        |
   C—                        C—

D-Glyceraldehyde         L-Glyceraldehyde
```

Since carbon No. 2 in glyceraldehyde has four different groups, namely, H, HC=O, OH, and H_2COH, which can be arranged in two different ways, it is called an asymmetric or chiral carbon atom. The aldoses containing three carbon atoms are called trioses.

There are two possible arrangements for each chiral atom, and there are four (2^2), eight (2^3), and sixteen (2^4) possible arrangements (isomers) for tetroses, pentoses, and hexoses, respectively. The general formula for describing the number of possible optical isomers is 2^n, where n is equal to the number of chiral atoms in a molecule.

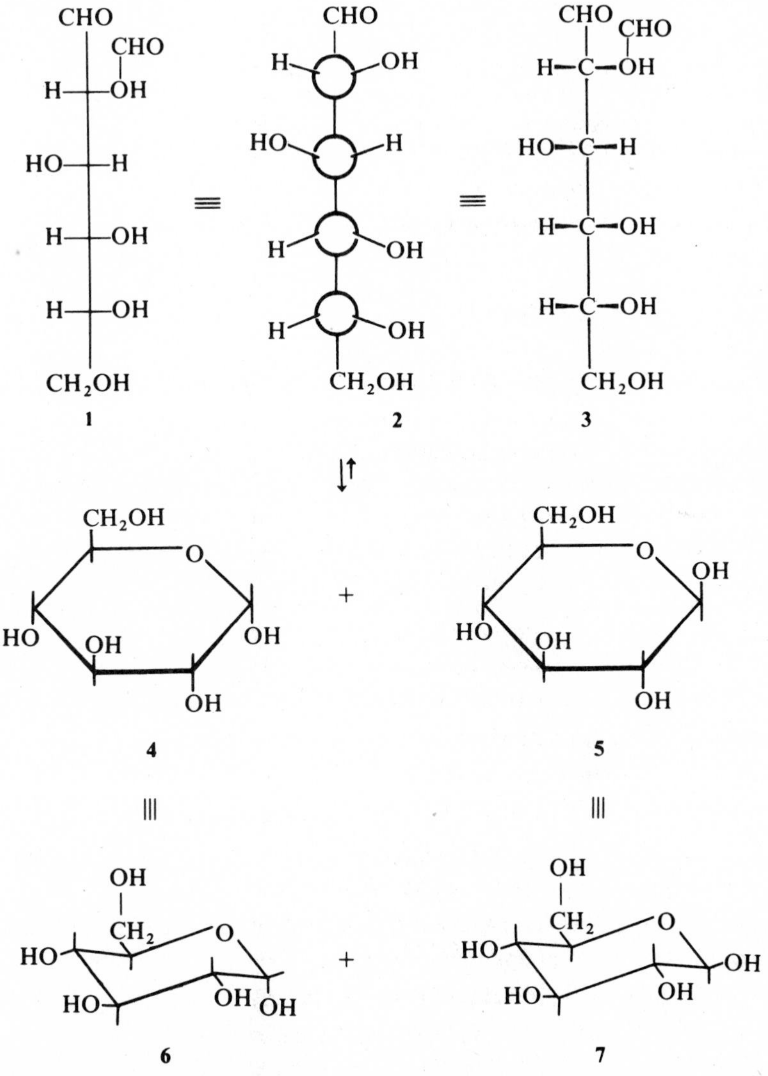

α—D—(+)—Glucopyranose β—D—(+)—Glucopyranose

Figure 5.1. Open (top) and cyclic (middle and bottom) formulas for glucose. The top left form is called a Fischer projection. Structures **4** and **5** are called Haworth structures, and structures **6** and **7** are called "chair" structures.

We will direct our attention to pentoses and hexoses and specifically to the D isomers, that is, those with the hydroxyl group on the right-hand side of the chiral carbon atom farthest away from the carbonyl group. The OH on carbon No. 5 in D-glucose is on the right-hand side, as shown by the skeletal molecular structure

$$
\begin{array}{ll}
1 & C{=}O \\
  & | \\
2 & C{-} \\
  & | \\
3{-}C & \\
  & | \\
4 & C{-} \\
  & | \\
5 & C{-} \\
  & | \\
6 & C{-} \\
\end{array}
$$

D-Glucose

Actually, the linear form shown here is in equilibrium with the cyclic or ring form (Carbon 6 is not chiral since it has two H's directly attached to it) shown in Figure 5.1.

The hydroxyl group on carbon No. 1 in the cyclic form of D-glucose, which is called the anomeric carbon atom, may be arranged in two different ways. The two forms are called α- and β-D-glucopyranose or simply the α and β forms of D-glucose. These glycopyranoses and all other low-molecular-weight aldoses and ketoses are called monosaccharides.

Condensation of two monosaccharides produces a disaccharide. The everyday compound we call sugar is a disaccharide with the common name of sucrose. Sucrose contains one unit of α-D-glucose and one of β-D-fructose. The C—O—C ether linkage connecting the two monosugar units is called a glycosidic linkage or bond. The geometry of the C—O—C linkage of α-D-glucose with β-D-fructose is called an alpha linkage whereas the C—O—C linkage formed through condensation of with the β-D-glucose is called a beta linkage.

5.3 STARCH

D-Glucose is the building block or repeating unit in the principal polysaccharides, starch and cellulose. These giant molecules serve as the reserve carbohydrates in plants. The reserve carbohydrates in humans and many other animals is a polysaccharide called glycogen.

Glycogen is synthesized rapidly in a process called glycogenesis and hydrolyzed with equal rapidity to D-glucose in a process called glycogenolysis. These polymerization (building up) and depolymerization (breaking down) processes are regulated by the hormone insulin, which is excreted by the pancreas.

Two molecules of α-D-glucopyranose may join together and produce a molecule of a disaccharide called α-maltose. The mechanism for this dimerization is, of course, more complicated, but this general statement will be adequate for our discussion. As shown in the following structural formula, the two α-D-glucopyranose monomer units in α-maltose are joined through an oxygen or acetal linkage between carbons 1 and 4:

α-Maltose

When this building-up process is continued in plants, a giant molecule or polymer called amylose is produced. This polymeric chain, which may contain a thousand or more maltose units, is relatively flexible and tends to form a spiral or helix, as shown in Figure 5.2.

An iodine molecule fits well in this helix and gives a characteristic blue color test for starch. The other form of starch, called amylopectin, is a highly branched structure

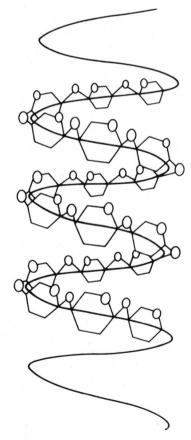

Figure 5.2. The loose, somewhat helical geometry of amylose, which is one of the two major components of starch.

in which branches or chain extensions are formed on the No. 6 carbon atoms of the repeating D-glucose units.

Natural starch is a polydisperse macromolecule consisting of different ratios of amylose and amylopectin, depending on its source. For example, the so-called waxy starches contain over 98% amylopectin. The polydisperse nature appears to be characteristic of natural structural and storage materials (such as glycogen and starch), but not of those macromolecules whose shape, size, and electronic nature are critical in carrying out specific biological functions (such as DNA and enzymes).

5.4   CELLULOSE

When two molecules of β-D-glycopyranose are joined together through carbon atoms 1 and 4, the product is called cellobiose. As shown in Figure 5.3, cellulose is a polymer made up of many cellobiose units.

The polymeric chains connected by alpha acetal linkages in starch are flexible and helical. These conformations result from intramolecular hydrogen bonding, that is, attractions between hydrogen and oxygen atoms in the same large molecule. C. Hanes employed the helical structure in an attempt to explain the water solubility of amylose

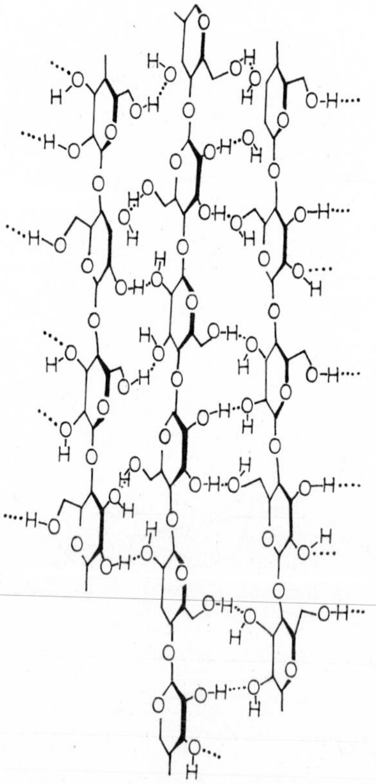

Figure 5.3. Puckered sheet structure of cellulose chains. The dots denote hydrogen bonding.

in 1937. R. Rouble confirmed the presence of this helical structure by x-ray diffraction studies in 1943.

In contrast, the beta acetal linkages in cellulose are less flexible, and the chains tend to remain extended and may be joined to each other by intermolecular hydrogen bonds, that is, attractions between hydrogen and oxygen atoms in adjacent molecules, as shown in Figure 5.3.

The intermolecular hydrogen bonds in cellulose are so strong that it is insoluble in water, whereas the linear chains of alpha-linked glucose units in amylose are water soluble. Further, because of the lack of intermolecular bonding, amylose is flexible in contrast to cellulose, which is rigid.

The difference between the alpha and beta linkages in these carbohydrate polymers shows up not only in varying physical properties, but also in their digestibility. Humans possess a gut enzyme that will break down alpha linkages specifically. Thus, we can eat potatoes and gain food value from them. However, humans cannot digest cellulose, but termites have enzymes that can digest this β-polysaccharide. Thus, they are able to feast on our timberland and wooden portions of our houses.

The rigidity of wood is due to hydrogen bonding between the cellulose molecules and lignin. Wood exposed to ammonia (NH_3), a base, will begin to degrade as a result of the rupture of the hydrogen bonds. The partially degraded cellulose can then be shaped. When the ammonia is washed away, new hydrogen bonding occurs, locking the molecules into a new form, similar to that which occurs when we "set" our hair. Longer exposure to bases disrupts the glycosidic bonding and causes a permanent loss of strength.

Acids act similarly but faster to disrupt the glycosidic (acetal) bonds. Thus, polysaccharides degrade to their original monosaccharide units (D-glucose) when heated with acids. Enzymes catalyze this degradation and provide both plants and humans with a source of D-glucose.

5.5 COTTON

Cotton is grown in semitropical regions throughout the world. This essentially pure cellulose has been harvested for many thousands of years in China, Egypt, and Mexico. There are several species of *Gossypium* plants, but *Gossypium hirsutum* is the most common cotton plant.

Large-scale production of cotton was hampered until the invention of the cotton gin by Eli Whitney in 1793. This machine, which consists of a rotating spiked cylinder, displaced the labor-intensive hand process for separating the cotton from the cotton seeds. The production of cotton was also increased by the invention of the mechanical cotton picker in the twentieth century. Cotton was "king" in the cotton-producing states for almost a century, but has now been partially displaced by synthetic fibers, such as polyesters and nylon. Over 85 million bales (16.7 million tons) of cotton are produced annually worldwide.

Cotton, which is a cellulosic fiber, readily absorbs water and is stronger when wet than dry. Cotton is a good source of cellulose and it is the source for most of the cellulose employed in the synthesis of rayon, cellulose acetate, and cellulose nitrate.

PAPER

It is believed that paper was invented by Ts'ai Lun in China in the second century A.D. Paper was first produced in the United States by William Rittenhouse in Germantown, Pennsylvania, in 1690. The original Chinese paper was a mixture of bark and hemp, but prior to the eighteenth century, much of the paper was made from rags. Paper was named after the papyrus plant, which is no longer used for the production of paper.

Modern paper is made from wood pulp (cellulose), which is obtained by the removal of lignin from debarked wood chips by use of chemicals, such as sodium hydroxide, sodium sulfite, or sodium sulfate. Newsprint and paperboard, which is thicker than paper, may contain some residual lignin. Lignin is the structural support and adhesive matter of the plant world. Wood contains cellulose bonded by at least 25% lignin.

The book you are reading, the newspaper, materials used to write notes on, and even some clothing are made of paper. If you rip a piece of ordinary paper (not your book, please!), you will note that paper consists of small fibers. Most of these cellulosic fibers are randomly oriented, but a small percentage of fibers are aligned in the direction in which the paper was produced from a watery slurry to a water-free sheet on a series of heated rolls. The papermaking process is sketched in Figure 5.4.

Wood and other woody products contain mostly cellulose and lignin. In the simplest papermaking process, the wood is chopped (actually torn) into smaller fibrous particles as it is pressed against a rapidly moving pulpstone. A stream of water washes the fibers away and dissolves much of the water-soluble lignin. The fibrous material is concentrated into a paste called pulp. The pulp is layered into thin sheets and rollers are employed both to squeeze out water and to assist in achieving paper of

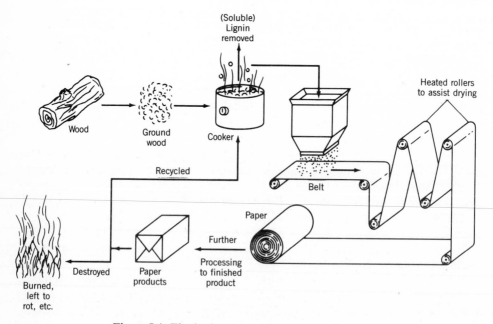

Figure 5.4. The basic process of paper manufacturing.

a uniform thickness. Paper produced by this mechanism is not very white or strong because the remaining lignin is somewhat acidic and causes hydrolytic breakup of the cellulose chains. Most of the paper utilized for newsprint is of this type or regenerated, reused paper.

The sulfate process, also called the kraft process (kraft comes from the Swedish word for strong since good strength paper is produced), is more commonly employed. The kraft process is also favored over the sulfite process because of environmental considerations, since the sulfite process employs more chemicals that must be disposed of—particularly mercaptans (RSH), which are quite odorous, similar to the compounds emitted by a frightened skunk. Present research involves reclaiming and recycling these pulping chemicals, and so far more than a 10-fold decrease in the amount of chemical used per volume of paper produced has been attained.

If pure cellulose pulp were used, the fiber mat formed would be largely water soluble with only surface polar and hydrogen bonding acting to hold the fibers together. White pigments such as clay and titanium dioxide (TiO_2) are added to help "cement" the fibers together and to fill the voids, thus producing a firm, white writing surface. Resins, surface-coating agents, and other special surface treatments (such as coating with polypropylene and polyethylene) are employed for paper products intended for special purposes such as milk cartons, ice cream cartons, roofing paper, extra strength uses, light building materials, and drinking cups. The cellulose supplies the fundamental structure and most of the bulk (about 90% of the weight) and strength for the paper product whereas the additives provide special properties needed for special applications.

As costs rise the interest in recycling paper also increases. Community recycling centers and paper drives organized by local Scout, church, or school groups help in this recycling process. Today about 20 to 25% of our paper products are being recycled. This percentage could be doubled if more emphasis was placed on the collection of waste paper. Unfortunately, recycled paper costs slightly more than virgin pulp. On the other hand, recycling can reduce the destruction of natural resources (by saving forests) and minimize the load on waste-disposal systems.

5.7 OTHER CARBOHYDRATE POLYMERS

Glycogen is a food-reserve polysaccharide of animals that also occurs in some yeasts and fungi. It is found in the liver and muscles of animals, and its structure is similar to that of amylopectin, but it has a larger number of chain branches on carbon No. 6.

Dextran is a poly-α-D-glucose linked throughout the Nos. 1 and 6 carbon atoms with occasional branching at the No. 3 carbon atom. Dextran is an amorphous solid that forms random coils in aqueous (water) solutions. Dextrans have been used as blood plasma volume extenders, in pharmaceuticals, to increase the viscosity of foodstuffs (such as ice cream), and as emulsion stabilizers.

Many plants and some species of seaweed contain polyuronides in which the methylol group (CH_2OH) in the repeating unit of the polymeric carbohydrate is replaced by a carboxylic acid group (COOH). These gums, such as alginic acid, agar, pectic acid, and carrageenan, are water soluble. Some polyuronides, such as galactomannans, carrageenans, agar, gum arabic, and alginates, are used as food additives in ice cream, pie fillings, gelled desserts, and salad dressings. Many are also used in the pharmaceutical industry as encapsulating materials and as emulsion stabilizers. Agar is used as a culture medium for bacteria. Pectic acid is widely used in the making of jams and jellies.

The shells of crustacea and some insects and fungi consist of celluloselike polymers in which the hydroxyl groups on carbon No. 2 of the repeating glucose unit are replaced by an acetylamino group ($NH(CO)CH_3$). This polymer, called chitin, has also been found in fossils that are over 500 million years old.

5.8   LIGNIN

Lignin functions as the structural support and cement material of the plant world. It constitutes about 25% of wood, and is thus one of the most abundant natural polymers. Since lignin is relatively inert and is insoluble in the plant, it is not easily isolated in a pure and undegraded state. However, it can be extracted from wood by dissolving it in a dilute solution of acetic acid (H_3COOH) and ethyl acetate ($H_3CCOOC_2H_5$) in the "ester process."

Because of the synthetic sequence that occurs in plants, lignin appears to have a two-dimensional sheet structure. Polysaccharides, such as cellulose and hemicellulose, are deposited as plant cell walls are formed. Lignin is then synthesized to fill the spaces between the polysaccharide fibers. This process constrains the "three-dimensional" growth of lignin. The structure of lignin is complex and variable, but it contains ethers, aromatic and aliphatic alcohol functions, ketones, and aromatic rings.

Lignin sulfonate is used as a dispersant and wetting agent and is an important additive in the preparation of oil well drilling muds. Lignin sulfonates are used in road binders, industrial cleaners, and adhesives. Lignin obtained by alkali extraction is used as an additive for cement, filler for rubber, and as a dye dispersant. Nevertheless, considerable lignin is burned as fuel at paper mills. It is currently available at 20–30 cents per pound, making it a potentially attractive feedstock for other uses.

5.9   OTHER NATURAL PRODUCTS FROM PLANTS

Many natural resins are fossil resins exuded from trees thousands of years ago. Exudates from living trees are called recent resins, and those obtained from dead trees are called semifossil resins. Humic acid is a fossil resin found with peat, brown coal, or lignite deposits throughout the world. It is a carboxylated phenolic-like polymer that is used as a soil conditioner, as a component of oil drilling muds, and as a scavenger for heavy metals.

Amber is a fossil resin found in the Baltic Sea regions, and sandarac and copals are found in Morocco and Oceania, respectively. Other fossil resins, called Manila, Congo, and Kauri, are named after their geographic source.

Frankincense and myrrh, which are mentioned in the *New Testament*, contain polyuronides. Bitumens, which were used by Noah for waterproofing the ark in the biblical story, occur as asphalt at Trinidad Lake (West Indies) and as gilsonite in Utah.

5.10   PHOTOSYNTHESIS

The sun is the source that winds the clock of life. Green plants absorb solar energy and converts it to carbohydrates in a process called photosynthesis. Photosynthesis is the process in which carbon dioxide (CO_2) combines with water (H_2O) to form glucose ($C_6H_{12}O_6$).

Photosynthesis begins with the absorption of light by chlorophyll in plants. The absorbed energy excites electrons in this green pigment to higher energy states. When the electrons return to their original "ground" state, the released energy is used to decompose water, produce a strong reducing agent, and energize a select phosphate ester from nucleic acid.

Reduction refers to processes whereby electrons are added in a chemical reaction, and oxidation refers to the process of giving up electrons:

$$2Fe^{3+} + 2e^- \rightarrow 2Fe^{2+} \quad \text{(reduction)}$$

$$H_2 \rightarrow 2H^+ + 2e^- \quad \text{(oxidation)}$$

The overall reaction, which is the sum of the two preceding reactions, is

$$H_2 + 2Fe^{3+} \rightarrow 2H^+ + 2Fe^{2+} \quad \text{(net reaction)}$$

Detailed discussion of this topic is found in most introductory chemistry books.

The synthesis of glucose is described by the equation

$$6CO_2 + 6H_2O \xrightarrow{\text{light}} C_6H_{12}O_6$$

Chlorophyll-*a*

The most important and common plant pigment is chlorophyll, which is a polymer with a structure similar to that of heme proteins, except that iron is replaced by magnesium. A number of compounds are grouped under the name of chlorophyll. The preceding structure is the "active" part of chlorophyll-*a*, which was confirmed by H. Fischer, R. Willstatter, and James B. Conant.

The importance of photosynthesis in the biosphere cannot be overstated; without photosynthesis, there would be no biosphere. The photosynthetic process plays a major role in maintaining the desirable amounts of CO_2 and O_2 in the atmosphere. We breathe in O_2 and emit CO_2; conversely, plants absorb CO_2 and H_2O and form carbohydrates and emit O_2. This natural cycle emphasizes the harmony or balance that is often apparent in nature and demonstrates the symbiotic relationship between the plant and animal kingdom.

5.11   POLYHYDROXYALKANOATES

Chemists at the Pasteur Institute in Paris produced biologically polyhydroxybutyrate (PHB, $-[O-CH(CH_3)CH_2CO]_n-$) in the 1920s by a controlled bacterial fermentation using *Alcaligenes autrophus*. This biodegradable polyester was produced commercially by W. R. Grace, but production was discontinued in the late 1950s. PHB (Biopol) is now being produced in India under a license from ICI, who, through its Marlborough Biopolymer subsidiary, is also producing a copolymer of hydroxybutyrate and hydroxyvalerate ($HOCH(C_2H_5)CH_2COOH$), which has better impact resistance than the homopolymer PHB. PHB and its copolymer have piezoelectric properties, that is, they generate an electric current when compressed.

GLOSSARY

Acetal: The oxygen linkage between monosaccharides in polysaccharides.

Aldo: Prefix for aldehyde compounds (—CHO).

Aldose: A compound containing hydroxyls and an aldehyde group, like glyceraldehyde.

Amino acid: Compound with the general formula

$$\begin{array}{c} H \\ | \\ H_2NCCOOH \\ | \\ R \end{array}$$

Amylopectin: A highly branched starch molecule.

Amylose: A linear starch molecule.

Anomeric: A carbon atom on which the hydroxyl groups may be arranged in two different ways, that is, alpha and beta.

Asymmetric: A molecule that is not symmetrical, that is, the atoms or groups around the carbon atom may be arranged in two different ways.

Bitumen: Asphalt and gilsonite resins.

Carbonyl:

$$\diagup_{\diagdown} C={O}$$

Chiral: An asymmetric carbon atom; derived from the Greek word *cheir*, meaning hard.

Chlorophyll: A green pigment in plants.

Dextro: Derived from the Latin word *dexter*, meaning right.

Dextrose: A trivial name for D-glucose.

Empirical formula: Simplest formula.

Fossil resin: Aged exudates from tropical trees.

Fructose: $C_6H_{12}O_6$, a ketohexose.

Glucose: $C_6H_{12}O_6$, an aldose with six carbon atoms and five hydroxyl groups.

Glyceraldehyde:

$$\underset{\underset{OH}{|}}{H_2C}-\underset{\underset{OH}{\overset{\overset{H}{|}}{C}}}{}-\overset{\overset{H}{|}}{C}=O$$

Glycogen: The reserve carbohydrate in animals.

Glycogenesis: The polymerization of glucose to glycogen.

Glycogenolysis: The hydrolysis of glycogen to form glucose (depolymerization).

Hexose: An aldose or ketose with six carbon atoms.

Humic acid: A carboxylated phenolic-like semifossil resin found in peat and lignite deposits.

Hydroxyl: —OH.

Insulin: A hormone that regulates glycogenesis and glycogenolysis.

Isomers: Molecules with the same empirical formulas.

Isomers, optical: Molecules with similar formulas but with arrangements in space that rotate the plane of optical light in equal and opposite directions.

Isomers, steric: Molecules with substituents arranged differently in space.

Levo: Derived from the Latin word *laevos*, meaning left.

Levulose: A trivial name for D-fructose.

Light, polarized: Light that is vibrating in a single plane.

Lignin: The noncellulosic material in wood.

Lignin sulfonate: The product of the reaction of lignin and sulfuric acid.

Maltose: The disaccharide made up of two molecules of D-glucose.

Monosaccharide: The simplest saccharide, that is, fructose and glucose.

Natural polymer: Giant molecules such as starch, cellulose, and proteins that occur in nature.

Oxidation: The loss of electrons by a molecule or ion.

Photosynthesis: The production of glucose by the catalytic combination of carbon dioxide and water.

Polydisperse: A mixture of polymers with different molecular weights.
Pyranose: A cyclic molecule consisting of five carbon atoms and one oxygen atom.
Recent resin: Exudate from live trees.
Reduction: The addition of electrons to a molecule or ion.
Semifossil resin: Exudate from dead trees.
Skeletal formula: A formula in which the hydrogen atoms are omitted.
Solar energy: Energy from the sun.

REVIEW QUESTIONS

1. What is a natural polymer?
2. What is the empirical formula for glucose ($C_6H_{12}O_6$)?
3. What optically active acid occurs in milk?
4. In the dextro–levo convention, what would your right hand be called?
5. How do dextrose (D-glucose) and levulose (D-fructose) differ chemically?
6. The simplest amino acid is glycine

$$\overset{\displaystyle H}{\underset{\displaystyle H}{(N_2HCCOOH)}}$$

   Is glycine optically active?
7. What is the shape of the pyranose molecule?
8. How many anomeric carbon atoms are present on D-glucose?
9. How many monosaccharide repeating units are there in maltose?
10. Which is more linear: amylose or amylopectin?
11. Is cellulose mono- or polydisperse?
12. What polymer is present in paper?
13. Which will have less residual solvent: a fossil resin or a recent resin?
14. What is the generic name for asphalt and gilsonite?
15. Define oxidation.
16. Define reduction.
17. Which can you digest: starch or cellulose?

BIBLIOGRAPHY

Brauns, F. E. (1952). *The Chemistry of Lignin.* New York: Academic Press.
Hicks, E. (1961). *Shellac.* New York: Chemical Publishing Co.
Hoiberg, A. V. (1964). *Bituminous Materials: Asphalt, Tars and Pitches.* New York: Wiley–Interscience.
Ott, E., Sperlin, H. M., and Grafflin, M. W. (1954). *Cellulose and Cellulose Derivatives.* New York: Wiley–Interscience.

Pearl, I. A. (1967). *The Chemistry of Lignin.* New York: Dekker.

Schnitzer, M., and Khan, S. U. (1972). *Humic Substances in the Environment.* New York: Dekker.

Seymour, R. B., and Carraher, C. E. (1981). *Polymer Chemistry: An Introduction.* New York: Dekker.

Whistler, R. L. (1988). In R. B. Seymour and H. F. Mark (Eds.), *Applications of Polymers* (Chap. 12). New York: Plenum.

Whistler, R. L., and BeMiller, J. N. (1953). *Industrial Gums.* New York: Academic Press.

Whistler, R. L., and Paschall, E. F. (1965). *Starch: Chemistry and Technology.* New York: Academic Press.

Whistler, R. L., and Smart, C. L. (1953). *Polysaccharide Chemistry.* New York: Academic Press

ANSWERS TO REVIEW QUESTIONS

1. A giant molecule found in nature.
2. CH_2O.
3. Lactic acid

$$(CH_3 \overset{\overset{\displaystyle H}{|}}{\underset{\underset{\displaystyle OH}{|}}{C}} - COOH)$$

4. Dextro-hand.
5. Dextrose is an aldohexose and levulose is a ketohexose.
6. No. It does not contain a chiral carbon atom (only three different groups on carbon no. 1).
7. It has a ring or cyclic structure.
8. One.
9. Two (it is a disaccharide).
10. Amylose.
11. Polydisperse.
12. Cellulose.
13. Fossil resin.
14. Bitumen.
15. Loss of electrons.
16. Gain of electrons.
17. Starch.

CHAPTER VI

Nature's Giant Molecules: The Animal Kingdom

6.1 INTRODUCTION

One of the strongest and most rapidly growing areas of polymer science is that of natural polymers. Our bodies are largely composed of polymers: deoxyribonucleic acid (DNA), ribonucleic acid (RNA), proteins, and carbohydrate polymers. These polymers are related to aging, awareness, mobility, strength, and so on, all character- istics that contribute to our being "alive and well." Many medical, health, and biological projects and advances are concerned with materials that are polymeric. There is an ever increasing emphasis on molecular biology, that is, science applied at the molecular level to biological systems. An understanding of natural polymers is advantageous to intelligent citizens and to those who desire to understand and contribute positively to advances in biology, but these will be discussed only briefly in this chapter.

Starch, cellulose, and lignin are the building blocks of the plant world, whereas proteins and nucleic acids serve a similar role in the animal kingdom. Natural rubber, resins, and gums are also polymeric and play an important role in our everyday activities. The shape and size of these natural polymers are critical to their ability to carry out their highly specialized functions.

We are witnessing a reemergence of the use of natural polymers in many new and old industrial applications, since natural polymers are renewable resources that nature continues to provide. There is no difference in the science and technology of natural and synthetic polymers, and manufacturing techniques suitable for application to synthetic polymers are normally equally applicable to natural polymers.

6.2   AMINO ACIDS

More than 200 amino acids exist, but only about 20 of them are necessary for the existence of animal life. Some of these amino acids can be synthesized in adequate quantities in the human body (Table 6.1). Each amino acid contains an amino

Table 6.1   The Twenty Amino Acids Commonly Found in Proteins[a]

Glycine	Gly	H—CH—COOH, NH$_2$
Alanine	Ala	CH$_3$—CH—COOH, NH$_2$
*Valine	Val	CH$_3$—CH—CH—COOH, CH$_3$ NH$_2$
*Leucine	Leu	CH$_3$CH—CH$_2$—CH—COOH, CH$_3$ NH$_2$
*Isoleucine	Ile	CH$_3$—CH$_2$—CH—CH—COOH, CH$_3$ NH$_2$
*Phenylalanine	Phe	C$_6$H$_5$—CH$_2$—CH—COOH, NH$_2$
Tyrosine	Tyr	HO—C$_6$H$_4$—CH$_2$—CH—COOH, NH$_2$
Serine	Ser	HO—CH$_2$—CH—COOH, NH$_2$
*Threonine	Thr	CH$_3$—CH—CH—COOH, OH NH$_2$

Table 6.1 (*Continued*)

Cysteine	Cys	$HS-CH_2-\underset{\underset{NH_2}{\mid}}{CH}-COOH$
*Methionine	Met	$CH_3-S-CH_2CH_2-\underset{\underset{NH_2}{\mid}}{CH}-COOH$
*Tryptophan	Trp	indole ring structure with $CH_2-\underset{\underset{NH_2}{\mid}}{CH}-COOH$
Proline	Pro	pyrrolidine ring structure H_2C-CH_2, H_2C $CH-COOH$, $N-H$
Asparagine	Asn	$NH_2-\underset{\underset{O}{\parallel}}{C}-CH_2-\underset{\underset{NH_2}{\mid}}{CH}-COOH$
Aspartic acid	Asp	$HO-\underset{\underset{O}{\parallel}}{C}-CH_2-\underset{\underset{NH_2}{\mid}}{CH}-COOH$
Glutamine	Gln	$NH_2-\underset{\underset{O}{\parallel}}{C}-CH_2CH_2-\underset{\underset{NH_2}{\mid}}{CH}-COOH$
Glutamic acid	Glu	$HO-\underset{\underset{O}{\parallel}}{C}-CH_2CH_2-\underset{\underset{NH_2}{\mid}}{CH}-COOH$
*Lysine	Lys	$NH_2-CH_2CH_2CH_2CH_2-\underset{\underset{NH_2}{\mid}}{CH}-COOH$
*Arginine	Arg	$NH_2-\underset{\underset{NH}{\parallel}}{C}-NH-CH_2CH_2CH_2-\underset{\underset{NH_2}{\mid}}{CH}-COOH$
*Histidine	His	imidazole ring structure with $CH_2-\underset{\underset{NH_2}{\mid}}{CH}-COOH$

*Amino acids not synthesized naturally by humans are called essential amino acids. These are denoted by an asterisk.

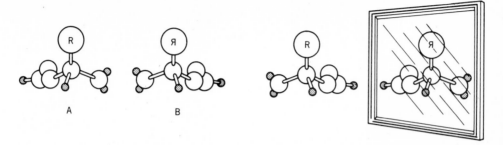

Figure 6.1. α-Amino acids. Representations illustrate optically active forms.

functional group (NH$_2$) and a carboxylic acid functional group (COOH). These two functional groups are attached to the same carbon atom, which is called the alpha carbon atom.

$$
\begin{array}{c}
\text{amine}\underbrace{}_{} \\
\text{(H}_2\text{N)}\!-\!\overset{\displaystyle R}{\underset{\displaystyle H}{C}}\!-\!\overset{\displaystyle O}{\overset{\displaystyle \|}{C}}\!-\!\text{OH} \quad\text{acid}
\end{array}
$$

first or alpha carbon atom

With the exception of glycine, in which the R group is a hydrogen atom, two optical isomers exist for each α-amino acid. As shown in Figure 6.1, these two forms are mirror images of each other. Similar phenomena are our right and left hands, which are approximately mirror images of each other (Figure 6.2). Structures A and B of Figure 6.1 are called optical isomers since they rotate the plane of polarized light in opposite directions. All the essential amino acids are of levo optical isomeric form, that is, the L form.

Since amino acids contain both acidic (RCOOH) and basic (RNH$_2$) groups in the same molecule, they are called zwitterions, after the Greek word *zwitter*, which means

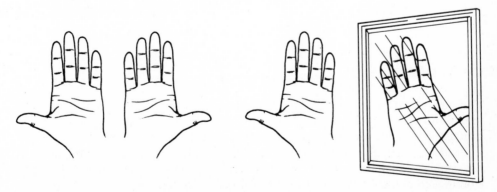

Figure 6.2. Left and right hands. Representations illustrate optically active forms.

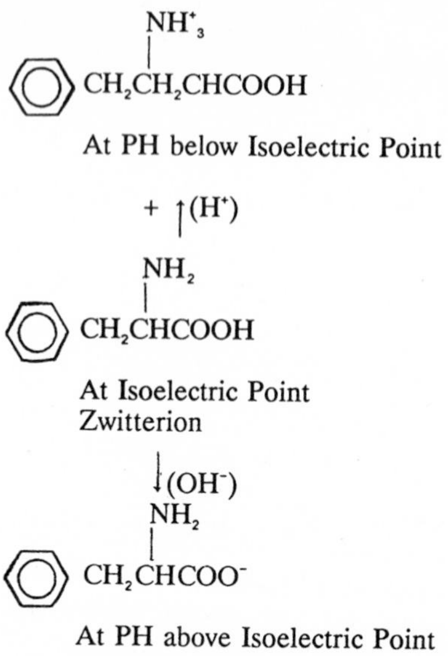

$$NH^+_3$$
$$|$$

⬡ $CH_2CH_2CHCOOH$

At PH below Isoelectric Point

$$+ \uparrow (H^+)$$

$$NH_2$$
$$|$$

⬡ $CH_2CHCOOH$

At Isoelectric Point
Zwitterion

$$\downarrow (OH^-)$$
$$NH_2$$
$$|$$

⬡ CH_2CHCOO^-

At PH above Isoelectric Point

Figure 6.3. Structure of an amino acid above and below its isoelectric point.

hybrid. Amino acids can have different electrical charges, depending on the pH of the solution.

The pH of a solution is a measure of acidity or basicity of that solution. A pH of 7 is neutral, that is, neither acidic nor basic. The solution is acidic at pH values less than 7, and the lower the pH value, the more acidic the solution. Basic solutions have pH values higher than 7, and the higher the value, the more basic the solution.

The pH at which the negative and positive charges are balanced (i.e., there is no movement toward either pole in an electric field) is characteristic for each amino acid and is called the isoelectric point. Shifts in the amount and form of the various amino acids are controlled in the body by variations in pH, thus allowing the body to effectively regulate the supply of specific amino acids. This assures essential supplies of specific amino acids for producing proteins for hair, skin, and so on. The changes in structure of phenylalanine (Phe) with changes in pH are shown in Figure 6.3.

6.3 PROTEINS

The name protein is derived from the Greek word *proteios*, meaning of first importance. G. V. Mulder coined this word in 1835 when he recognized that these nitrogen-containing organic compounds were essential for all life processes.

Proteins are copolymers made up of 20 different amino acids. Many lower animals can synthesize all of these amino acids, but as noted in Table 6.1, humans must obtain 8 of these from their diet. Fortunately, soybeans, which have served as a staple food for centuries, contain all the essential amino acids.

6.4   PROTEIN STRUCTURE

The structures of naturally occurring L-leucine and D-leucine are shown for comparative purposes:

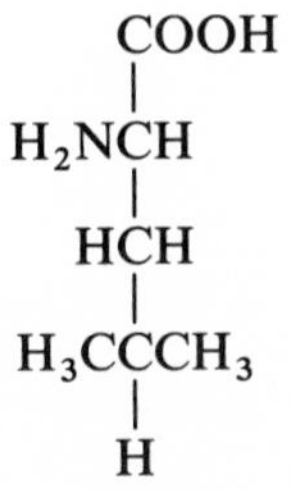

$$
\begin{array}{c}
\text{COOH} \\
| \\
\text{H}_2\text{NCH} \\
| \\
\text{HCH} \\
| \\
\text{H}_3\text{CCCH}_3 \\
| \\
\text{H}
\end{array}
$$

L-Leucine (Leu)
(naturally occurring)

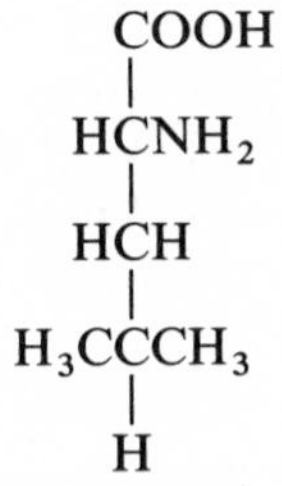

$$
\begin{array}{c}
\text{COOH} \\
| \\
\text{HCNH}_2 \\
| \\
\text{HCH} \\
| \\
\text{H}_3\text{CCCH}_3 \\
| \\
\text{H}
\end{array}
$$

D-Leucine (Leu)
(does not occur naturally)

The amino acids

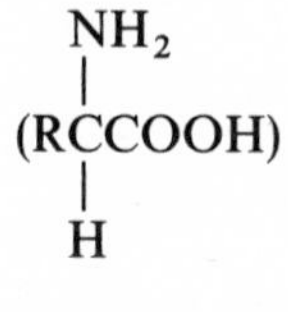

$$
\begin{array}{c}
\text{NH}_2 \\
| \\
\text{(RCCOOH)} \\
| \\
\text{H}
\end{array}
$$

in proteins are arranged in characteristic head to tail order and joined through peptide linkages

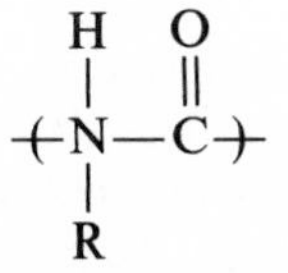

$$
\begin{array}{cc}
\text{H} & \text{O} \\
| & \| \\
+\text{N}—\text{C}+ \\
| \\
\text{R}
\end{array}
$$

Peptide (or Amide) linkage

A dipeptide (or dimer) is produced when two amino acids are joined together. The low-molecular-weight polymers containing relatively few amino acid residues or mers are called polypeptides. The high-molecular-weight polymers or macromolecules are also polypeptides but are usually called proteins.

It is of interest that the peptide linkages that connect the various amino acid units to form proteins are structurally similar to those connecting the synthetic nylons:

$$\begin{array}{ccc} R & H & O \\ | & | & \| \\ \text{---C---N---C---} \\ | \\ H \end{array} \qquad\qquad \begin{array}{ccc} & H & O \\ & | & \| \\ \text{---R---N---C---} \end{array}$$

Peptide linkage
for proteins   Amide linkage
for nylons

We can represent the structures of proteins by using the symbols for the amino acids. The dipeptide made from the reaction between valine and glycine through reaction of the carboxylic acid function on the valine with the amine group on the glycine is represented as Val·Gly and given the name valylglycine. If the amine function of the valine were reacted with the acid function of glycine then the representations are reversed, giving Gly·Val the name of glycylvaline. In each case the amino acid whose carboxyl group is involved in the formation of the amide linkage is placed first. This convention is biochemical shorthand for representing complex structures.

$$\begin{array}{c} H \quad O \quad H \qquad\qquad O \\ | \quad\; \| \quad\; | \qquad\qquad \| \\ H_2N\text{---}C\text{---}C\text{---}N\text{---}CH_2\text{---}C\text{---}OH \\ | \\ CH \\ /\;\;\backslash \\ H_3C \qquad CH_3 \end{array}$$

$$\begin{array}{c} O \quad H \quad H \quad H \\ \| \quad\; | \quad\; | \quad\; \| \\ H_2N\text{---}CH_2\text{---}C\text{---}N\text{---}C\text{---}C\text{---}OH \\ | \\ CH \\ /\;\;\backslash \\ H_3C \qquad CH_3 \end{array}$$

Val·Gly
Valylglycine   Gly·Val
Glycylvaline

When a protein is boiled in hydrochloric acid, hydrolysis of many of the amide repeating units occurs, and a mixture of dimers, monomers, trimers, and so on is produced.Because of the extreme importance of proteins, techniques have been developed to allow the sequence of amino acids to be reconstructed from such data. Molecular weight measurements and electron and x-ray diffraction measurements provide information on the actual shape and size of the protein. This process is tedious, and hence it has been used for only a few of the more important and simpler proteins.

Lysome vasopressin is composed of nine amino acid units. It is excreted by the pituitary gland, is employed clinically as a hypertensive agent, and was the first naturally occurring hormonal polypeptide synthesized in the laboratory. The Nobel prize was awarded in 1955 to V. du Vigneaud for this synthesis. The chemical and abbreviated structure for bovine vasopressin is

Cys———┐
↓   │
Tyr   │
↓   │
Phe   S
↓   │
Gly   S
↓   │
Asn   │
↓   │
Cys———┘
↓
Pro
↓
Arg
↓
Gly—NH$_2$

Bovine vasopressin

Proteins can be divided according to function, as noted in Table 6.2, or according to overall shape as fibrous or globular. Fibrous proteins can be likened to a strand of rope—they are largely linear and nonbranched. Globular proteins exhibit a great deal of twisting and turning, with the overall shape dependent on the specific sequence of amino acid units, and are often "held in shape" through cross-links that "lock in" a specific geometry.

Table 6.2   Protein Classification According to Function

Function	Example	Description
A. Structural proteins	Collagen	Major component of connective tissue in animals, including bones, cartilage, and tendons.
	Keratins	Comprise most protective coverings of animals: hair, hoofs, claws, feathers, beaks, nails.
B. Regulator proteins		
1. Enzymes	Chymotrypsin	Involved in digestive process, cleaves polypeptides excreted by pancreas.
	Lysozyme	Involved in digestion, cleaves polysaccharide chains; found in many natural sources such as egg whites.
2. Hormones	Bradykinin	Regulates blood pressure; in blood plasma.
	Insulin	Required for normal glucose metabolism.
C. Transport proteins		
	Hemoglobin	Responsible for oxygen transport from lungs to the cells and for removal of waste carbon dioxide from cells; found in red blood cells.
	Myoglobin	Responsible for binding oxygen, which it obtains from hemoglobin, and storing it until needed; found in muscle tissue.

Linus Pauling received the Nobel prize in 1954 for his work on protein structure. The sequence of amino acid units joined by peptide linkages in the polypeptide chain is called the primary structure. Because of essentially free rotation around the covalent bonds in this chain, the macromolecules may assume an infinite number of shapes or conformations. However, Pauling showed that certain conformations are preferred because of intramolecular and intermolecular hydrogen bonding forming the so-called secondary structure.

The shape, size, and specific sequence of amino acid units composing a protein determine the specific function of the protein. For example, fibrous proteins are typically found as connective tissue, including tendons, bones, and cartilage, whereas globular proteins such as hemoglobin and myoglobin are often involved with transport. Because the structure of proteins is so important, great effort has been made to describe the structure of specific proteins in an effort to understand particular structure–property relationships. Four classifications are employed to describe the structure of a protein.

Primary structure

Primary structure describes the specific sequence of amino acid units composing the protein. Typically, only the primary bonding is considered when describing the primary structure of a protein. The diagrams of methionine enkephalin (Figure 6.4) and beef insulin (Figure 6.5) show primary structures.

Secondary structure

As the number of amino acid units increases, interaction between the amino acid units within a single chain becomes possible. The major "driving force" fixing preferred geometrics is secondary bonding forces, which are primarily hydrogen bonds. Two major secondary structures are observed, that is, the helix (Figure 6.4) and the sheet (Figure 6.5). Within these two major categories there exist variations. For instance, the particular carbon atoms involved with bonding and the number of amino acids within a complete circle of the spiral lead to names such as alpha helix and beta helix. Within such a straightforward division of secondary structure, further variations can arise. Thus, wool consists of helical protein chains connected to give a "pleated" sheet.

Tertiary Structure

Discussions related to the tertiary structure of a protein focus on the overall folding, that is, the turning of the protein chains. Although it is important to remember that such chains may exist in a helix or sheetlike secondary structure, the tertiary structure concerns only the overall, gross shape of the protein chain. Both secondary and primary forces in the form of cross-links are important in determining the tertiary structure.

Quaternary Structure

Protein chains can group together to give a larger, more intricate structural arrangement described as a quaternary structure. Thus, hemoglobin is composed of four protein chains, and each protein chain is described in terms of a tertiary structure and the sum are described in terms of a quaternary structure. Just as there are

Figure 6.4. Alpha helix conformation of proteins.

primary, secondary, tertiary, and quaternary structures for biological giant molecules, there are analogous structures for the somewhat simpler, more regular synthetic polymers. Illustrations are given in Figure 6.6.

Protein chains can also interact with other protein chains. Here we will consider only one type of such interactions, those forming multiple helices. Nature employs the triple helix as a building block; it has greater strength than a simple helix composed of a single protein chain, greater flexibility than simple sheet proteins, and results from the intertwining of three alpha helices. Disulfide (—S—S—) bonds act as cross-links and hydrogen bonding provides the cohesive forces holding the protein chains together. This triple helix is also called a protofibril.

Human hair is an example of a macrofibril, with the various triple helices, microfibrils, and finally the macrofibril held together by hydrogen bonding and sulfide cross-links. Moths excrete an enzyme that breaks such disulfide cross-links in the

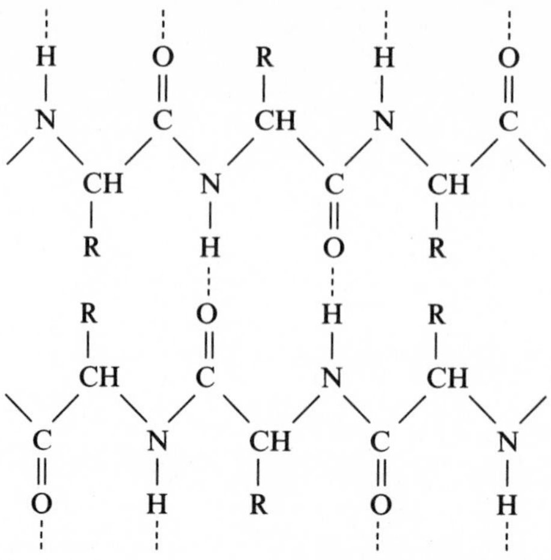

$$
\begin{array}{ccccc}
\text{H} & \text{O} & \text{R} & \text{H} & \text{O} \\
| & \| & | & | & \| \\
\text{N} & \text{C} & \text{CH} & \text{N} & \text{C} \\
\diagdown\diagup & & & \diagdown\diagup & \diagdown \\
\text{CH} & \text{N} & \text{C} & \text{CH} & \\
| & | & \| & | & \\
\text{R} & \text{H} & \text{O} & \text{R} &
\end{array}
$$

Figure 6.5. Beta arrangement or pleated sheet conformation of proteins.

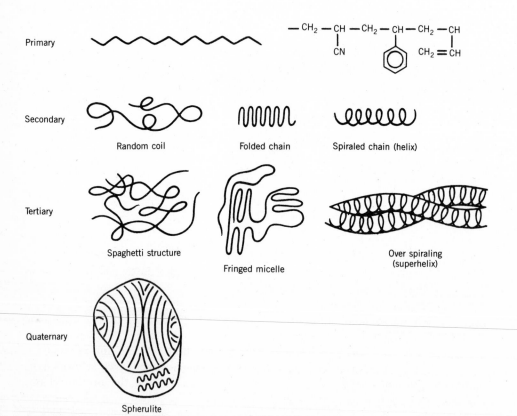

Figure 6.6. Schematic representations of various structural levels for linear synthetic giant molecules.

deterioration of wool. These cross-links are also cleaved and reformed in the cold waving of hair.

The preferred structure of fibrous proteins, such as hair, wool, and silk, is a sheet in which the hydrogen atoms on the nitrogen atom of the amide group (CONH) in one chain are attracted to the oxygen atoms of the amide group on an adjacent chain. In contrast, the preferred structure of nonfibrous proteins is a right-handed alpha helix, which results from intramolecular hydrogen bonding, that is, within the same molecule. Polymers of both L and D amino acids form helices with 3.6 units per turn. However, the winding direction is right for the helices of the L polymer and left for the helices of the D polymer.

Fibrillar proteins, such as keratin of the hair and nails, collagen of connective tissue, and myosin of the muscle, are strong, water-insoluble polymers. In contrast, globular proteins, such as enzymes, hormones, hemoglobin, and albumin, are usually weaker and more water soluble polymers.

6.5 ENZYMES

In addition to serving as food and structures for the animal kingdom, proteins, in the form of enzymes (from the Greek word *enzymos*, which means soup dough), perform specific functions by acting as catalysts in biological reactions. These specific proteins are responsible for reactions varying from eye movement, to the maintenance of body temperature, to the production of blood cells, to the digestion of food.

The "lock and key" concept proposed by Nobel laureate E. Fischer in the early 1900s is the most widely accepted theory for the specificity of enzymatic reactions. According to this theory, the chiral atoms in the amino acids provide a geometric pattern that permits specific reactions to occur while the reactant is locked in place. The products are released rapidly after the extremely fast reaction occurs, and the mechanism is then repeated numerous times. Nobel laureate J. Sumner isolated the crystalline enzyme urease in 1926. However, the leading organic chemists and biochemists of that era maintained that it was impossible to isolate crystalline proteins. It is now believed that specificity of an enzyme molecule is related to its interaction with three groups in the substrate. Many enzymes have approximately rounded overall shapes. Since the spherical shape requires less energy than rod- or coil-shaped polymers, it facilitates easy transport of enzymes.

6.6 WOOL

Wool from sheep was woven into fabrics by the ancient inhabitants of Egypt, Nineveh, Babylon, and Peru. The first factory in America using water power to weave wool was established at Hartford, Connecticut, in 1788. About 40 thousand tons of wool is used annually in the United States, and the worldwide consumption is 3 million tons annually.

The polymer chains in wool consist of parallel polypeptide alpha helices joined by disulfide (S—S) bonds. When wool is ironed with a wet cloth, immersed in an alkaline detergent, or subjected to tension in the direction of the helical axis, the hydrogen bonds parallel to the axes and the disulfide linkages are broken, and the structure can be elongated to nearly 100% of its possible length.

The opening and closing of these disulfide cross-linking groups allows the curling and uncurling of wool and of human hair. For humans, the wavesetting lotions sodium bisulfite ($NaHSO_3$) and ammonium thioglycolate ($HSCH_2COONH_4$) in hot- and cold-hair waving, respectively, open disulfide linkages and disrupt hydrogen bonds. The hair is curled and a neutralizer is added to reform the disulfide linkages, thus setting the hair in a curled or straight manner as desired.

6.7 SKIN PROTEIN

Elastin, a structural protein of skin, is somewhat similar to α-keratin in that it is composed of polypeptide chains linked covalently to form a network. In this case, the cross-linking results from the reaction of four lysine side chains from four adjacent polypeptide chains.

In contrast to the protein chains in wool, under moderate loads; elastin will stretch in a reversible, rubberlike fashion. The tetradentate nature (four reactive lysine groups) of the desmosine cross-links presumably aids in the re-formation of coiled polypeptide chains upon the release of tension and thus contributes to the elastic properties of skin protein.

6.8 SILK

Sericulture, that is, the culture of the silkworm (*Bombyx mori*), and the weaving of the silk filaments produced by the mulberry silkworm were of prime importance over 5000 years ago. In 2640 B.C., the Empress HSi-Ling-Shi developed the process of reeling by floating the cocoons on warm water. This process and the silkworm itself were monopolized by China until about A.D. 550 when two missionaries smuggled silkworm eggs and mulberry seeds from China to Constantinople (Istanbul). The crystalline silk fiber is three times as strong as wool.

Because of its high cost, only a small amount of silk (55 thousand tons) is woven worldwide annually. Both the discrete fibers of wool and the continuous filaments of silk are made up of macromolecules in which the repeating units consist of about 20 different amino acids.

6.9 NUCLEIC ACIDS

Nucleic acids, which are found in the nucleus of all living cells, are responsible for the synthesis of specific proteins and are involved in the generic transmission of characteristics from parent to offspring. Nucleic acids were discovered in 1867 by J. Miesher, who isolated these materials from the remnants of pus from cells. Since nucleic acid was found in the nucleus of cells, he called this material "nuclein," but this name has been changed to nucleic acid. It should be noted that nucleic acids are also found in the cytoplasm as well as in the nucleus of cells.

P. Levene, who discovered D-deoxyribose in 1929, showed that D-riboses were present in pure nucleic acid. Avery, MacLeod, and McCarty showed that deoxy-nucleic acid (DNA) was the basic genetic component of chromosomes in 1944. Nobel laureate Alexander R. B. Todd synthesized adenosine diphosphate (ADP) and adenosine triphosphate (ATP) in 1947. These compounds are not only important components of nucleic acids but are also involved in biological energy transfer.

The terms DNA and RNA are derived from the specific sugar moiety present. Deoxyribose is the sugar present in deoxyribose nucleic acids (DNAs), whereas ribose

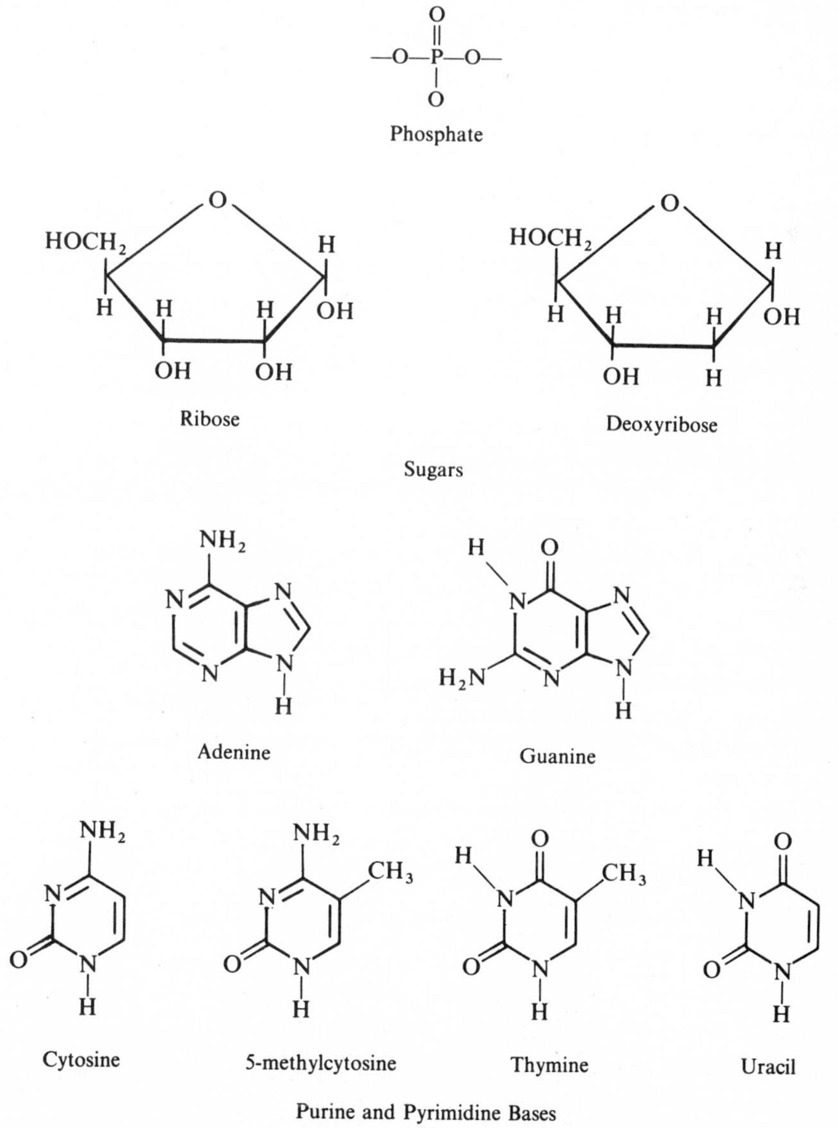

Figure 6.7. Chemical structures of components of nucleic acids.

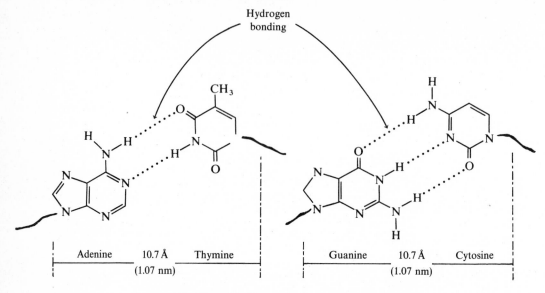

Figure 6.8. Allowable base pairs in nucleic acids (DNA).

sugar is present in ribose nucleic acids (RNAs). The difference between the two sugars is the absence of one of the hydroxyl (OH) groups in the deoxyribose.

The three components of a nucleotide are a purine or a pyrimidine base, a pentose, and phosphoric acid (Figure 6.7). As already noted, the difference between DNA and RNA is the pentose component and the particular bases that are present. Both DNA and RNA contain adenine, guanine, and cytosine, whereas RNA also contains uracil and DNA also contains thymine and 5-methylcytosine. A third difference is the tendency of RNA to be single-stranded and not to possess a regular helical structure, whereas DNA can be double-stranded and typically forms a regular helical structure.

The purine and pyrimidine bases, called adenine (A), guanine (G), cytosine (C), and thymine (T), are held together by hydrogen bonds in the parent cell. The pyrimidine molecules are smaller than the purine molecules, and one of each of these bases can fit between the strands in the DNA double helix, as shown in Figures 6.8 and 6.9. This formation of base pairs may be remembered from the mnemonic expression Gee-CAT. The pentose molecules are joined together through the phosphate units and the base portions are present as substituents on the pentose molecule. A portion of a DNA chain is shown in Figure 6.10.

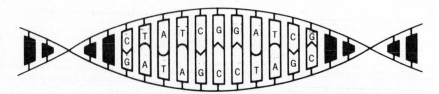

Figure 6.9. A schematic representation of the double helix of DNA.

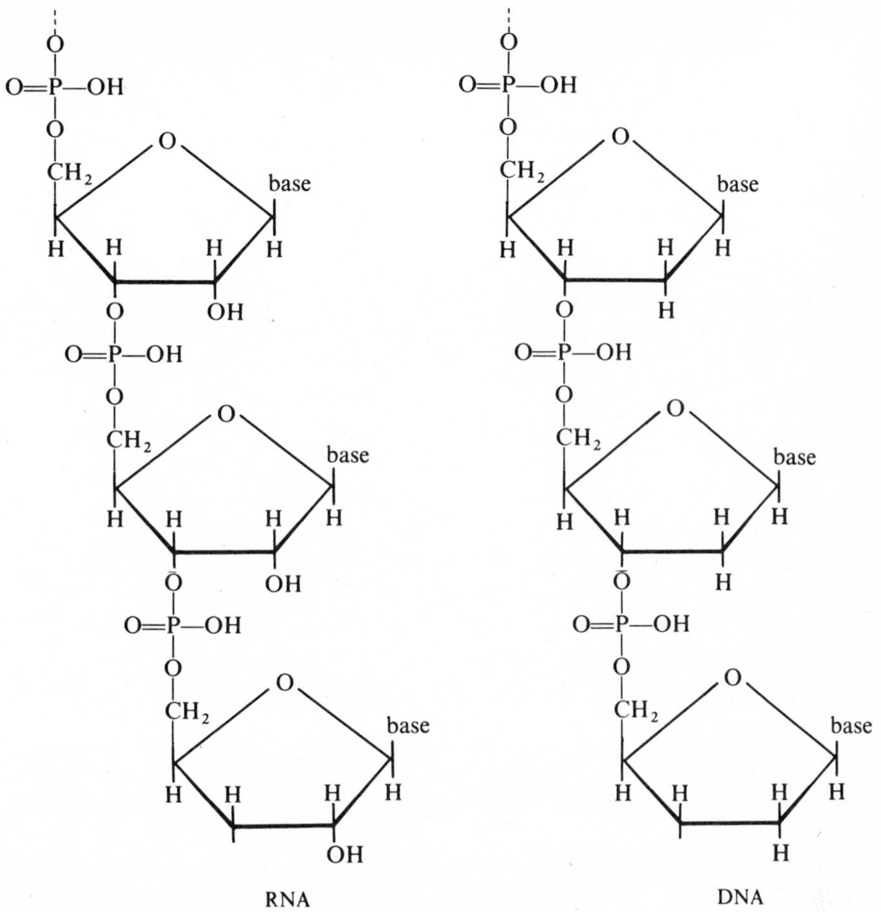

Figure 6.10 Representative structural units of a segment of RNA (left) and DNA (right).

Nobel laureates James Watson and Francis Crick correctly postulated the double-stranded helical structure in 1953; this was confirmed by x-ray diffraction in 1973. The complete biochemical synthesis of a biologically active DNA of a virus was accomplished by utilizing two enzymes discovered by Nobel prize winner Arthur Kornberg in 1967. The enormity of such a DNA molecule is easier to understand when one considers that a single DNA molecule laid lengthwise would be about a centimeter in length.

6.10   THE GENETIC CODE

DNA controls the synthesis of RNA, which in turn controls the synthesis of proteins. The general steps can be outlined as follows: First DNA synthesizes smaller RNA molecules called messenger RNA (m-RNA). The DNA determines the sequences of bases that will be presented in the messenger RNA through pairing with a partially untwisted portion of a DNA chain.

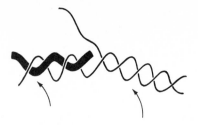

Another kind of RNA called transfer RNA (t-RNA) serves as the "traffic officer" for amino acids; it selects only certain amino acids and transports these amino acids to the m-RNA template. The molecular weight of transfer RNA is much lower than that of m-RNA, and it is more soluble and more mobile in the cell fluids. There are at least 20 different forms of t-RNA (soluble RNA), each being specific for a given amino acid.

The anticodons, which consist of a specific sequence of bases, allow the t-RNA to bind with specific sites, called the codons of m-RNA. The order in which amino acids are brought by t-RNA to the m-RNA is determined by the sequence of codons. This sequence constitutes what is referred to as a genetic message. Individual units of that message (individual amino acids) are designated by triplets of nucleic acid units.

The m-RNA can be visualized as a template with indentions that have specific spatial (geometric, steric) and electronic characteristics.

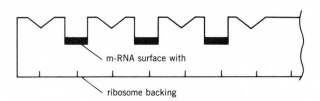

m-RNA surface with

ribosome backing

Briefly, instructions to replicate a DNA molecule can be visualized as a long sentence containing about 10 letters, one for each nucleic acid unit. There are only four letters in this alphabet—A, C, G, and T—but since the t-RNA operates on a 3-base code, 4^3 or 64 different sequences are possible. This number is far more than is needed to recall only 20 amino acids and more than one sequence may be available to recall a specific amino acid. Sequences are also available for the messages of "start here" and "end here." After much experimentation, each of these 64 sequences has been identified with a specific amino acid or other message.

Each cell contains a chromosome, which in turn contains a DNA molecule. Reproduction occurs on signal when a new cell is needed as described in the foregoing.

6.11   GENETIC ENGINEERING

A gene is a section of DNA in a chromosome. Genetic engineering consists mainly of uniting pieces of spliced DNA from different sources to produce a recombinant DNA. The potential applications of genetic engineering sometimes appear in fiction, where the possibility of cloning human beings has fascinated novelists, playwrights, and comic book writers. The term clone comes from the Greek word *klon*, meaning a cutting used to propagate a plant. Cell cloning is the production of identical cells from a single cell. Similarly, gene cloning is the production of identical genes from a single gene. During gene cloning, genes from different organisms are often joined together to form one artificial molecule known as a recombinant DNA.

The chemical reactions used in gene cloning are analogous to those encountered in elementary organic chemistry. The first of these, the "cutting" of a gene, is actually the hydrolysis of DNA. The second type of reaction involves joining DNA molecules together by a dehydration reaction. Most procedures require a vector (plasmids) and restriction enzymes. Vectors are the name given to the material that carries the recombinate DNA into a cell so that it will be accepted in the new environment and allowed to reproduce. Today the usual vectors for bacterial cells are plasmids, which are small, free-floating ringlets of DNA that are present in most cells and carry genetic information concerning resistance to antibiotics.

Restriction enzymes cut double-stranded DNA at predictable places, leaving the DNA with uneven ends containing a short segment of only one of the strands. These ends are called sticky ends since they can combine with other sticky ends through a process in which the bases are paired. These restriction enzymes cut DNA strands only at locations where there is a palindrome base structure. A palindrome is an arrangement of letters that reads the same way forward and backward, such as

DOG, OTTO, GOD

or, for the four bases (*Gee-CAT*),

GAATTAAG

would be a palindrome as would

AGATTAGA

Thus for a paired sequence

we see a palindrome in the pair where the top pair reads in reverse direction to the lower pair. A restriction enzyme may then split this pair as

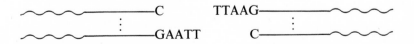

producing two sticky ends—one containing the uncoupled TTAA portion and on the lower strand an AATT uncoupled portion.

A second DNA sample can now be "cut" by employing the same restriction enzyme to produce complementary sticky ends:

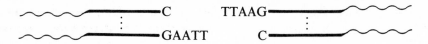

These DNA portions from the two different genes are mixed along with another enzyme (DNA lipase) that will chemically bind the complementary sticky ends:

The new recombinate DNA then enters a cell, carrying with it the new information.

New techniques and sequences are continually being developed and new applications are being studied. The goal of genetic engineering is to couple desired characteristics. For instance, if the synthesis of a specific protein is desired but its natural production occurs slowly and only as a trace material, a genetic engineer might couple a gene portion that dictates production of that specific protein with a gene portion that rapidly reproduces itself.

GLOSSARY

Adenosine phosphate: A purine base.

Amino group: —NH_2.

Anticodon: A specific sequence of purine and pyrimidine bases that permits t-RNA to bond with a specific set.

Bonds, secondary: Bonds based on van der Waals forces, that is, London dispersion bonds, dipole–dipole interactions, and hydrogen bonds.

Carboxyl group: —COOH.

Chiral atom: A carbon atom with four different groups.

Conformation: Different shapes of a molecule.

D-Ribose: A five-carbon carbohydrate (pentose).

Deoxynucleic acid: DNA.

Deoxyribose: A ribose in which a hydrogen atom replaces a hydroxyl group.

Desmosine: A skin protein.

Disulfide bond: S—S.

DNA: Deoxyribonucleic acid.

Elastin: A skin protein.

Enzyme: A biological catalyst.

Essential amino acid: Amino acids required for good health.

Fibrillar protein: Hairlike, water-insoluble protein structures resulting from intermolecular hydrogen bonding.

Genetic engineering: Any artificial process used to alter the genetic composition.

Globular protein: Water-soluble protein structures with intramolecular hydrogen bonds.

Gly: Abbreviation for glycine; other abbreviations for amino acids are shown in Table 6.1.

Glycine: The first member of the α-amino acid homologous series ($CH_2(NH_2)COOH$).

Hydrophilic: Water-loving.

Hydrophobic: Water-hating.

Insulin: A hormone that controls glycogenesis and glycogenolysis.

Intermolecular bonding: Secondary bonding between atoms in two different molecules.

Isoelectric point: The pH value at which the negative and the positive charges on an amino acid are equal. This point is characteristic for each amino acid.

Isomers: Molecules with identical structures, that is, the same formulas.

Kwashiorkor: Chronic disease caused by deficiency of essential amino acids.

L-Amino acid: Levo isomer that is the typical isomer for naturally occurring amino acids.

Lock and key concept: Theory of enzymatic activity that requires a specific fitting of a reactant in the enzyme structure that permits a specific reaction to occur.

Lysine: An essential amino acid (see Table 6.1).

Milk of magnesia: $Mg(OH)_2$.

Molarity: A measure of the number of moles present in a liter of solution. A 1 molar solution of sodium hydroxide (NaOH) contains 40 g of sodium hydroxide.

Mole: 6.023×10^{23} particles.

Molecular biology: Science applied at the molecular level in biological systems.

Mutation: A mistake in coding transfers.

Nucleotide: The repeating unit in nucleic acids.

Optical isomer: An isomer that rotates the plane of polarized light.

Pellagra: A chronic disease caused by a deficiency of lysine.

Peptide linkage:

$$
\begin{array}{cc}
H & O \\
| & \| \\
\!\!\!\!-\!N\!-\!C\!- \\
| & \\
R & \\
\end{array}
$$

pH: An acidity scale in which 7.0 is the neutral point. pH values less than 7.0 are acidic, and those greater than 7.0 are alkaline.

Polypeptide: Name commonly used for low-molecular-weight amino acid polymers.

Polysome: A complex formed from RNA and ribosomes.

Primary structure: Structure resulting from primary bonding of atoms in proteins.

Prosthetic group: A nonprotein group, such as glucose, joined to a protein molecule.

Protein: A polymer made up of repeating units of α-amino acids.

Purine: A heterocyclic molecule with two fused rings.

Pyrimidine: A heterocyclic molecule consisting of one nitrogen-containing ring.

R: An alkyl group ($H(CH_2)_n$-).

Ribonucleic acid: RNA.

RNA: Ribonucleic acid.

m-RNA: Messenger RNA.

t-RNA: Transfer RNA.

Secondary structure: Conformations in proteins resulting from intermolecular and intramolecular attractions.

Sericulture: The culture of the silkworm and weaving of silk filaments.

Sodium hydroxide: NaOH.

Tertiary structure of proteins: Three-dimensional shape of protein molecules based on foldings of protein chains.

Tufting: Splitting of fibers.

Zwitterion: A polar molecule made up of an anion and a cation. In amino acids, the anion is the carboxyl ion ($—COO^-$) and the cation is the ammonium ion ($—NH_3^+$).

REVIEW QUESTIONS

1. What does DNA stand for?

2. What does RNA stand for?

3. How many different repeating units are present in proteins?

4. What does R stand for?

5. How does glycine differ from all other amino acids?

6. What does Gly stand for?

7. Are naturally occurring amino acids D or L?

8. What is the name given to a molecule like an amino acid that contains both an acidic and a base group?

9. What is the pH of distilled water?

10. Is the pH of an acid less than or greater than 7.0?

11. Define isoelectric point.

12. Give two major properties of an enzyme.

13. What is the repeating linkage in a protein?

14. What is the term used for hydrogen bonds between atoms in the same molecule?

15. Is the primary structure in proteins dependent on configurations or conformations?

16. Is the secondary structure in proteins dependent on configurations or conformations?

17. What type of hydrogen bonds are present in fibrillar proteins?

18. What type of hydrogen bonds are present in globular proteins?

19. Which is water repellent: hydrophilic or hydrophobic?

20. What is silk culture called?

21. What is the difference between D-ribose and D-deoxyribose?

22. What is the repeating unit in nucleic acid called?

23. The mnemonic expression Gee-CAT stands for what bases?

24. What is the general name for the bases in DNA?

BIBLIOGRAPHY

Bohenski, R. C. (1987). *Modern Concepts of Biochemistry.* Newton, MA: Allyn & Bacon.

Conn, E. E., *et al.* (1987). *Outlines of Biochemistry.* New York: Wiley.

Davidson, J. N. (1972). *The Biochemistry of Nucleic Acids.* New York: Wiley.

Dickerson, R. E., and Geis, I. (1969). *The Structure and Action of Proteins.* New York: Harper & Row.

Kiopple, K. D. (1966). *Peptides and Amino Acids.* New York: Benjamin.

Kochetkov, N. K., and Budovskil, E. I. (1971). *Organic Chemistry of Nucleic Acids.* New York: Plenum.

Merrifield, R. B. (1963). Solid phase peptide synthesis. *Journal of the American Chemical Society,* **85**, 2149.

Perotz, M. F. (1962). *Proteins and Nucleic Acids.* Amsterdam: Elsevier.

Prave, P., *et al.* (1987). *Fundamentals of Biotechnology.* New York: VCH Publishers.

Scheraga, A. A. (1961). *Protein Structure.* New York: Academic Press.

Schulz, G., and Schirmer, R. (1979). *Principles of Protein Structure.* New York: Springer–Worley.

Walter, P., Melen, J., and Nofer, J. (1976). *Peptides.* Ann Arbor, MI: Ann Arbor Science Publishers.

ANSWERS TO REVIEW QUESTIONS

1. Deoxyribonucleic acid.

2. Ribonucleic acid.

3. About 20 different amino acid residues (repeating units).

4. In general, R can represent almost any moiety that is unspecified.

5. It does not contain a chiral carbon atom, therefore has no optical isomers.

6. Glycine.

7. L (levo).

8. Zwitterion.

9. 7.0.

10. Less than 7.

11. The pH at which the negative charge equals the positive charge in an amino acid. This is the neutral point.

12. Enzymes are catalysts and are specific.

13. The peptide linkage

$$\begin{array}{ccc}
\overset{\displaystyle H}{\underset{\displaystyle |}{}} & & \overset{\displaystyle R}{\underset{\displaystyle |}{}}\;\;\overset{\displaystyle H}{\underset{\displaystyle |}{}} \\
+\text{N}\!-\!\text{CO}+ & \text{or} & +\text{C}\!-\!\text{N}\!-\!\text{CO}+ \\
& & \underset{\displaystyle H}{|}
\end{array}$$

14. Intramolecular hydrogen bonds.

15. Configurations, that is, primary bonds.

16. Conformations.

17. Intermolecular hydrogen bonds.

18. Intramolecular hydrogen bonds.

19. Hydrophobic.

20. Sericulture.

21. D-Ribose contains one more oxygen atom than D-deoxyribose.

22. Nucleotide.

23. Guanine, cytosine, adenine, and thymine.

24. Purine and pyrimidine bases.

CHAPTER VII

Derivatives of Natural Polymers

7.1 INTRODUCTION

In addition to their use as food, shelter, and clothing, polymeric carbohydrates, proteins, and nucleic acids are also important commercial materials.

7.2 DERIVATIVES OF CELLULOSE

C. F. Schönbein accidentally produced explosive cellulose nitrate in 1846 when he used his wife's cotton apron to wipe up a spilled mixture of nitric acid (HNO_3) and sulfuric acid (H_2SO_4) and dried the wet apron by heating it on the stove. Gun cotton

117

or cellulose trinitrate is produced when concentrated acids are used, but the more polar and less expensive cellulose dinitrate is obtained when dilute solutions of the acids are used. A segment of the polymer chain of cellulose dinitrate has the structure

Cellulose dinitrate

For all of these polyfunctional materials, substitution occurs such that a repeating unit may contain, for example, one, two, or three substituents for each repeat unit. Thus cellulose dinitrate will contain, on the average, two nitrate substituents per unit but some units will contain three, some two, some one, and some zero nitrate groups.

Nitrated cellulose, in a mixture of ethanol (C_2H_5OH) and ethyl ether (($C_2H_5)_2O$), called collodion, is still used as a liquid court plaster (Nu/Skin). It was modified by J. N. Hyatt to produce a plasticized cellulose dinitrate, which he called celluloid.

A plasticizer is a nonvolatile compound that enhances the segmental motion of the polymer chains. Hyatt, who was a printer and inventor of the roller bearing, used camphor to plasticize cellulose dinitrate. He added pigments to this plasticized polymer to produce billiard balls and for his efforts won a prize of $10,000. These balls had previously been carved out of the tusks of elephants. A competitive flexibilized product called Parkesine was produced by A. Parkes in England in 1861. In spite of these inventions, over 100,000 elephants were killed annually for their tusks in the nineteenth century.

Brush handles, baby rattles, shirt fronts, and collars were made from celluloid and phonograph records were made by Thomas A. Edison from shellac in the nineteenth century. However, because of the combustibility of celluloid and the high cost of shellac, the use of these plastics was limited.

The fact that cellulose nitrate is highly flammable and explosive should not be surprising since most highly nitrated compounds, such as TNT and dynamite (glyceryl nitrate), are highly flammable and explosive.

Trinitrotoluene
(TNT)

Glyceryl trinitrate
(Nitroglycerin)

When all the hydroxyl groups in cellulose are nitrated, the material is known as nitrated cotton or gun cotton. When a mixture of gun cotton and lesser nitrated cellulose is mixed with ethanol (ethyl alcohol, CH_3CH_2OH) and diethyl ether ($CH_3CH_2OCH_2CH_3$), it produces a jellylike mass. This can be rolled into strips and dried to a consistency of dry gelatin. When ignited in small quantities in the open, this smokeless powder burns readily without exploding. Under the confinement of a rifle or gun cylinder, decomposition after ignition is rapid and much heat and gas are evolved. From the following balanced equation, it can be shown that from a 1-g (about the size of a small pea) charge of gun cotton, gas at about room pressure would occupy a volume of one quart. This expanding gas propels the shell from the gun. Since there is no solid residue left by the decomposition of gun cotton, the barrel chamber is left clean.

$$\longrightarrow 5xCO + 7xCO_2 + 4xH_2 + 3H_2O + 3N_2$$
$$\text{or for } x = 1 \rightarrow 5CO + 7CO_2 + 4H_2$$
$$+ 3H_2O + 3N_2$$

P. Schutzenberger reacted cellulose to produce cellulose triacetate in 1869, but the large-scale use of this cellulosic derivative was delayed because expensive solvents, such as chloroform, were required to dissolve this cellulose derivative. G. Miles solved this dilemma in 1902 by removing one of the acetyl groups from each repeating unit to produce cellulose diacetate in a process called partial saponification, which is simply soap making. The chemical equations for acetylation of cellulose and partial deacetylation by saponification are

acetic anhydride   Cellulose triacetate

Cellulose diacetate

Henry and Camile Dreyfus developed the fiber-making process for cellulose acetate. This fiber, which is now called acetate rayon, has been produced in the United States by Celanese Corporation (now Hoechst-Celanese) since 1924. Cellulose triacetate is employed as the fabric for many tricot fabrics and sportswear. Today's triacetate is shrink and wrinkle (permanent press modified) resistant and easily washed.

J. Brandenberger made films of cellulose acetate in 1908 when he sprayed a solution of cellulose acetate on a tablecloth in an attempt to waterproof it. He found that he could peel the cellulose acetate from the cloth to produce a thin, transparent film. Because it was less flammable, cellulose acetate film was used as a replacement for celluloid film. Cellulose acetate is also used as a lacquer, called "dope," which was used to coat cloth for World War I airplanes. It is still used as a fingernail polish.

7.3   CELLOPHANE

Transparent film, such as cellophane, is produced by passing a solution of cellulose xanthate, called viscose, through a slit and precipitating this sheet in an acid solution. The xanthation reaction, with one hydroxyl group (OH) in the repeating units of cellulose, is shown in the following abbreviated equation, in which Cell is used as the abbreviation for cellulose.

$$\text{Cell—OH} + \text{NaOH} \xrightarrow{-H_2O} \text{Cell—O, Na}^+$$

$$\text{Cell—O, Na}^+ + CS_2 \longrightarrow \text{Cell—O—C(=S)—S}^-, \text{Na}^+$$

$$\underset{\text{Cellulose xanthate}}{\text{Cell—O—C(=S)—S}^-, \text{Na}^+} + H_2SO_4 \longrightarrow \underset{\substack{\text{Sodium} \\ \text{acid} \\ \text{sulfate}}}{NaHSO_4} + \underset{\substack{\text{Carbon} \\ \text{disulfide}}}{CS_2} + \underset{\text{Cellophane}}{\text{Cell—OH}}$$

The annual production of cellophane film was almost 500 million pounds in the 1960s, but because of its high cost compared to other film products, its annual production in the United States has decreased to about 150,000 pounds.

7.4   RAYON

Although Robert Hooke suggested the possibility of making "artificial silk" in 1665, the first man-made fiber was not produced until two centuries later by Hilaire de Chardonnet. These regenerated cellulosic filaments were produced by passing collodion through small, uniformly sized holes, called spinnerets, evaporating the ethanol and ether solvent, and then removing the nitrate groups from the cellulose nitrate filaments (Figure 7.1).

Three principal methods have been employed in the synthesis of rayon. These are the viscose process, the cuprammonium process, and the acetate process. Viscose and

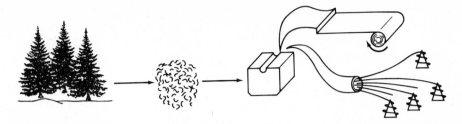

Figure 7.1. Cellophane and viscose rayon are produced from cotton or wood pulp that is matted, shredded, dissolved in basic solutions, and pumped through a slit (top) or spinneret (bottom) into an acid bath, where the cellophane or rayon precipitates to form sheets and threads.

cuprammonium rayons have similar chemical and physical properties. Both are dyed easily and lose their strength when wet (disruption of hydrogen bonds). Acetate rayon is easily softened when ironed and boiling water decreases its luster, but it is a finer texture and more readily dyed than viscose or cuprammonium rayons.

7.5 DERIVATIVES OF STARCH

Starch obtained from rice, wheat, potatoes, and corn is used worldwide as a source of food. Starch obtained from corn and potatoes is used industrially for adhesives and sizing for paper and textiles in the United States. Most starches contain about 25% amylose and 75% amylopectin. However, waxy corn is almost 100% amylopectin. Over 125 million tons of corn are used annually as a source of amylopectin in the United States.

The hydroxyl groups on each glucose residue in starch undergo the same reactions as the hydroxyl groups in cellulose. At high degrees of substitution (DS), starch becomes a thermoplastic. However, unplasticized starch triacetate tends to be brittle, and the higher esters are soft and weak.

A wide variety of graft copolymers of starch have been made, largely through addition of vinyl monomers such as acrylonitrile and methyl acrylate to free radical sites generated by irradiation or chemical treatment (Figure 7.2). The so-called "Super Slurper" that absorbs hundreds of times its own weight of water is a graft copolymer.

Soluble starch is produced by treating starch with 7.5% hydrochloric acid for 1 week at room temperature. The adhesiveness of starch is improved by oxidation with hydrogen peroxide (H_2O_2) or sodium hypochlorite (NaOCl) (bleach). Partially degraded starch (dextrin) is used as an adhesive on postage stamps. Starch nitrate is used as an explosive.

7.6 LEATHER

Leathermaking, which was the first man-made polymer reaction, involves the introduction of cross-links into the protein molecule. The tanners who produced this most historic of materials were some of our earliest craftsmen. That this ancient

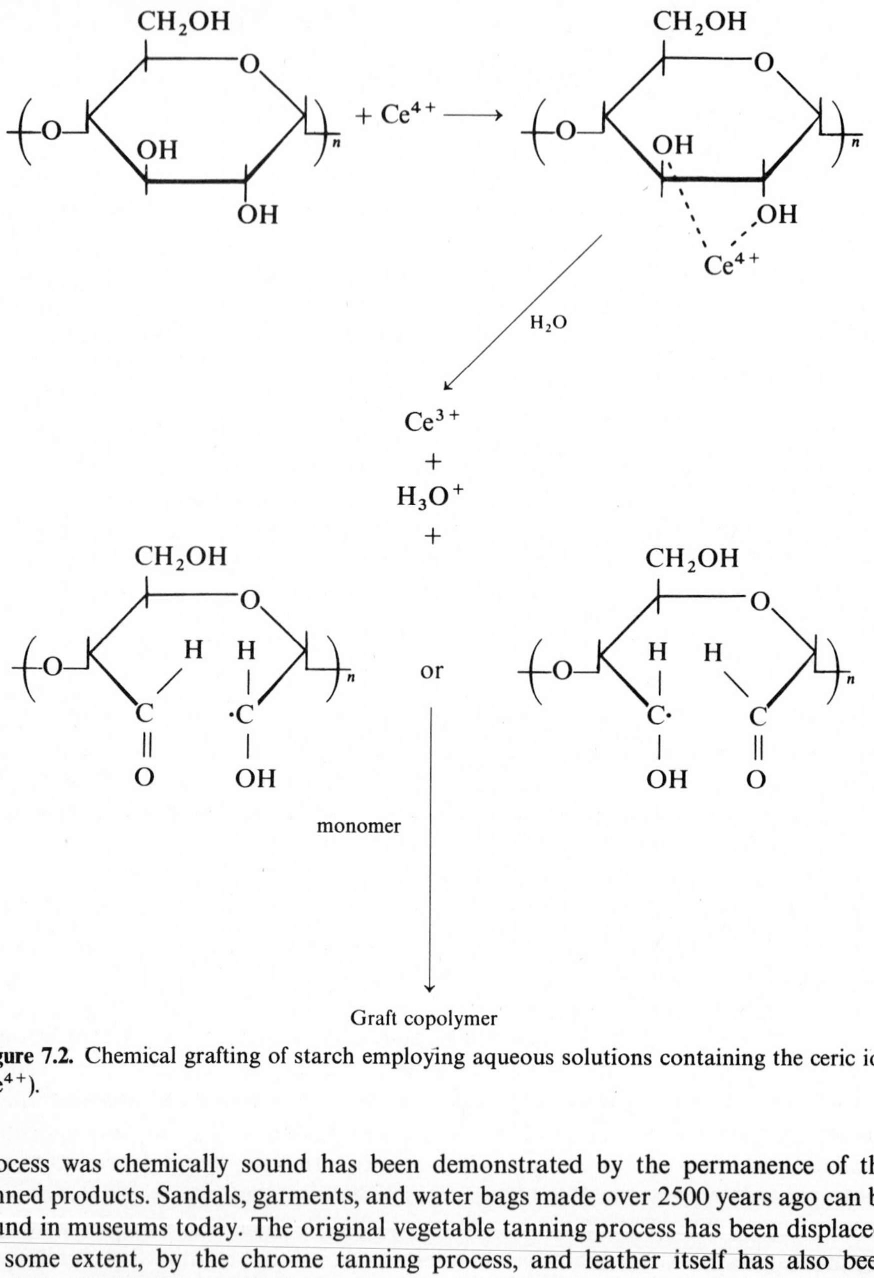

Figure 7.2. Chemical grafting of starch employing aqueous solutions containing the ceric ion (Ce^{4+}).

process was chemically sound has been demonstrated by the permanence of the tanned products. Sandals, garments, and water bags made over 2500 years ago can be found in museums today. The original vegetable tanning process has been displaced, to some extent, by the chrome tanning process, and leather itself has also been partially displaced by synthetic sheeting, but the vegetable tanning art is still practiced widely throughout the world.

7.7   REGENERATED PROTEINS

Proteins, such as those in animal blood, hides, and milk, are readily available and have been used as food and adhesives for many years. The proteinaceous gluten in wheat flour has been used as an adhesive in dough formation for centuries. The first animal glue factory was built in Holland in 1690, and a comparable plant was built in the United States in 1808.

Casein is obtained by acidification or by the addition of rennet to skimmed milk. E. Childs patented casein as a plastic in 1885, and W. Krische and A. Spitteler used this proteinaceous material to produce a commercial moldable plastic (artificial horn, Galalith) in 1897. Zein from maize and protein from soybeans and peanuts have been used to a limited extent for molded plastics, but these products are not competitive with general-purpose synthetic plastics.

In addition to food and adhesives, the widest use of proteins is for regenerated fibers. A. Ferretti produced casein fiber (Lanital) by forcing an alkaline solution of casein through small holes (spinnerets) into an acid bath. Other casein fibers are called Aralac and Merinova.

Regenerated protein fibers are also made from soybeans, peanuts (Ardil), corn (Zein), and chicken feathers. All of these regenerated protein fibers must be immersed in formalin to prevent bacterial attack. Unlike polyamide fibers, such as nylon, some of these bacteria-resistant polyamide fibers, called by the generic name Azlon, are attacked by moths.

7.8   NATURAL RUBBER

In 1839, Charles Goodyear cross-linked natural rubber by heating it with a small amount of sulfur. Since the rubbery chains are held together by just a few sulfur cross-links, the "vulcanized" product is flexible and is said to have a low cross-linked density. His brother, Nelson Goodyear, produced hard rubber or ebonite by heating rubber with a large amount of sulfur to produce a plastic product with high cross-linked density.

The cross-linking of rubber by the formation of sulfur cross-links is a complex chemical reaction. However, Charles Goodyear carried out his first vulcanization reactions on his wife's kitchen stove and many competitors followed his simple procedure and ignored his patents. He spent several hundreds of thousands of dollars contesting infringements on his patents, and because of his indebtedness, he spent considerable time in a debtor's prison. Notably, his attorney in "The Great India Rubber Patent Infringement Case" in 1852 was Daniel Webster. The court ruled in favor of Goodyear, but he died in 1860 before he could earn enough money from his royalties to pay his debts.

Charles Goodyear was the first to convert a linear elastomer to a cross-linked elastomer, and Nelson Goodyear was the first to convert an elastomer to a thermoset. The invention of the pneumatic tire by Dunlop in 1888 and the fountain pen by Waterman in 1884 were dependent on the inventions of Charles and Nelson Goodyear, respectively.

7.9   DERIVATIVES OF NATURAL RUBBER

Vulcanized rubber, of course, is a derivative, but this product is just called rubber by English-speaking consumers. Nevertheless, there are many other derivatives of natural rubber that were produced prior to the development of the synthetic polymer industry.

In 1927, Fisher isomerized rubber in the presence of acids, such as *p*-toluene sulfonyl chloride ($H_3CC_6H_4SO_2Cl$). These thermoplastics, or "thermoprenes," were used as adhesives for bonding rubber to metal. The most widely used product, called Pliolite, was produced by heating rubber in the presence of chlorostannic acid (H_2SnCl_6). This cyclized rubber continues to be used to a limited extent, but because of high costs it has been displaced by synthetic adhesives and coatings.

Since rubber is an unsaturated polymer, it adds chlorine readily. This chlorination reaction was known for many years but was not commercialized until 1925, when Peachey patented the chlorinated thermoplastic product under the trade name of Duroprene and Alloprene. Chlorinated rubber was produced in Germany under the trade name of Tornesit and later in the United States under the trade name of Parlon. Chlorinated rubber has been used as a protective coating for swimming pools.

Rubber hydrochloride (Pliofilm) is produced by the reaction of natural rubber with gaseous hydrogen chloride (HCl). Films of rubber hydrochloride, cast from solutions of this derivative in benzene, have been used as transparent packaging materials.

7.10   MODIFIED WOOL

Permanent-press wool is machine-washable, retains imparted creases or pleats, and requires minimum ironing after washing and drying. In the CSIRO solvent–resin–steam process, a reactive polyurethane is applied to the garment, which is pressed and hung in a steam oven. The polymer is cured on the garment in the required shape. The International Wool Secretariat's permanent-press process employs a thiol-terminated prepolymer.

Keratins have a high cystine content, approximately 500 μmol/g for wool. These cystine residues cross-link with adjacent protein chains via disulfide bonds, thus restricting their conformational motion. Treatment of wool with reducing agents converts these disulfide cystine groups to two cysteine residues:

$$—NHCHCH_2—SS—CH_2CHCO— \rightarrow —NHCHCH_2SH + HSCH_2CHCO$$
$$\vert\vert\vert\vert$$
$$—CONH——CONH—$$

7.11   JAPANESE LACQUER

Japanese lacquer is a naturally occurring, phenolic coating material obtained from the lacquer tree *Rhus vernicifera*. Its use has been dated back 5000–6000 years to continental China. The sap consists of urushiol (65–70%), a polysaccharide plant gum (5–7%), a copper-bearing lactose (<1%), and water (20–25%). On drying, Japanese lacquer forms a cellular structure about 0.1 μm in diameter, with gum as the cell wall.

7.12   NATURAL POLYMERS THROUGH BIOTECHNOLOGY

Poly(3-hydroxybutyrate) (PHB) is produced by bacteria. It occurs in soil bacteria, estuarine microflora, blue-green algae, and microbially treated sewage. A good laboratory source is *Alcaligenes eutrophus*. These microorganisms can be grown in large tanks on a variety of substrates, including natural sugars, ethanol, or gaseous mixtures of carbon dioxide and hydrogen. Large quantities of PHB accumulate as discrete intercellular granules, which may constitute up to 80% of the cell's dry weight.

7.13 OTHER PRODUCTS BASED ON NATURAL POLYMERS

Gutta-percha and balata, which are trans isomers of polyisoprene, have been used as cable coatings and molding resins. Since their chemical formulas are identical to the cis isomer of polyisoprene, they undergo all of the reactions listed for natural rubber. However, none of these derivatives has been produced commercially.

Shellac is the secretion of the lac insect (*Laccifer lacca*), which feeds on the sap of trees in southeastern Asia. Thomas Edison used this product for molding gramophone records. Moldings of mica-filled shellac have also been used for electrical insulators. Alcohol solutions of shellac continue to be used as coatings, but oleoresinous coatings are more widely used.

The art of decorative painting was developed in prehistoric times, and the first successful human nonvocal communication was probably through paintings on cave walls. Varnishes based on natural materials, such as beeswax, were used over 3000 years ago.

All oil paints consist of finely ground pigments, a solvent or thinner, and a resin-forming component called a binder. The solution of the binder in the solvent is called the vehicle. Some modern paints contain natural resins, such as copals, kauri, manila, and others. However, these natural resins as well as the oleoresinous paints are being displaced by coatings based on synthetic resins, which are generally superior and more economical than the natural products.

GLOSSARY

Acetate rayon: Cellulose acetate fibers.

Alum: Aluminum sulfate.

Ardel: Peanut protein fiber.

Azlon: Protein fibers.

Binder: Resin-forming component in paint.

Cachuchu: South American Indian name for rubber.

Casein: Milk protein.

Cellophane: Regenerated cellulose film.

Celluloid: Cellulose nitrate plasticized by camphor.

Cellulose: A polycarbohydrate with repeating units of D-glucose joined by beta acetal linkages.

Cellulose acetate: A product of the reaction of cellulose and acetic anhydride.

Cellulose nitrate: A product of the reaction of nitric acid and cellulose (erroneously called nitrocellulose).

Cellulose xanthate: The reaction product of sodium cellulose and carbon disulfide (Cell-CSS$^-$, Na$^+$).

Chlorinated rubber: The reaction product of rubber and chlorine.

Cinnabar: HgS.

Collodion: A solution of cellulose nitrate in an equimolar mixture of ethanol (C_2H_5OH) and ethyl ether (($C_2H_5)_2O$).

Cyclized rubber: A derivative of rubber with a cyclic (ring) structure.

Dextrin: Partially degraded starch.

Dope: A solution of cellulose acetate in acetone.

Ebonite: Hard rubber.

Formalin: 37% aqueous solution of formaldehyde.

Galalith (milkstone): Molded casein articles.

Gluten: Wheat flour protein.

Hematite: Fe_2O_3.

Hydrogen peroxide: H_2O_2.

Lanital: Casein fiber.

Lapis lazuli: Egyptian blue.

Leather: Cross-linked protein produced by a tanning process.

Linear elastomer: An elastomer with a continuous polymer chain.

Malachite green: $CuCO_3 \cdot Cu(OH)_2$.

Methylol group: $HOCH_2-$.

Ocher: Hydrated iron oxide pigment.

Orpiment yellow: As_2S_3.

Plasticizer: A flexibilizing additive.

Pyrolusite: MnO_2.

Rayon: Regenerated cellulose fiber.

Rubber hydrochloride: The reaction product of rubber and gaseous hydrogen chloride.

Saponification: Hydrolysis of an ester to produce an alcohol and an acid.

Schweitzer's solution: $Cu(NH_3)_4(OH)_2$.

Shellac: A natural resin obtained from the excreta of coccid insects that feed on twigs of trees in southeastern Asia.

Sizing: A coating on paper or textiles.

Sodium hypochlorite: NaOCl.

Spinnerets: Small, uniform-sized holes used in fiber production.

Starch: A polysaccharide with repeating units of D-glucose joined by alpha acetal linkages.

Tanning: The cross-linking of animal skins by tannic acid.

Thermoprene: Cyclized rubber.

Thermoset polymer: Cross-linked polymer.

Titanium dioxide: TiO_2.

Vehicle: Solution of a binder in a solvent.

Viscose: A solution of cellulose xanthate.

Zein: Corn protein.

REVIEW QUESTIONS

1. Why is the term nitrocellulose incorrect?

2. What is the solvent present in collodion?

3. What is the effect of camphor on cellulose nitrate?

4. Why is cellulose acetate less flammable than cellulose nitrate?

5. What is the original meaning of saponification?

6. What is the principal difference between rayon and acetate rayon?

7. Would you expect a viscose rayon plant to have a bad odor? Why?

8. What is the difference between rayon and cellophane?

9. What is the principal difference between amylose starch and cellulose?

10. Which has the higher degree of polymerization (DP): starch or dextrin?

11. What is the oldest man-made cross-linked polymer?

12. What famous chemist gave the name of rubber to *Hevea braziliensis*?

13. Why is casein no longer used as a molded plastic?

14. Which has the higher cross-linked density: soft vulcanized rubber or ebonite?

15. Is thermoprene elastic?

16. Is chlorinated rubber elastic?

17. What is the principal use for rubber hydrochloride?

18. What name is given to the resinous component of a paint?

19. What is the most widely used pigment?

20. What pigments are red, white, and blue?

BIBLIOGRAPHY

Abraham, H. (1960–1963). *Asphalts and Allied Substances* (6th ed.). Princeton, NJ: Princeton University Press Van Nostrand.

Alexander, S. H. (1975). The chemistry and technology of bituminous coatings (Chap. 45). In J. K. Craver and R. W. Tess (Eds.), *Applied Polymer Science*. Washington, D.C.: American Chemical Society.

Carraher, C. E., and Tsuda, M. (1980). *Modification of Polymers*, ACS Symposium Series. Washington, DC: American Chemical Society.

Cowan, J. C. (1975). Chemistry and technology of drying oils (Chap. 36). In J. K. Craver and R. W. Tess (Eds.), *Applied Polymer Science*. Washington, DC: American Chemical Society.

Curtis, L. G., and Crowley, U. D. (1988). Cellulose esters (Chap. 43). In R. W. Tess and G. A. Poehlein (Eds.), *Applied Polymer Science*. Washington, DC: American Chemical Society.

Floyd, D. E., and Wittcoff, H. (1975). Chemistry and technology of polyamide resins in coatings (Chap. 47). In J. K. Craver and R. W. Tess (Eds.), *Applied Polymer Science*. Washington, DC: American Chemical Society.

Goodman, M. (1976). Biopolymers—New frontier. *Polymer Preprints*, **17**, 235.

Hicks, E. (1961). *Shellac.* New York: Chemical Publishing Co.

Jen, Y., and McSweeney, E. E. (1975). Chemistry and technology of tall oil and naval stores (Chap. 51). In J. K. Craver and R. W. Tess (Eds.), *Applied Polymer Science.* Washington, DC: American Chemical Society.

Laudise, M. A. (1985). Dimerized fatty acids (Chap. 40). In R. W. Tess and G. W. Poehlein (Eds.), *Applied Polymer Science.* Washington, DC: American Chemical Society.

Mantell, C. L., Kopf, C. W., Curtis, J. L., and Rogers, E. M. (1942). *The Technology of Natural Resins.* New York: Wiley.

McGregor, and Greenwood, C. T. (1980). *Polymers in Nature.* New York: Wiley.

Morton, M. (1973). *Rubber Technology.* New York: Van Nostrand–Reinhold.

Ott, E., Spurlin, H. M., and Grafflin, M. W. (1955). *Cellulose and Cellulose Derivatives* (Vol. 5). New York: Interscience.

Seymour, R. B. (1978). Acetylation of cellulose solutions. *Journal of Polymer Science—Chemistry Edition,* **16**, 1.

Seymour, R. B., and Johnson, E. J. (1977). Solutions of cellulose in organic solvents (Chap. 19). In F. W. Harris and R. B. Seymour (Eds.), *Structure–Solubility Relationships in Polymers.* New York: Academic Press.

Whistler, R. L., and BeMiller, J. N. (1966). *Industrial Gums.* New York: Benjamin.

Wint, R. F., and Shaw, K. G. (1985). Nitrocellulose, ethylcellulose, and related esters (Chap. 44). In R. W. Tess and G. W. Poehlein (Eds.), *Applied Polymer Science.* Washington, DC: American Chemical Society.

Zelinski, R. P. (1975). Chemistry and technology of rubber (Chap. 28). In J. K. Craver and R. W. Tess (Eds.), *Applied Polymer Science.* Washington, DC: American Chemical Society.

ANSWERS TO REVIEW QUESTIONS

1. Nitrate (NO_3) groups and not nitro (NO_2) groups are present.

2. An equimolar mixture of ethanol and ethyl ether.

3. It serves as a flexibilizer or plasticizer for the stiff cellulose nitrate.

4. Acetate groups are less flammable (less explosive) than nitrate groups.

5. Soap making, that is, the alkaline hydrolysis of animal fat to produce glycerol and sodium salts of fatty acids (soap).

6. Rayon is a regenerated cellulose fiber; cellulose acetate is an acetyl derivative of cellulose.

7. Yes, carbon disulfide (CS_2) is odiferous.

8. The physical form—rayon is a fiber, cellophane is a film; both are regenerated cellulose.

9. The arrangement of the acetal linkages. Cellulose has beta acetal linkages and amylose starch has alpha acetal linkages between the D-glucose repeating units.

10. Starch; dextrin is degraded starch.

11. Leather.

12. Joseph Priestley.

13. Too expensive, subject to degradation, and properties are inferior to those of many other thermoplastics. It is used as an adhesive.

14. Ebonite—the maximum cross-linked density is approached when large amounts of sulfur are reacted with natural rubber.

15. No. It is a plastic used as a coating and as an adhesive.

16. No. It is a hard thermoplastic used as a coating.

17. As a packaging film (Pliofilm).

18. Binder (resin-forming component).

19. Titanium dioxide (white pigment).

20. Pb_3O_4, TiO_2, and lapis lazuli. (There are many others.)

CHAPTER VIII

Physical and Chemical Testing of Polymers

8.1 TESTING ORGANIZATIONS

The selection of general-purpose polymers has sometimes been the result of trial and error, misuse of case history data, or questionable guesswork. However, since polymeric materials must be functional, it is essential that they be tested using meaningful use-oriented procedures. Both the designer and the user should have an understanding of the testing procedure used in the selection of a polymeric material for a specific end use. They should know both the advantages and the disadvantages of the testing procedure, and designers should continue to develop additional empirical tests.

Fortunately, there are many standards and testing organizations whose sole purpose is to assure the satisfactory performance of materials. The largest standards organization is the International Standards Organization (ISO), which consists of members from 89 countries and many cooperative technical committees. There is also the American National Standards Institute (ANSI) and the American Society for Testing and Materials (ASTM), which publishes its tests on an annual basis. Other important reports on tests and standards are published by the National Electrical Manufacturing Association (NEMA), Deutsches Institut fur Normenausschuss (DIN), and the British Standards Institute (BSI).

8.2   EVALUATION OF TEST DATA

Unlike the physical data compiled for metals and ceramics, the data for polymers are dependent on the life span of the test, the rate of loading, temperature, preparation of the test specimen, and so on. Some of these factors, but not all, have been taken into account in obtaining the data listed in tables in subsequent chapters of this book. Published data may vary for the same polymer fabricated on different equipment or produced by different firms and for different formulations of the same polymer or composite. Hence, the values cited in the tables are usually labeled "Properties of Typical Polymers."

Many tests used by the polymer industry are adaptations of those developed previously for metals and ceramics. None is so precise that it can be used with 100% reliability. In most instances, the physical, thermal, and chemical data are supplied by the producers, who are expected to promote their products in the marketplace. Hence, in the absence of other reliable information, positive data should be considered as upper limits of average test data and an allowance of at least $\pm 5\%$ should be assumed by the user or designer.

8.3   HEAT DEFLECTION TEST

The heat deflection standard, which is now called Deflection Temperature of Plastics under Flexural Load (DTUL) (ANSI/ASTM D648-72/78), is a result of "round-robin" testing by all interested members of the ASTM Committee D20. This standard was accepted several decades ago. As shown by the numbers after D648 in the test designation, it was revised and reapproved in 1972 and reapproved in 1978, respectively.

The DTUL test measures the temperature at which an arbitrary deformation occurs when plastic specimens are subjected to an arbitrary set of testing conditions. The standard molded test span measures 127 mm in length, 13 mm in thickness, and 3–13 mm in width. The specimen is placed in an oil bath under a 0.455 or 1.820 MPa load in the apparatus shown in Figure 8.1, and the temperature is recorded when the specimen deflects by 0.25 mm.

Since crystalline polymers, such as nylon 6,6, have a low heat deflection temperature value when measured under a load of 1.820 MPa (264 psi), this test is often run at 0.460 MPa (66 psi). However, the 1.82 MPa load is standard for composites.

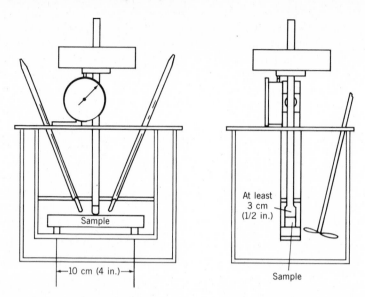

At least
3 cm
(1/2 in.)

Sample

Sample

←10 cm (4 in.)→

Figure 8.1. Apparatus for heat deflection under load (1.820 or 0.460 MPa) test.

The results of this test must be used with caution. The established deflection is extremely small, and in some instances may be, in part, a measure of warpage or stress relief. The maximum resistance to continuous heat is an arbitrary value for useful temperatures, which is always below the DTUL value.

8.4   COEEFICIENT OF LINEAR EXPANSION

Since it is not possible to exclude factors such as changes in moisture, plasticizer, or solvent content, or release of stresses with phase changes, ANSI/ASTM D696-79 provides only an approximation of the true thermal expansion. The values for thermal expansion of unfilled polymers are high, relative to that of other materials of construction, but these values are dramatically reduced by the incorporation of fillers and reinforcements.

   In this test, the specimen, measuring between 50 and 125 mm in length, is placed at the bottom of an outer dilatometer tube and below the inner dilatometer tube. The outer tube is immersed in a bath and the temperature is measured. The increase in length (ΔL) of the specimen as measured by the dilatometer is divided by the initial length (L_0) and multiplied by the increase in temperature to obtain the coefficient of linear expansion (α). The formula for calculating this value is

$$\alpha = (\Delta L/L_0)T$$

8.5   COMPRESSIVE STRENGTH

Compressive strength, or the ability of a specimen to resist a crushing force, is measured by crushing a cylindrical specimen in accordance with ASTM-D695.

Direction of applied force (stress)

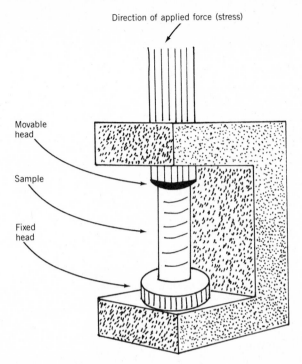

Movable
head

Sample

Fixed
head

Figure 8.2. Apparatus for measurement of compression-related properties.

The test material is mounted in a compression tool as shown in Figure 8.2, and one of the plungers advances at a constant rate. The ultimate compression strength is equal to the load that causes failure divided by the minimum cross-sectional area. Since many materials do not fail in compression, strengths reflective of specified deformation are often reported.

8.6 FLEXURAL STRENGTH

Flexural strength or crossbreaking strength is the maximum stress developed when a bar-shaped test piece, acting as a simple beam, is subjected to a bending force perpendicular to the bar (ANSI/ASTM D790-71/78). An acceptable test specimen is one that is at least 3.2 mm in depth and 12.7 mm in width and long enough to overhang the supports, but the overhang should be less than 6.4 mm on each end.

The load should be applied at a specified crosshead rate and the test should be terminated when the specimen bends or is deflected by 0.05 mm/min. The flexural strength (S) is calculated from the following expression in which P = the load at a given point on the deflection curve, L = the support span, b = the width of the bar, and d = the depth of the beam. Figure 8.3 shows a sketch of the test and the expression for calculating flexural strength is $S = PL/bd^2$.

One may use the following expression in which D is the deflection to obtain the maximum strain (r) of the specimen under test:

$$r = 6Dd/L$$

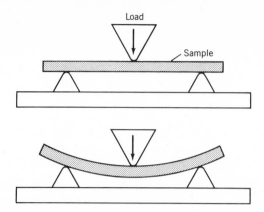

Figure 8.3. Sketch of effect of load on test bar in ASTM test 790.

One may also obtain data for flexural modulus, which is a measure of stiffness, by plotting flexural stress (S) versus flexural strain (r) during the test and measuring the slope of the curve obtained.

8.7   MELTING POINT

As is the case with the glass transition temperature, melting will be observed to occur over a temperature range since it takes time for the chains to unfold. If the temperature is raised very slowly, a one to two degree range will be observed. The determination of the melting point requires only visual observation of when melting occurs as the sample is heated.

8.8   WATER ABSORPTION

Water absorption can be determined through weight increase when a dried sample is placed in a chamber of specified humidity and temperature (ANSI/ASTM D570-63(1972)).

8.9   IMPACT TEST

Impact strength may be defined as toughness or the capacity of a rigid material to withstand a sharp blow, such as that from a hammer. The information obtained from the most common test (ANSI/ASTM D256-78) on a notched specimen (Figure 8.4) is actually a measure of notch sensitivity of the specimen.

In the Izod test, a pendulum-type hammer, capable of delivering a blow of 2.7 to 21.7 J, strikes a notched specimen (measuring 127 mm × 12.7 mm × 12.7 mm with a 0.25mm notch), which is held as a cantilever beam. The distance that the pendulum travels after breaking the specimen is inversely related to the energy required to break the test piece, and the impact strength is calculated for a 25.4-mm test specimen.

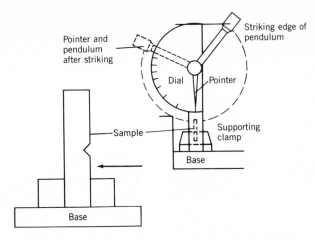

Figure 8.4. Notched Izod impact test (ASTM D256).

8.10   TENSILE STRENGTH

Tensile strength or tenacity is the stress at the breaking point of a dumbbell-shaped tensile test specimen (ANSI/ASTM D638-77). The elongation or extension at the breaking point is the tensile strain. As shown in Figure 8.5, the test specimen is 3.2 mm thick and has a cross-section of 12.7 mm. The jaws holding the specimen are moved apart at a predetermined rate and the maximum load and elongation at break are recorded. The tensile strength is the load at break divided by the original cross-sectional area. The elongation is the extension at break divided by the original gauge length multiplied by 100. The tensile modulus is the tensile stress divided by the strain. As an alternative to reporting the tensile strength, one may determine the slope of the tangent to the initial portion of the elongation curve.

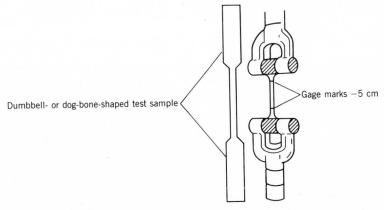

Figure 8.5. Tensile test showing the dog-bone specimen clamped in the jaws of an Instron tester.

8.11   HARDNESS TEST

Hardness is the resistance of a material to local deformation. This test utilizes an indentor, which may be a sharp pointed cone in the Shore D Durometer test or a ball in the Rockwell test.

The Shore durometer is a spring-loaded indentor with a scale that shows the extent of indentation, with 100 being the hardest rating on the scale. The Rockwell tester measures the indentation of a loaded ball, usually on the R scale. The diameter of the ball used in the R scale is 12.7 mm (Figure 8.6).

8.12   DENSITY (SPECIFIC GRAVITY)

Specific gravity is simply the density (mass per unit volume) of a material divided by the density of water. In cgs units the density of water is about 1.00 g/cc at room temperature. Thus, at room temperature the density and specific gravity values are essentially the same. Specific gravity is often used because it is unitless, whereas a density, although commonly given in cgs units, can be given in other weight per volume units such as pounds per quart.

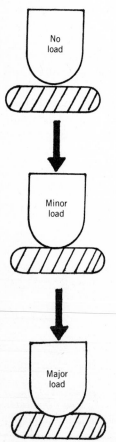

Figure 8.6. Illustration of Rockwell hardness test equipment.

8.13   GLASS TRANSITION TEMPERATURE

Qualitatively, the glass transition temperature corresponds to the onset of short-range (typically one- to five-atom chains) coordinated motion. Actually, many more (often 10 to 100) atoms may attain sufficient thermal energy to move in a coordinated manner at T_g.

The glass transition temperature (ASTM D-3418) is the temperature at which there is an absorption or release of energy as the temperature is raised or lowered. T_g may be determined using any technique that signals an energy gain or loss.

It must be emphasized that the actual T_g of a sample is dependent on many factors, including pretreatment of the sample and the method and conditions of determination. For instance, the T_g for linear polyethylene has been reported to be from about 140 to above 300 K. Calorimetric values for polyethylene centralize about two values, 145 and 240 K; thermal expansion values are quite variable within the range of 140 to 270 K; NMR values occur between 220 and 270 K; and mechanical determinations range from 150 to above 280 K. The method of determination and the end property use should be related. Thus, if the area of concern is electrical, then determinations involving dielectric loss are appropriate.

Whether a material is above or below its T_g is important in describing the material's properties and potential end use. Fibers are composed of generally crystalline polymers that contain polar groups. The polymers composing the fibers are usually near their T_g to allow flexibility. Cross-links are often added to prevent gross chain movement. An elastomer is cross-linked and composed of essentially nonpolar chains; the use temperature is above its T_g. Largely crystalline plastics may be used above or below their T_g. Coatings or paints must be used near their T_g so that they have some flexibility but are not rubbery. Adhesives are generally mixtures in which the polymeric portion is above its T_g. Thus the T_g is one of the most important physical properties of an amorphous polymer.

8.14   RESISTANCE TO CHEMICALS

The resistance of polymers to chemical reagents has been measured as described in ANSI/ASTM D543-67/78, which covers 50 different reagents. In the past, the change in weight and appearance of the immersed test sample have been reported. However, this test has been updated to include changes in physical properties as a result of immersion in test solutions.

Most high-performance polymers are not adversely affected by exposure to nonoxidizing acids and alkalies. Some are adversely affected by exposure to oxidizing acids, such as concentrated nitric acid, and all amorphous linear polymers will be attacked by solvents with solubility parameters similar to those of the polymer. Relatively complete tables showing resistance of polymers to specific corrosives have been published.

REVIEW QUESTIONS

1. Why is the coefficient of linear expansion important to know for materials used in aircraft?

2. Why is it important to know the rate of addition of load in the compressive strength test?

3. Why is it important to establish standard test conditions?

4. If the heating rate of a sample was low, would you expect the melting point obtained to be lower or higher than a melting point obtained when heating the sample faster?

5. Compare the impact test with the test for flexural strength.

6. What is the density of a piece of plastic that weighs 30 g and that occupies a volume of 20 cc?

7. Arrange the following in order of increased density: wood that floats on water, a piece of heavy plastic that sinks when placed in water, and a paper clip made of metal.

BIBLIOGRAPHY

Alfrey, T. (1948). *Mechanical Behavior of Polymers.* New York: Interscience.

American Society for Testing and Materials. *1976 Book of ASTM Standards,* Parts 26, 34, 36, Plastics. Philadelphia: American Society for Testing and Materials.

Braun, D., Cherdron, H., and Kern, W. (1972). *Techniques of Polymer Synthesis and Characterization.* New York: Wiley–Interscience.

Carraher, C. E. (1977). Resistivity measurements. *Journal of Chemical Education,* **54,** 576.

Carraher, C. E., Sheats, J., and Pittman, C. U. (1978). *Organometallic Polymers.* New York: Academic Press.

Craver, C. D. (1983). *Polymer Characterization.* Washington, DC: American Chemical Society.

McCaffery, E. M. (1970). *Laboratory Preparation for Macromolecular Chemistry.* New York: McGraw–Hill.

Mittall, K. L. (1983). *Physicochemical Aspects of Polymer Surfaces.* New York: Plenum.

Nielsen, L. E. (1966). *Mechanical Properties of Polymers.* New York: Reinhold.

Painter, P. C., Coleman, M. M., and Koenig, J. L. (1982). *The Theory of Vibrational Spectroscopy and Its Application to Polymeric Materials.* New York: Wiley.

Seanor, D. (1982). *Electrical Properties of Polymers.* New York: Academic Press.

Seymour, R. B. (1982). *Plastics vs. Corrosives.* New York: Wiley.

Seymour, R. B., and Carraher, C. E. (1984). *Structure–Property Relationships in Polymers.* New York: Plenum.

Seymour, R. B., and Steiner, R. H. (1955). *Plastics for Corrosion Resistant Applications.* New York: Reinhold.

Turi, E. A. (1981). *Thermal Characterization of Polymeric Materials.* New York: Academic Press.

Ward, I. M. (1983). *Mechanical Properties of Solid Polymers.* New York: Wiley–Interscience.

Williams, J. G. (1984). *Fracture Mechanics of Polymers.* New York: Halstead.

Zachariades, A. E., and Porter, R. S. (1983). *The Strength and Stiffness of Polymers.* New York: Dekker.

ANSWERS TO REVIEW QUESTIONS

1. Temperatures for aircraft operation can vary greatly, thus it is important that a good match exists between bonded materials in the aircraft.

2. Some materials will act differently dependent on the rate of load application.

3. So that comparison of test results are more reliable.

4. Less—since the slower heating rate will allow the chains a longer time to unfold.

5. See Sections 8.6 and 8.9. In the impact test the load is more rapidly applied.

6. $D = 30\,\text{g}/20\,\text{cc} = 1.5\,\text{g/cc}$.

7. Wood, plastic, clip.

CHAPTER IX _____

Fibers

9.1 INTRODUCTION

Before the advent of man-made fibers, clothing was made from natural fibers, that is, plant fibers (cotton, hemp, and jute), and animal fibers (fur and hair; wool and silk). The modification of natural fibers began in the early 1800s and continues today.

Fibers are threadlike strands with a length (l) to thickness (d) ratio of at least $100 : 1$. Fibers are usually spun into yarns, which are made into textiles, which are then fabricated into finished products such as rugs, clothing, and tire cords. In this chapter, the basic concepts and processing techniques of fibers will be described. Individual fiber groups will also be discussed.

Fiber properties such as high tensile strength and high modulus are characteristic of polymers having good molecular symmetry, which allows the chains to be closely associated with one another in order to enhance dipolar and hydrogen bonding interactions between the chains.

Even though polystyrene chains may exhibit high symmetry, polymers such as polystyrene are not fibers, because the forces associated with inter- and intrachain

140

attractions have low energy. Likewise, a nylon that is produced by the condensation of 2-*tert*-butylterephthalic acid and 1-phenyl-1-*n*-butylhexamethylenediamine is not a good fiber because of the bulky butyl and phenyl groups, which prevent close chain association. Branched nylon 6,6 is also a poor fiber since the branching also prevents close chain association. These structures are shown in Figure 9.1.

In contrast, wool, silk, and cellulose contain both the essential strong secondary bonding forces and unit symmetry, which are necessary for good fibers. Polymers such as nylon 6,6 also exhibit the required molecular symmetry and good bonding. In fact, nylon was the first man-made fiber that performed as well as its natural proteinaceous counterparts, wool and silk.

Fibers generally exhibit the following characteristics: they are thermoplastics (able to be molded or shaped through application of heat), abrasion resistant (withstand surface wear), resilient (spring back when deformed), strong, and relatively nonabsorbent. In addition, synthetic fibers are usually resistant to mildew, rot, and moths. Some polymers, such as nylon 6,6, may be used as both fibers and plastics.

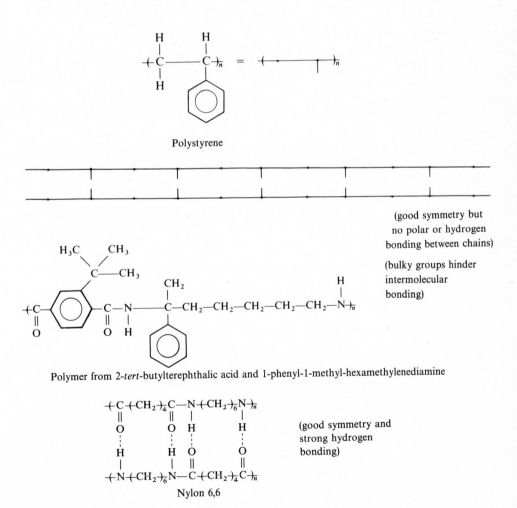

Figure 9.1. Fibrous and nonfibrous structures.

It is important to note that fiber producers make the fibers but usually do not make the finished products. Thus, the fiber producers, like the farmers who grow our food, depend on others for the processing of their products.

Staining is an interesting problem. Most items that we worry about, such as tomato paste, coffee, and foods, are hydrophilic. They must be hydrophilic to be metabolized by our bodies, which are "walking water containers," that is , proteins, nucleic acids, carbohydrates, and water are themselves hydrophilic.

The second major group of staining materials is human excretions such as perspiration, which are also hydrophilic. Thus, fabrics such as nylons and polyesters that hydrogen bond and interact favorably with these wastelike materials are typically easily stained. Even so, rugs and clothing made from nylons and polyesters feel softer to the touch and "breathe" when worn and thus offer advantages that more than offset the "staining problem." Staining is often resisted by application of a surface coating that is somewhat hydrophobic, but its major role is in preventing the staining material from penetrating the inner fibers of the fabrics. Scotchgard® is one of the many available surface treatments.

Hydrophobic (water-repellant) fabrics, such as polyolefin textiles, do not accept stains readily, and thus the staining material is readily washed away. However, such fabrics must undergo additional treatments before they are soft to the touch and even then may be inferior in this regard to nylons and polyesters.

9.2 PRODUCTION TECHNIQUES

Fibers can be produced as continuous filaments, as staple yarns, as the staple (short fibers) itself, or as filament yarn, depending on the processing and intended end use. Most natural fibers are of the staple variety and are made into staple yarn. These fiber types are illustrated in Figure 9.2.

Most man-made fibers (including regenerated cellulose and cellulosic derivatives) are formed by forcing a solution of the polymer through tiny holes (called spinnerets).

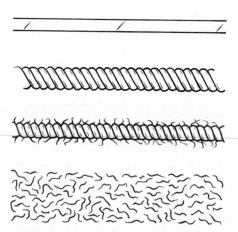

Figure 9.2. Illustrations of fiber types. Top to bottom: continuous monofilament, filament yarn, staple yarn, and staple.

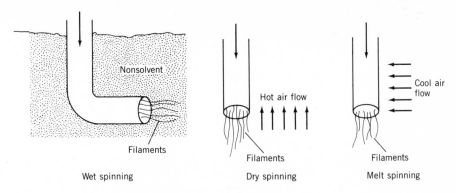

Figure 9.3. Types of spinning processes.

Some fibers, such as rayon, are produced by wet spinning, in which the filaments from the spinnerets are passed through chemical baths that insolubilize the soluble polymer. The solvent is evaporated by passing warm air by the spinneret as the polymer solution exits. Other fibers, such as cellulose acetate, are produced by evaporation of the solvated filament as it passes through the spinneret. In the melt spinning technique, the feedstock, such as pellets of nylon 6,6, are melted, forced through a spinneret, and cooled to form continuous nylon fibers. These processes, which should not be confused with the process of "spinning" of fibers to form yarn, are illustrated in Figures 9.3 and 9.4.

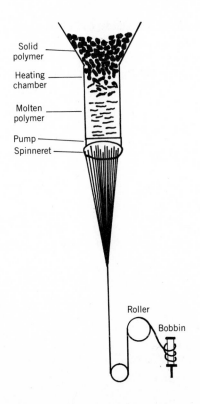

Figure 9.4. Diagram of the melt spinning process for the production of thread and filaments.

Filaments of polymers from spinnerets are then stretched to reduce their diameter, orient the polymer chains, and permit the fibers to arrange themselves along the pull axis. A wide variety of fiber sizes and strengths can be achieved by adjusting the pull rate and spinneret size. Polypropylene fibers are also produced by a unique fibrillation process in which strips of the polymer film are twisted and stretched.

The denier of a filament or yarn is the weight of 9000 m expressed in grams. The lower the denier, the lighter and finer the yarn. For example, 15 denier filaments are often used in women's hosiery, whereas 840 denier filaments are used in tires.

Monofilaments, which are obtained by continuous "pulls" from the spinneret, are employed for a variety of uses, such as fishline. The monofilaments can also be woven into items such as sheer curtains or knitted to form products such as hosiery. Continuous strands of two or more filaments can be twisted or braided together to form filament yarns.

Large groups of continuous untwisted filaments are called "tow" and are often cut or broken to produce short segments, that is, staple fibers. Natural fibers such as wool and cotton are also staple fibers. The staple fiber can be twisted or spun or used without textile spinning as filling in mattresses, comforters, pillows, and sleeping bags.

Yarns spun from staple are more irregular than filament yarns, since the short ends of the fibers projecting from the yarn surface produce a fuzzy effect. Spun yarns are also more bulky than filament yarns of the same weight. Therefore, they are often used for porous, warm fabrics and for the production of nonsmooth fabric surfaces.

Textured filament yarns are made by twisting (throwing) the yarn in a designated manner. New filament yarns are being produced by untwisting, false twisting, deknitting, knitting, and crimping. Different bulk and stretch properties provide new fabrics for the fashion designer.

Both natural and synthetic fibers must be dyed for cosmetic effects. Cellulose fibers are often scoured with alkaline solutions, treated with an agent to prevent mildew, soaked in copper aqueous or mercury salt solutions, and treated with water repellents such as quaternary ammonium compounds such as $R_4N^+Cl^-$. Wool must be scoured and is often made crease and wrinkle resistant by immersing in aqueous solutions of melamine–formaldehyde prepolymers.

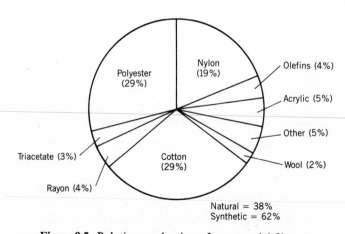

Figure 9.5. Relative production of commercial fibers.

Cotton and other fibers or textiles are dyed by immersing in a solution or dispersion of the selected dye. Direct dyes, such as azo dyes, are used on cotton and rayon; reactive dyes, such as those with chlorotriazinyl groups, and triarylmethane dyes are used on polyamides; and bis azo disperse dyes and basic dyes are used on polyester and acrylic fibers. Fibers can be colored at various steps during their production. Fibers that are colored before spinning are called spun-dyed, dope-dyed, and solution-dyed. Fibers can also be dyed after being made into finished textiles.

The characteristics of two or more staple fiber types can be achieved by blending the fibers together prior to spinning into yarn. Various types of monofilament or filament yarns may also be combined and twisted together to form a "combination" filament yarn. The relative production of various commercial fibers is shown in Figure 9.5.

9.3   NYLONS

W. Carothers and J. Hill produced low-melting polyester fibers in the duPont laboratories in 1932 but shelved this project in favor of the development of the more heat-resistant polyamide fibers. The first polyamide fiber, poly-ω-aminononanoic acid (or nylon-9) $\pend{NH(CH_2)_8CO}_m$, had a higher melting point (380°F) than the original polyester fibers but was softened at the temperature used for ironing.

These investigators then synthesized polyhexamethylenedipamide, which had a melting point of 505°F. This pioneer synthetic fiber was produced commercially by duPont at Seaford, Delaware, in 1939. Carothers recognized that high-molecular-weight polymers could not be synthesized unless the reactants were extremely pure. Accordingly, he made a salt from the diamine and diacid, removed the impurities by crystallization, and then heated the purified salt to produce a polymer, which he called a "super polyamide." This polymer is now called nylon 6,6 because both reactants have six carbon atoms. The equation showing that the degree of polymerization (DP) is a function of the purity of the reactants ($DP = (1 - P)^{-1}$) is called the Carothers equation. P is defined as the extent of reaction, which must be at least 0.995 for polymer formation.

$$H_2N(CH_2)_6NH_2 + HOOC(CH_2)_4COOH \rightarrow H_2N(CH_2)_6NH_3^+, {}^-OOC(CH_2)_4COOH$$

$$nH_2N(CH_2)_6NH_3^+, \bar{O}OC(CH_2)_4COOH \xrightarrow{\Delta}$$

$$\left[\underset{}{\pend{NH-(CH_2)_6-NH-\overset{\displaystyle O}{\overset{\|}{C}}-(CH_2)_4-\overset{\displaystyle O}{\overset{\|}{C}}}_n} \right] + nH_2O$$

Nylon 6,6
(polyhexamethyleneadipamide)

Carothers also found from his earlier work with polyesters that for the reaction to produce long chains water must be removed, resulting in a shift of the equilibrium to favor larger chains. This was accomplished using what was called a molecular still, actually a hot plate assembly fitted with vacuum to help in the removal of water. This is an example of applying old knowledge to new problems.

Before nylon could be produced for public consumption, scientists needed to find large, inexpensive sources of the reactants—hexamethylenediamine and adipic acid. The duPont Company scientists devised a scheme for producing these two reactants from coal, air, and water. The process was laborious and later scientists developed procedures to make these two reactants from agricultural by-products such as rice hulls and corn cobs. Several chemical equations are involved in these synthetic procedures:

$$\text{Benzene} \xrightarrow[\text{catalyst}]{H_2} \text{Cyclohexane} \xrightarrow[\text{catalyst}]{O_2} \text{Cyclohexanol} + \text{Cyclohexanone} \xrightarrow[\Delta]{HNO_3}$$

Benzene
(from petroleum)   Cyclohexane   Cyclohexanol   Cyclohexanone

$$HO-\overset{O}{\underset{\|}{C}}+CH_2)_{\overline{4}}\overset{O}{\underset{\|}{C}}-OH$$

Adipic acid

$$\text{Tetrahydrofuran} + 2CO + H_2O \xrightarrow[\text{catalyst}]{\Delta} HO-\overset{O}{\underset{\|}{C}}+CH_2)_{\overline{4}}\overset{O}{\underset{\|}{C}}-OH$$

Tetrahydrofuran
(from farm by-product)

Adipic acid

$$\Delta \,\Big|\, NH_3$$

$$H_2N+CH_2)_{\overline{6}}NH_2 \xleftarrow[\Delta,\ \text{pressure}]{\text{catalyst}} H_2N-\overset{O}{\underset{\|}{C}}+CH_2)_{\overline{4}}\overset{O}{\underset{\|}{C}}-NH_2 \longleftarrow$$

Hexamethylenediamine   Adipamide

Nylon is one of our most important polymers and it can be formed into sheets, rods, fibers, bristles, tubes, and coatings and used as a molding powder. Nylon 6,6 still accounts for the majority of nylons synthesized, with fiber applications including uses in dresses, lace, parachutes, tire cord, upholstery, underwear, carpets, ties, suits, socks, fishing line, bristles for brushes, and thread (including surgical thread).

Other nylons have also been developed. Nylon-6 (polycaprolactam, Perlon), which was patented by P. Schlack in 1937, is produced on a large scale in Europe, but only on a moderate scale in the United States.

$$\left[NH(CH_2)_5-\overset{}{\underset{\underset{O}{\|}}{C}} \right]_n$$

Nylon-6

Nylon 4,6 (Stanyl), which is more hydrophilic than nylon 6,6, and nylon-12 are also available commercially.

Aromatic nylons, prepared from the condensation of terephthalic acid and aliphatic diamines, have high melting points and are tough and strong. These nylons, which are called aramids, are second-generation fibers and are utilized in the construction of radial tires, bulletproof clothing, and fiber-reinforced composites. The principal aramid fibers are poly-*p*-benzamide (Kevlar) and polyphenyleneisophthalamide (Nomex). Over 10,000 tons of Kevlar are produced annually.

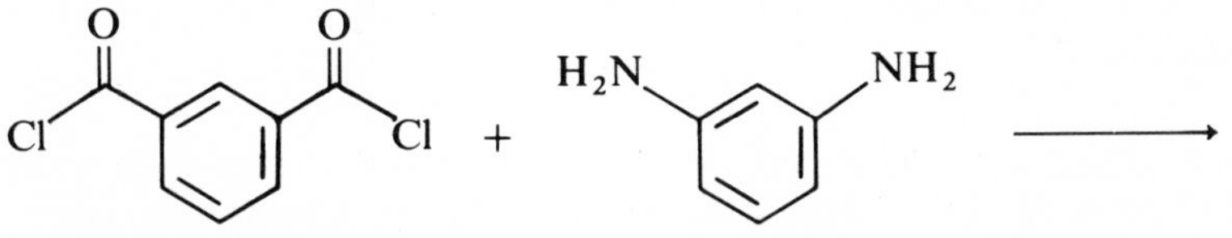

Phthaloyl dichloride

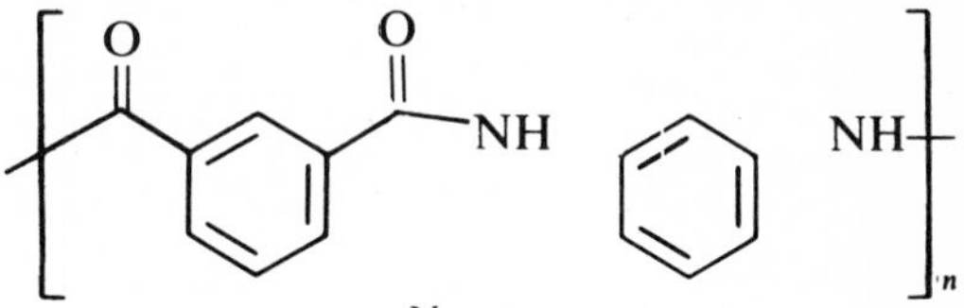

Nomex

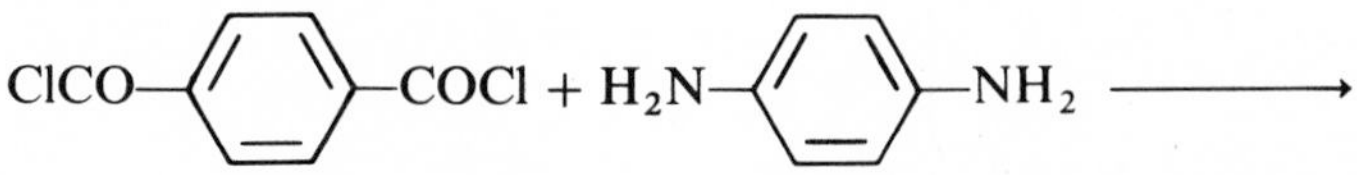

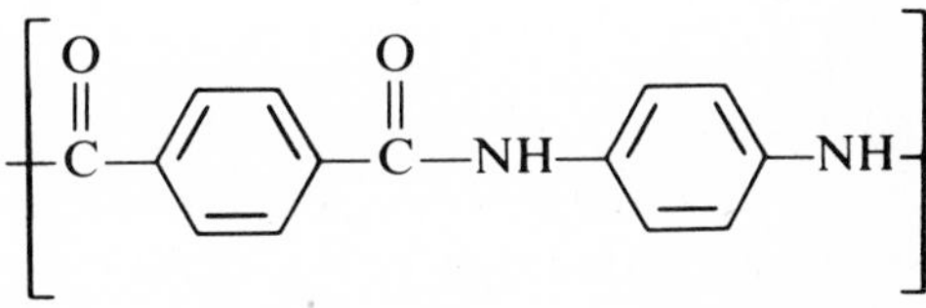

Kevlar

A number of other nylon fibers have also been made. Quiana fibers were developed as a synthetic silk material and are now employed in the production of blouses, dresses, and shirts. Quiana fibers are spun from the nylon made from condensing dodecanedioic acid and di-4-aminocyclohexylmethane. Interestingly, it has a lower melting point (205°C) than nylon 6,6 but a higher glass transition temperature (135°C). The high T_g enables fabrics made from Quiana to resist wrinkles and creasing during laundering.

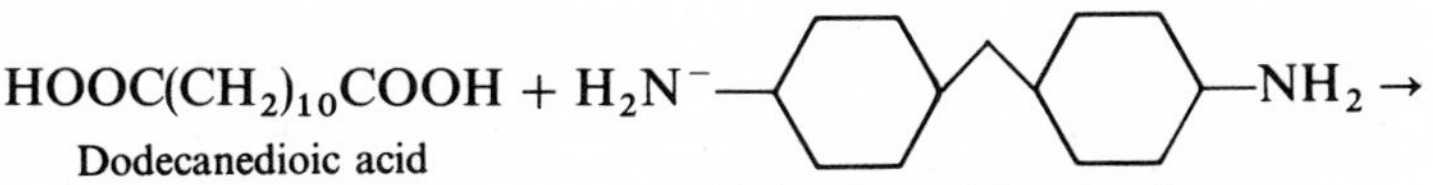

$$HOOC(CH_2)_{10}COOH + H_2N-$$

Dodecanedioic acid

Di-4-aminocyclohexylmethane

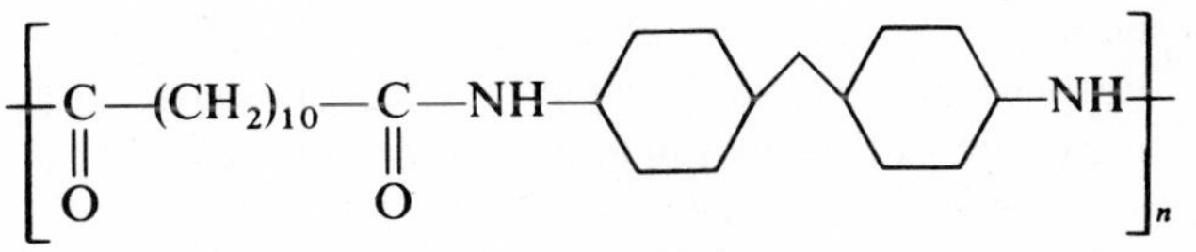

Quiana

9.4 POLYESTERS

J. R. Whinfield and J. T. Dickson substituted phthalic acid for Carother's adipic acid and produced a relatively high melting (507°F) polyester fiber in 1940. This synthetic fiber was produced in England in 1940 and in the United States by duPont in 1945. Polyester fibers are now produced by a number of companies and are sold under a variety of trade names, such as Dacron, Kodel, Terylene, and Trevira.

The most popular polyester is obtained by the condensation of ethylene glycol (the major ingredient in antifreeze) and terephthalic acid. These two reactants can be made from a number of petrochemical feedstocks. As shown by the following equation, terephthalic acid is produced by the catalytic oxidation of p-xylene.

Ethylene glycol is produced by the hydrolysis of ethylene oxide, which is obtained by the catalytic oxidation of ethylene produced from natural gas or petroleum crude. Both xylene and ethylene can also be made from coal and oil shale and ethylene has been made by the destructive distillation of alcohol, corn, wheat, and other natural, renewable products. Thus, the source of the starting material is variable.

Polyesters are manufactured as films, plastics, and fibers. Polyester fibers are mainly used in making fabrics such as carpets, clothing, upholstery, and underwear. The fibers are also employed in the construction of tire cord. Polyester fabrics are easy to care for and resist mildew, rot, and fading. Most of the outerwear garments are permanent-press textiles.

The formula for the repeating unit in most polyester fibers is

$$\left[\!-OCH_2CH_2O-\overset{\overset{\displaystyle O}{\|}}{C}-\!\!\bigcirc\!\!-\overset{\overset{\displaystyle O}{\|}}{C}-\!\right]_n$$

Poly(ethylene terephthalate)(PET)

Figure 9.6 shows ball-and-stick models of poly(ethylene terephthalate).

9.5   ACRYLIC FIBERS

In the 1940s, chemists at Monsanto and duPont dissolved polyacrylonitrile in dimethylacetamide (DMAC) and produced unique fibers by passing these solutions through spinnerets and evaporating the solvent. Modacrylic fibers, such as copolymers of vinylidene chloride, are related but more easily dyed than are copolymers of acrylonitrile and other monomers. These were produced commercially in the 1950s.

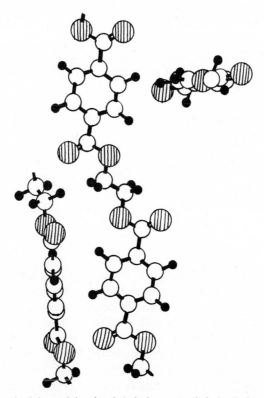

Figure 9.6. Ball-and-stick models of poly(ethylene terephthalate) showing various views.

Fibers from both polyacrylonitrile and copolymers derived from acrylonitrile are classified jointly as acrylic fibers. The Textile Fiber Products Identification Act divides these fibers into two categories: acrylic fibers are those containing at least 85% by weight of acrylonitrile, and modacrylic fibers contain less than 85% but at least 35% acrylonitrile.

$$
n \quad \underset{H}{\overset{H}{\diagdown}} C = C \underset{CN}{\overset{H}{\diagup}} \quad \rightarrow \quad \left(C - C \right)_n
$$

Acrylonitrile   Polyacrylonitrile

$$
x \quad \underset{H}{\overset{H}{\diagdown}} C = C \underset{CN}{\overset{H}{\diagup}} \; + \; y \quad \underset{H}{\overset{H}{\diagdown}} C = C \underset{Cl}{\overset{Cl}{\diagup}} \quad \rightarrow \; \left(C - C \right)_x \left(C - C \right)_y
$$

Acrylonitrile   Vinylidene chloride   Copolymer

These fibers are long-lasting, dry rapidly, and resist fading, wrinkling, and mildew. They are used in making carpets, sportswear, sweaters, and blankets as well as many other products. As in the case of nylon and polyesters, acrylics are also being employed as coatings and plastics.

The monomer acrylonitrile is synthesized at an annual rate in the United States of 700,000 tons by the ammoxidation of propylene, the latter being a product obtained from petroleum.

$$
\underset{H_3C}{\overset{H}{\diagdown}} C = C \underset{H}{\overset{H}{\diagup}} \; + NH_3 + O_2 \xrightarrow{\text{catalyst}} \; \underset{CN}{\overset{H}{\diagdown}} C = C \underset{H}{\overset{H}{\diagup}}
$$

Propylene   Acrylonitrile

9.6   GLASS FIBERS

Glass fibers may be thinner than human hair and may look and feel like silk. These flexible glass fibers are stronger than steel and will not burn, stretch, or rot. The ancient Egyptians used coarse glass fibers for decorative purposes.

Edward Libbey, an American glass manufacturer, exhibited a dress made from fiberglass and silk at the Columbian Exposition in Chicago in 1893. Fiberglass was made in Germany as a substitute for asbestos during World War I and, because of the toxicity of asbestos, is being used again for this purpose. The Owens Illinois Glass Company and the Corning Glass Works developed practical methods for making fiberglass commercially in the 1930s.

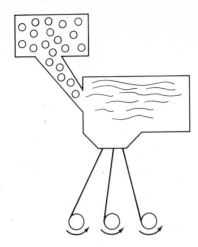

Figure 9.7. Illustration of one procedure for producing fiberglass.

Fiberglass, which is an inorganic polymeric fiber, is made from the same raw materials as those used to make ordinary glass. As shown in Figure 9.7, glass marbles are melted in special furnaces and molten glass passes through spinnerets (bushings) at the furnace bottom. The hot filaments are wound on a spinning drum, which stretches and orients the glass filaments. The principal use of fiberglass is as a reinforcement for polyester and epoxy resin.

Fiberglass wool is produced by exposing the hot filaments to high-pressure steam jets to produce fibers that are gathered together to form a white wool-like mass, which is used for insulation.

Fiberglass can also be woven into a fabric to make tablecloths and curtains. The end products can be dyed, do not wrinkle or soil easily, and need no ironing after washing. Textile material is also used for electrical insulation. In bulk form, it is used for heat and sound insulation and for air filters. The insulation properties are a result of the high bulk property of fiberglass whereby still air also acts as a good buffer to thermal changes.

Fiberglass is also employed in the manufacture of reinforced plastics that are strong yet lightweight. Car bodies, ship hulls, building panels, aircraft parts, and fishing rods are popular examples. The fibers can be woven, matted together, or used as individual strands depending on the nature and price of the final product. About 400,000 tons of fiberglass are produced annually in the United States.

9.7   POLYOLEFINS

A number of olefinic polymers are being used as fibers. The most popular is polypropylene fiber, which is used in the production of outdoor–indoor carpets, cordage, and upholstery. Polypropylene fibers are low melting (below 355°F) and are degraded by sunlight. Yet, because of their resistance to soiling and ease in cleaning, stabilized polypropylene fibers are widely accepted as carpeting for heavily traveled, dirt-attracting areas, such as store and home entrances, patios, and swimming pool areas. The formulas for polyolefin and polypropylene are

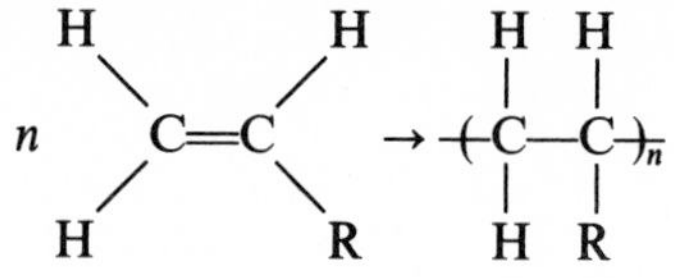

Polyolefin

where R=H, CH_3, or other hydrocarbon group.

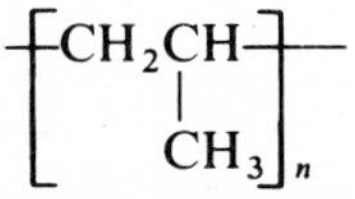

Polypropylene (PP)

The strength of polypropylene filaments (Spectra) has been increased dramatically by chain extension, that is, stretching solvent-swollen filaments.

9.8   POLYURETHANES

Spandex (Lycra) is an elastomeric fiber popularized by "cross-your-heart" bra commercials. It was introduced by duPont in 1958. Lycra is a segmented copolymer, with each segment contributing its own properties to the whole material. The soft segment is composed of flexible macroglycols whereas the rigid segment is formed from 4,4-diisocyanatodiphenylmethane (MDI) and hydrazine (Figure 9.8).

Macroglycols can be made from polyesters, polyethers, and polycaprolactones. The key is that the end groups are both alcohols (hydroxyls). The macroglycols are generally short chains with DPs around 40. The hydroxyl end groups are reacted with an excess of the MDI diisocyanate to form urethane linkages with isocyanate end groups. These in turn are further reacted with hydrazine to form urealike and urethane linkages, giving segmented elastomeric fibers, that is, Lycra. This process illustrates the ability of scientists to design molecules with specific, desired properties. Lycra is used extensively in the manufacture of foundation garments, swimsuits, and running and exercise suits.

9.9   OTHER FIBERS

There are a number of other fibers being produced for specialty applications. For example, polyimides have good thermal stability and are employed where resistance to high temperature is required.

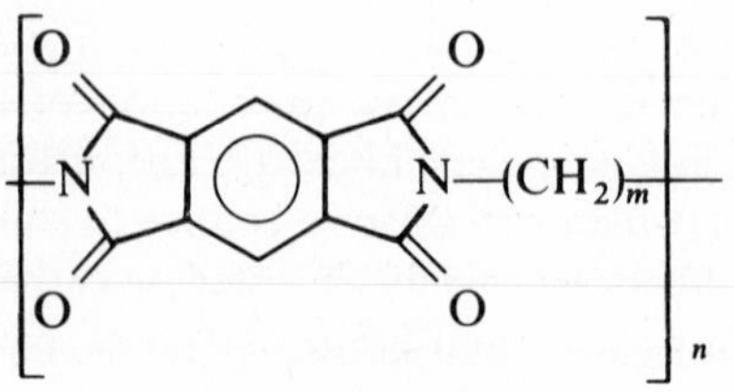

Polyimide

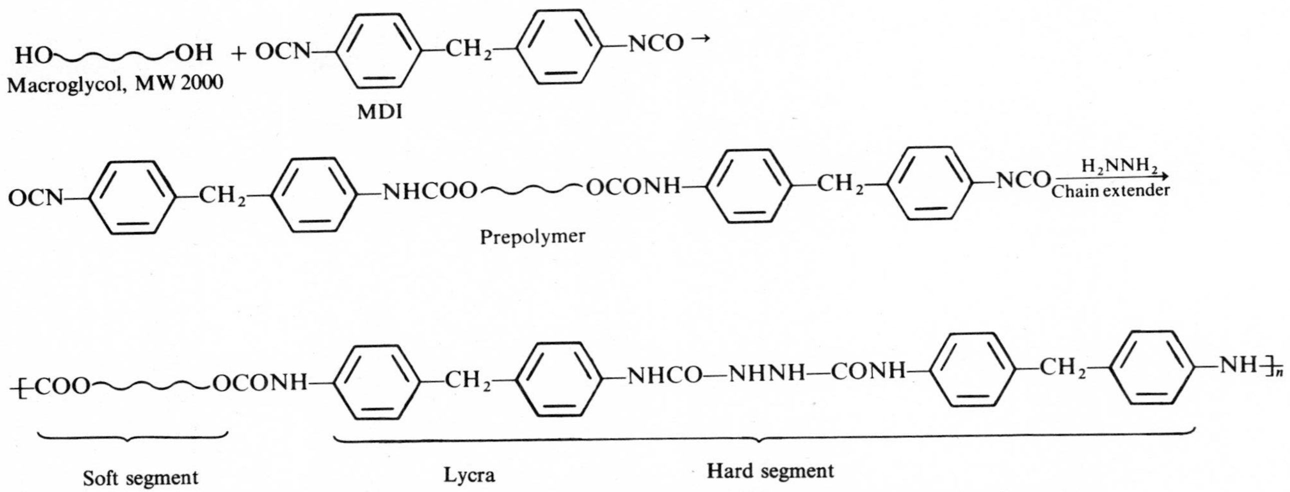

HO⌇OH + OCN⌬—CH₂—⌬NCO→
Macroglycol, MW 2000   MDI

OCN⌬—CH₂—⌬NHCOO⌇OCONH⌬—CH₂—⌬NCO $\xrightarrow[\text{Chain extender}]{H_2NNH_2}$

Prepolymer

+COO⌇OCONH⌬—CH₂—⌬NHCO—NHNH—CONH⌬—CH₂—⌬NH+ₙ

Lycra

⎧‾‾‾‾‾⎫   ⎧‾‾‾‾‾‾‾‾‾‾‾‾‾‾‾‾‾‾‾‾‾‾‾⎫
Soft segment   Hard segment

Figure 9.8. Synthetic outline for segmented elastomeric fibers, Spandex (Lycra).

153

Carbon fibers (also called graphite fibers) are used for applications where great strength, rigidity, and light weight are required. Graphite fibers are used to reinforce polymers in the construction of lightweight, highly durable bicycles, car bodies, golf club shafts, fishing rods, firemen's suits, and aircraft for space exploration. The Voyager, which flew around the world without refueling in 1986, was constructed from graphite and aramid-reinforced plastics.

As shown by the following equation, carbon fibers are produced by the pyrolysis of polyacrylonitrile. The polyacrylonitrile cyclizes at about 570°F and remains stable up to about 1290°F. This product, originally reported by Hautz in 1950, was dubbed black orlon after duPont's trade name for its light-colored polyacrylonitrile fiber. Graphite fibers are also produced from pitch and other materials.

Polyacrylonitrile → ("Black" orlon)

Boron filaments are produced by a unique process in which boron produced by the reaction of boron trichloride and hydrogen is deposited on a tungsten or graphite filament.

The physical properties of typical fibers are compared in Table 9.1.

Table 9.1   Physical Properties of Typical Fibers

Polymer	Tenacity (g/denier)	Tensile Strength (kg/cm^2)	Elongation (%)
Cellulose			
Cotton	2.1–6.3	3000–9000	3–10
Rayon	1.5–2.4	2000–3000	15–30
High-tenacity rayon	3.0–5.0	5000–6000	9–20
Cellulose diacetate	1.1–1.4	1000–1500	25–45
Cellulose triacetate	1.2–1.4	1000–1500	25–40
Protein			
Silk	2.8–3.0	3000–6000	13–31
Wool	1.0–1.7	1000–2000	20–50
Vicara	1.1–1.2	1000–1000	30–35
Nylon 6,6	4.5–6.0	4000–6000	26
Polyester	4.4–5.0	5000–6000	19–23
Polyacrylonitrile	2.3–2.6	2000–3000	20–28
Saran	1.1–2.9	1500–4000	20–35
Polyurethane (Spandex)	0.7	630	575
Polypropylene	7.0	5600	25
Asbestos	1.3	2100	25
Glass	7.7	2100	3.0

GLOSSARY

Acrylic fiber: Fiber containing more than 85% repeating units of acrylonitrile ($-[CH_2CH(CN)]-$).

Adipic acid: $HOOC-[CH_2]_4-COOH$.

Aminononanoic acid: $H_2N-[CH_2]_8-COOH$.

Ammoxidation: Oxidation in the presence of ammonia (NH_3).

Black orlon: Pyrolyzed Orlon.

Borazole: Boron nitride ($B\equiv N$).

Bulky group: Large substituent on a polymer chain.

Carothers, W.: The inventor of nylon 6,6.

Composite: A mixture of polymer and an addition such as fiberglass.

Copolymer: A macromolecule consisting of more than one repeating unit in the chain.

Dacron: Trade name for polyester fibers.

Denier: The weight in grams of 9000 m of a fiber; the finer the fiber, the lower the denier.

Dry spinning: The production of filaments by evaporation of the solvent from the solution, which was exuded from the spinneret.

Ethylene glycol: $HO-[CH_2]_2-OH$.

Fiber: Strong strands of the polymer with a length to diameter ratio of at least 100 to 1.

Fiberglass: Glass in the form of fine fibers.

Filament: A very long fiber.

Hexamethylenediamine: $H_2N-[CH_2]_6-NH_2$.

Hydrophobic: Water hating or water repellent.

Kodel: Trade name for polyester fibers.

Lycra: Trade name for spandex fibers.

Melt spinning: The production of filaments by cooling a molten exudate after it leaves the spinneret.

Modacrylic fiber: Acrylic fiber containing less than 85% repeating units of acrylonitrile ($-[CH_2CH(CN)]-$).

Nylon: A generic term for synthetic polyamides.

Nylon-6: Polycaprolactam.

$$\left[NH(CH_2)_5 - \overset{\displaystyle }{\underset{\displaystyle O}{C}} \right]_n$$

Nylon 6,6: The reaction product of hexamethylenediamine ($H_2N(CH_2)_6NH_2$) and adipic acid ($HOOC(CH_2)_4COOH$).

$$\left[NH(CH_2)_6NH - \overset{\displaystyle }{\underset{\displaystyle O}{C}} - (CH_2)_4 - \overset{\displaystyle }{\underset{\displaystyle O}{C}} \right]_n$$

Olefin: A hydrocarbon that is a member of the alkene homologous series $(H(CH_2)_nCH{=}CH_2)$.

Orlon: An acrylic fiber.

Polyamide: A polymer produced by the condensation of a diamine (H_2NRNH_2) and a dicarboxylic acid $(HOOCRCOOH)$.

Polyester: A copolymer produced by the condensation of a dicarboxylic acid $(HOOCRCOOH)$ and a diol $(HOROH)$.

Polyimide: A heat-resistant heterocyclic polymer.

Polypropylene: $[-CH(CH_3)CH_2-]_n$.

Polystyrene: $-[CH_2-CH(C_6H_5)-]_n$.

Propylene: $HC(CH_3){=}CH_2$.

Pyrolysis: Thermal degradation.

Spandex: An elastic fiber consisting of sequences of repeating units of polyesters and polyurethanes.

Spinnerets: Small, uniform holes used for the extrusion of filaments.

Spun-dyed fiber: Fiber that is dyed before the spinning process.

Staple fiber: A short fiber.

Tenacity: Tensile strength of a fiber expressed as g/denier. If a 100-denier yarn fails under a 300-g load, its tenacity is 300/100 or 3 g/denier.

Tensile strength: The maximum stress that a material can withstand without failure when stretched.

Textured yarn: Twisted filament yarn.

Thermoplastic: A linear or branched polymer that can be softened by heat and cooled to reform the solid.

Throwing: Twisting of filaments.

Tow: Several twisted filaments gathered together.

Urethane linkage:

$$\overset{\displaystyle H \quad O}{\underset{\displaystyle (-N-CO-)}{\displaystyle |\ \ \ \ ||}}$$

Yarn: Spun fibers used in weaving of fabrics.

Wet spinning: The production of filaments by passing the polymer solution, exuded from the spinneret, into a bath to insolubilize the polymer.

REVIEW QUESTIONS

1. What is the minimum length to diameter ratio for a substance to be classified as a fiber?

2. What atoms are involved in hydrogen bonding in nylons and proteins?

3. Which is stronger: hydrogen bonding or dipole–dipole interactions?

4. Besides hydrogen bonding, what else is characteristic of fiber molecules?

5. In the nylon nomenclature, which number is the number of carbon atoms in the diamine?

6. Which is more hydrophobic: cellulose or nylon 6,6?

7. How would you convert a long filament into a staple fiber?

8. How could you convert cellulose staple fiber to a long filament?

9. Is rayon produced by wet or dry spinning?

10. Are acrylic fibers produced by dry or melt spinning?

11. Name a fiber that is produced by melt spinning.

12. How are polypropylene fibers produced?

13. Which has the lower denier value: a nylon fishline or nylon fiber used to manufacture hosiery?

14. How do the polyester fibers produced originally by Carothers differ from today's polyester fibers, such as Dacron?

15. What type of spinning is used to produce glass fibers?

16. What is the precursor for some of the high-grade graphite fibers?

BIBLIOGRAPHY

Economy, J. (1976). *New and Specialty Fibers*. New York: Wiley.

Gaylord, N. W. (1974). *Reinforced Plastics*. Boston: Cahners Books.

Hall, A. J. (1974). *The Standard Book of Textiles*. London: Butterworth.

Lenz, R. W., and Stein, R. S. (1973). *Structure and Properties of Polymer Fibers*. New York: Plenum.

Mark, H. F., Atlas, S. M., and Cernia, C. (1967). *Man-Made Fibers: Science and Technology*. New York: Interscience.

Moncrieff, R. W. (1975). *Man-Made Fibers*. London: Butterworth.

Parrott, N. J. (1973). *Fiber Reinforced Materials Technology*. New York: Van Nostrand–Reinhold.

Settig, M. (1972). *Polyamide Fiber Manufacture*. Park Ridge, NJ: Noyes Corp.

Seymour, R. B., and Deanin, R. B. (1987). *Polymeric Composites: Their Origin and Development*. Utrecht, Holland: VNU Science Press.

ANSWERS TO REVIEW QUESTIONS

1. 100 to 1.

2. Hydrogen, nitrogen and oxygen.

3. Hydrogen bonding.

4. Structural symmetry.

5. The first integer.

6. Nylon 6,6.

7. By cutting it into small sections.

8. By converting it to rayon (regenerated cellulose).

9. Wet spinning.

10. Dry spinning.

11. Nylon.

12. By fibrillation and melt spinning.

13. Hosiery fiber.

14. Carothers produced an aliphatic polyester with a low softening point. Today's polyester fibers are aromatic polyesters with high softening points.

15. Melt spinning.

16. Acrylic fiber.

CHAPTER X

Rubber (Elastomers)

10.1 EARLY HISTORY

Athletes of the Mayan civilization used a ball made from rubber for their national game, called tlachti, over 1000 years ago. This game resembled modern basketball in that the ball was thrown through a circular stone hole. However, this game differed from the modern sport since only one goal was scored in each game and the members of the losing team could be executed.

The Indians of Mexico called the rubber tree *ule* and the rubbery product was called *ulei*. Some of the South American Indians called the tree *heve* but others called it *caaochu* or "weeping wood." The name caoutchouc is still in use in France, but *Hevea braziliensis* is the more widely used term for natural rubber (NR). A comparable elastomeric product is also present in the domesticated rubber plant (*Ficus elastica*), the guayule shrub (*Parthenium argentatum*), goldenrod (*Solidago*), and the dandelion (*Koksaghyz*). *Ficus elastica* was used unsuccessfully as a source of rubber in Malaysia, and *Castilloa elastica* and *Castilloa ulai* were used as the original sources of rubber in

Brazil, but *H. braziliensis* from Indonesian plantations is now the principal source of natural rubber.

The American Indians made waterproof boots and containers by dipping in the rubber latex. Latex, which is the liquid exuded by the rubber tree, is an aqueous emulsion of rubber. The name rubber was given to the sticky elastomeric material by Joseph Priestley, who used it to erase pencil marks from paper.

The use of natural rubber was limited by its characteristic stickiness. However, MacIntosh made a cloth sandwich from a solution of rubber in naphtha in 1823, and this type of construction is still used for waterproofing garments. However, there was little use for rubber until Charles Goodyear vulcanized (cross-linked) the crude product by heating it with sulfur. This discovery and other accidental discoveries are called serendipity after a name coined by Walpole. This author described three princes of Serendip or Sri Lanka who were seeking potential princesses but accidentally made many apparently more valuable discoveries.

The only major source of rubber in the nineteenth century was the wild rubber tree from Brazil, Central America, the west coast of Africa, and Madagascar. However, this supply was insufficient to meet the demands brought on by the introduction of the automobile, each of which required four pneumatic tires.

Since Brazil prohibited the export of rubber seeds or seedlings, H. A. Wickham smuggled 70,000 rubber seeds hidden in banana leaves and brought them to England in 1876. The 1900 seedlings that germinated and survived were used to start the rubber plantations in Malaya late in the nineteenth century. The first year's production of four tons of plantation rubber was small compared to the production of 50,000 tons of wild rubber obtained in 1900. However, the source of wild rubber continued to decrease with further exploitation, but over 1 million tons of plantation rubber were produced annually just prior to World War II.

A small amount of wild rubber is still obtained from Brazil, but over 90% of today's natural rubber supply is obtained from plantations of about 14 million acres in Indonesia, Malaysia, Thailand, Sri Lanka, India, Vietnam, Cambodia, and Sarawak. The latex from the rubber plant contains 36 to 40% of rubber.

10.2   GENERAL PROPERTIES OF ELASTOMERS

The individual polymer chains of elastomers are held together by weak intermolecular bonding forces, that is, London dispersion forces, which allow rapid chain slippage when a moderate pulling force is employed. Cross-links, which are introduced during vulcanization, permit rapid elongation of the principal sections, to a point where the chains are stretched to their elastic limit. Any additional elongation causes primary bond breakage (Figure 10.1). The cross-links, which are the boundaries for the principal sections, permit the rubber to "remember" its original shape, that is, the original orientation of the particular chains.

10.3   STRUCTURE OF NATURAL RUBBER (NR)

In 1826, Faraday used carbon–hydrogen analysis to show that rubber was a hydrocarbon with the empirical formula of C_5H_8. Subsequently, it was shown that the pyrolysis of natural rubber produced isoprene, which had the skeletal formula of

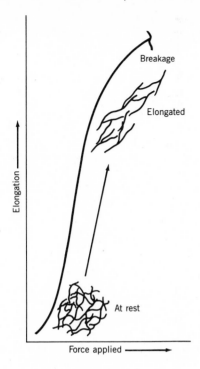

Figure 10.1. Illustration of typical applied-force–elongation behavior of rubber.

$C{=}C(C){-}C{=}C$. In the early part of the twentieth century, Harries added ozone (O_3) to natural rubber and showed that this elastomer consisted of repetitive units of 2-methyl-2-butene, $(C{-}C(C){=}C{-}C)_n$, where n was equal to several hundred.

These high-molecular-weight molecules are giant molecules. If we could see molecules of rubber (without the cross-links), they would look like strands of cooked spaghetti, and the polymer chains would be entangled much like the spaghetti strands. However, since these chains are in constant motion at room temperature, a can of worms serves as a more realistic model.

It is of interest to point out that the butene units in natural rubber have a cis arrangement, that is, the carbon–carbon chain extensions are on the same side of each ethylene unit ($C{=}C$). Thus, the skeletal chain of *Hevea* rubber would look like

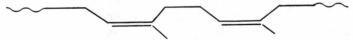

In contrast, another naturally occurring polymer of isoprene called balata or gutta-percha has a trans arrangement, as shown by

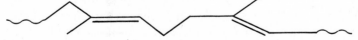

Ball-and-stick models of *Hevea* rubber and gutta-percha are shown in Figures 10.2 and 10.3. The trans arrangements permit these chains to fit closely together so that gutta-percha is a hard plastic, in contrast to the flexible cis polyisoprene. Staudinger received the Nobel prize for his interpretation of the correct structure of rubber and other macromolecules.

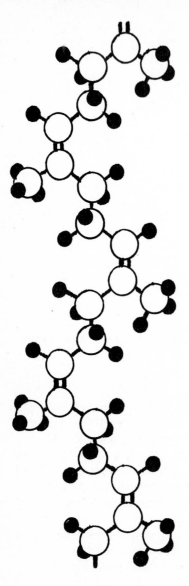

Figure 10.2. Ball-and-stick model of *Hevea* rubber (*cis*-1,4-polyisoprene).

10.4 HARVESTING NATURAL RUBBER

The *Hevea* tree grows best in hot, moist climates in acidic, well-drained soils. The cultivated rubber tree grows to 60 to 70 feet tall. The rubber latex flows through a series of tubes in the tree's cambium layer, that is, the outer wood layer directly beneath the bark. The latex oozes out when this layer is pierced.

Botanists continue to work on improving the tree and through grafting and breeding have grown trees that produce 1000% more rubber than the wild *Hevea* trees. Further work is being done on obtaining rubber from other plants, such as guayule, which will grow in the American southwestern desert area and can be harvested mechanically. Thus, even though natural rubber has been utilized by

Figure 10.3. Ball-and-stick model of gutta-percha (*trans*1,4-polyisoprene).

humans for several thousand years and cultivated for over a hundred years, research continues on the improvements of tree yields and the development of alternative sources.

Rubber tappers cut a narrow diagonal groove in the bark about 4 feet from the ground with a long, curved knife called a gouge (Figure 10.4). A U-shaped metal spout with a small cup to catch the latex is attached at the bottom of the cut. Trees are tapped for 25 to 30 years commencing when the young tree is 5 to 7 years of age.

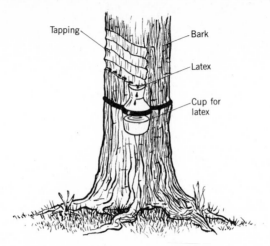

Figure 10.4. Illustration of how latex is collected from rubber trees.

Crude rubber is obtained by coagulation of the latex by the addition of formic acid (HCOOH) or acetic acid (H_3CCOOH). Some of the latex is not coagulated but is concentrated in machines called separators, which are similar to the cream separators employed in dairies. This rubber latex is used to make articles such as surgeon's gloves, condoms, tubing, elastic thread, and foam-backed carpeting.

10.5   STYRENE–BUTADIENE RUBBER (SBR)

Hofman synthesized isoprene in Germany in 1909, and the English chemists Matthews and Strange and the German chemist Harries converted methylisoprene to a rubbery product in 1910 by use of sodium metal. Over 2500 tons of this type of synthetic rubber were produced in Germany during World War I. Kaiser Wilhem equipped his Mercedes-Benz with synthetic rubber tires in 1912 and was impressed with their utility. However, since the methyl rubber was not reinforced by carbon black, these tires were not satisfactory when used on heavier equipment by the German army in World War I.

$$H_2C=\underset{\underset{\text{Isoprene}}{}}{\overset{\overset{\displaystyle H}{|}}{C}}-\underset{\underset{}{}}{\overset{\overset{\displaystyle CH_3}{|}}{C}}=CH_2 \qquad\qquad H_2C=\underset{\underset{\underset{\text{Methylisoprene}}{}}{\overset{\displaystyle CH_3}{|}}}{C}-\underset{\underset{\overset{\displaystyle |}{CH_3}}{}}{\overset{}{C}}=CH_2$$

In the late 1920s, Tschunker and Bock patented a method for producing a copolymer of 1,3-butadiene and styrene in an aqueous emulsion. The synthetic rubber molecule, which was called Buna-S, contained repeating units from both butadiene and styrene in a ratio of about $3:1$. Most of the SBR now produced contains about 20% of the 1,2 configuration, 20% of the *cis*-1,4, and 60% of the trans-1,4 con-figuration. The precise structure of SBR is varied, but it is reproducible. The irregular structure prevents the chains from close contact with one another and promotes rapid slippage of chain segments past one another.

The name Buna-S was derived from the first letters of butadiene (Bu) and styrene (S) and the chemical symbol for sodium (Na). Metallic sodium was employed to initiate the first polymerization of dimethylbutadiene. The German chemists obtained U.S. patents in which this novel polymerization process was described in detail. This synthetic rubber (Buna-S) was initially synthesized on an industrial scale by Germany's I. G. Farbenindustrie in 1933.

Prior to the bombing of Pearl Harbor in 1941, the Germans had an annual production capacity of 175,000 tons of Buna-S, and the Russians had an annual production capacity of 90,000 tons of sodium-catalyzed polybutadiene rubber. In contrast, annual American production of synthetic rubber, prior to 1942, was less than 10,000 tons, and most of this was specialty oil-resistant rubber that was not suitable for the manufacture of pneumatic tires.

Nevertheless, the production of Buna-S was duplicated in the United States during World War II, and the product was called GRS (Government rubber styrene). Over 50 GRS plants were constructed and operated in North America during the early 1940s, and the annual production of GRS reached 700,000 tons before the end of World War II.

After the war ended, the U.S. synthetic rubber production facilities were acquired by private industry, and the name for this synthetic rubber was changed by the

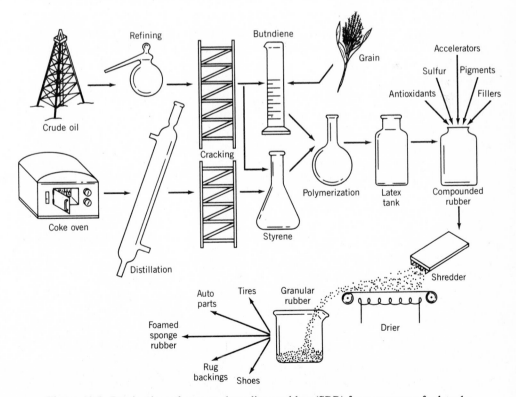

Figure 10.5. Production of styrene–butadiene rubber (SBR) from common feedstocks.

American Society for Testing Materials (ASTM) from GRS to SBR. Although this rubber was not as good as natural rubber, it was readily available and produced from inexpensive petroleum feedstocks as shown in Figure 10.5.

10.6   OTHER SYNTHETIC ELASTOMERS

Although the bulk of synthetic rubber is of the SBR variety, several other elastomers have been synthesized for special-purpose applications. These rubber products generally are more costly than natural and SBR rubber, but their special properties justify their higher costs.

In addition to synthesizing Buna-S, Tschunker and Bock also patented a process for the aqueous emulsion copolymerization of butadiene and acrylonitrile. They called this oil-resistant elastomer Buna-N. This copolymer is now produced under the ASTM name NBR.

$$\underset{H}{\overset{H}{\diagdown}}C=C\underset{C\equiv N}{\overset{H}{\diagup}}$$

Acrylonitrile

Another oil-resistant elastomer, called neoprene, was synthesized by Carothers and Collins and patented by duPont in 1931. This polymer is designated as CR by ASTM. It is of interest to note that the explosive vinylacetylene is the precursor of the monomer chloroprene. As shown by the following equation, the less explosive chloroprene was synthesized by the addition of hydrochloric acid (HCl) to vinylacetylene *in situ*.

$$n\,H_2C{=}CH{-}C{\equiv}CH \xrightarrow{\text{HCl}} n\,H_2C{=}CH{-}\underset{\underset{Cl}{|}}{C}{=}CH_2 \longrightarrow$$

Vinylacetylene   Chloroprene

$$\underset{\underset{Cl}{|}}{-(CH_2{-}CH{=}C{-}CH_2)_n}$$

Polychloroprene
(neoprene)

Isobutylene was polymerized at very low temperatures ($-150°F$) in the presence of aluminum chloride ($AlCl_3$) in Germany in the early part of the twentieth century. This polymer was sticky and could not be used as a vulcanizable elastomer.

$$n\,H_2C{=}C\underset{CH_3}{\overset{CH_3}{\diagdown}} \longrightarrow -(CH_2{-}\underset{\underset{CH_3}{|}}{\overset{\overset{CH_3}{|}}{C}})_n$$

Isobutylene   Polyisobutylene

Sparks and Thomas overcame this deficiency by introducing unsaturated linkages in the polymer chain. This was accomplished by the copolymerization of the isobutylene with a relatively small amount of isoprene so that a segment of the polymer chain would contain repeating units with the structures

$$
\begin{array}{ccc}
& H \quad CH_3 & \\
& | \quad\quad | & \\
+\!\!-\!C\!-\!C\!-\!\!- &) & \\
& | \quad\quad | & \\
& H \quad CH_3 &
\end{array}
\qquad \text{and} \qquad
\begin{array}{c}
H \quad CH_3 \; H \quad H \\
| \quad\quad | \quad\;\; | \quad\; | \\
+\!\!-\!C\!-\!C\!=\!C\!-\!C\!-\!\!+ \\
| \quad\quad\quad\quad\quad | \\
H \quad\quad\quad\quad\;\; H
\end{array}
$$

This elastomer, which is called butyl rubber, is designated as IIR by ASTM.

Ziegler and Natta received the Nobel prize in 1963 for the development of catalyst systems that were useful for the polymerization of ethylene and related unsaturated hydrocarbons. These Ziegler–Natta catalyst systems, containing the cocatalysts titanium trichloride ($TiCl_3$) and aluminumtriethyl ($Al(C_2H_5)_3$), were used subsequently for the polymerization of isoprene by chemists in several American rubber companies in 1955. This polyisoprene elastomer is designated as IR by ASTM and has been called by the incongruous trade name Natsyn, which stands for natural synthetic rubber.

The structure of the polymer chain in IR is similar to that of natural rubber and consists largely of cis repeating units. IR has many physical properties that are as good as those of the natural product. *cis*-Polybutadiene, which is produced by the polymerization of 1,3-butadiene by Ziegler–Natta or butyllithium catalysts, costs less than polyisoprene and has replaced the latter to a large extent.

$$
\begin{array}{cc}
CH_3 & \\
| & \\
n\,CH_2\!=\!C\!-\!CH\!=\!CH_2 & \\
\text{Isoprene} &
\end{array}
\qquad\qquad
\begin{array}{c}
H_3C \qquad\qquad\quad H \\
\;\backslash \qquad\qquad\quad / \\
C\!=\!C \\
/ \qquad\qquad\quad \backslash \\
+\!(CH_2 \qquad\quad CH_2\!-\!)_n \\
\text{Polyisoprene}
\end{array}
$$

The first American synthetic elastomer was synthesized by Patrick in 1927. Since the product was an organic polysulfide, it was called Thiokol. The prefix thio is the Greek word for sulfur. The ol suffix was used because the original objective was to produce a permanent antifreeze, that is, ethylene glycol.

$$
2n\,Cl\!-\!(CH_2)_2\!Cl + 2n\,NaS_x Na \longrightarrow \text{---}\!(CH_2)_2 S_x\!-\!(CH_2)_2 S_x\text{---}_n
$$

Ethylene dichloride   Sodium   Thiokol
                       polysulfide

Heat-resistant elastomers, called silicone rubbers, and designated as SI by ASTM, are produced by a repetitive condensation of dimethyldichlorosilane in the presence of water. The backbone of this polymer consists of siloxane units (—Si—O—Si—O—), which are characterized by a high bond strength.

$$
\begin{array}{ccc}
CH_3 & & CH_3 \\
| & & | \\
n\,Cl\!-\!Si\!-\!Cl & \xrightarrow{\;H_2O\;} & +\!(Si\!-\!O\!-\!)_n \\
| & & | \\
CH_3 & & CH_3
\end{array}
$$

The methyl groups on the chain act like oil or paraffin and provide nonstick and water-repellent properties in these elastomers.

Neither HDPE nor polypropylene (PP) is elastomeric. However, the copolymer of ethylene and propylene (EP) is an amorphous copolymer with elastomeric characteristics. A commercial vulcanizable elastomer (EPDM) is produced when a diene is added to ethylene and propylene before polymerization by a Ziegler–Natta catalyst.

Polyphosphazenes are useful as elastomers over an unusually broad temperature range. Polydichlorophosphazene, $\pm N{=}P(Cl_2)\pm_{\overline{m}}$, is unstable in humid atmospheres, but stable elastomers are produced when the chlorine substituents are replaced by phenoxy groups ($-OC_6H_5$) or other organic materials.

Elastomeric polyurethanes are produced by the reaction of a flexible polyester or polyether diol (HO—R—OH) with a diisocyanate (OCNRNCO).

10.7   PROCESSING OF ELASTOMERS

With only a few exceptions, the general steps involved in processing natural and synthetic rubber are the same. The exact steps vary according to the polymer utilized and its intended application.

The technicians involved in compounding rubber are called compounders but no chemical compounds are formed prior to the vulcanization or curing process. The additives are introduced to the solid elastomer on a two-roll wringerlike rubber mill or in an intensive mixer, such as a Banbury mixer.

The chief ingredients of compounded rubber are (a) sulfur, (b) accelerators, (c) pigments, (d) antioxidants, (e) reclaimed (recycled) rubber and (f) fillers, such as carbon black. Each of these additives performs a special function. Sulfur is added early in the processing and is largely responsible for the formation of cross-links in the vulcanization process. Accelerators (catalysts) are added to enhance the rate of vulcanization. Reinforcing pigments, such as carbon black, make rubber stronger and more resistant to wear. Antioxidants protect the rubber against chemical changes and the harmful effects of air, moisture, heat, and sunlight. Reclaimed rubber is recycled rubber that has been pretreated to make it compatible with the rubber mix.

The rubber mixture (compound) is shaped by (a) calendering, (b) extrusion, (c) molding, or (d) dipping. Calendering means rolling the rubber material into sheets. In extrusion, tube machines push the soft rubber material through different sized holes, similar to pushing toothpaste out of the tube. Extruded products include hoses, inner tubes, rubber stripping for use on refrigerator doors, and automobile windshields. Extruded products are vulcanized after they have been formed.

Most products are molded and vulcanized in the same step. Molded products include rubber tires, mattresses, and hard-rubber articles such as rubber hammers, gaskets, fittings, shoe soles, and heels.

Dipping is used to make products such as rubber gloves and balloons from liquid latex. Forms, typically made of glass, metal, or ceramic, are dipped into vats of latex. Repeated dipping increases the thickness of the product.

During vulcanization, the heat causes the sulfur to combine with the rubber to form cross-links. Generally, the more sulfur added, the harder the resulting rubber. Ebonite is formed from a compound containing one-third sulfur. Today, many other vulcanizing agents are used, but sulfur is still the major vulcanizing agent because of its low cost and availability.

Thermoplastic elastomers that may be used in place of cross-linked elastomers are not vulcanized but are simply molded, like thermoplastics in a heated mold. Rubbery products, such as solid tires and bumpers, are produced by reaction injection molding (RIM) in which the reactants are injected into the mold, where the polymerization reaction takes place.

Lattices of elastomers are processed like water-borne coatings. The curing or cross-linking agents and accelerators, stabilizers, and pigments are added as aqueous dispersions prior to curing.

Sponge rubber can be made from either dry rubber or latex. For dry rubber, chemicals such as sodium bicarbonate ($NaHCO_3$) are added that will form gas when heated during the vulcanization process and produce bubbles. Foam rubber, which is used for upholstery and foam strips for surgical use, is obtained by "whipping" air into the latex.

10.8   TIRES

The pneumatic (air-filled) tire was invented in 1845 by a Scottish engineer, Robert Thomson, but it was not strong enough for regular use. In 1888, John Dunlop, a Scottish veterinarian, developed usable rubber tubes for his son's tricycle.

The early automobile tires were single air-filled rubber tubes. Although they made movement easier and the ride smoother, they tended to develop many leaks. A two-piece tire consisting of a flexible, thin "inner tube" and a tough "outer tube" was developed in the early 1900s. The modern tubeless tire was developed in 1948. The basic parts of a tire are shown in Figure 10.6.

Modern tires are built on a slowly rotating roller called a drum. Initially, an inner liner of soft rubber is wrapped about the drum. A rubberized cord fabric is then laid, typically perpendicular to the initial inner liner. This fabric is called ply. A four-ply tire has four layers of cord fabric; a two-ply tire has two layers of cord. Next, layers of fibers (normally nylon, aramid (aromatic nylon), rayon, or fiberglass) or steel are laid. These layers are called belts. Most tires contain two belts.

An inner and outer ridge, called a bead, is then constructed. Each bead contains steel wire strands wound together into a hoop and covered with hard rubber. The ends of each ply are wrapped about the bead, forming the connective points of the tire to the wheel rim. Rubber sidewalls and outer tread material are then added. The individual parts, sidewalls, beads, plies, belts, and inner liner, are connected in a process called stitching.

The tire parts are now "permanently" cemented together in a process called vulcanization. The tire is removed from the drum and placed in a mold (curing press) that has the appropriate tread pattern. The mold acts like a large waffle iron: the "raw" tire is inserted in the mold, the mold is closed, heat is applied, and an inner air bag is filled with steam. The filled air bag pushes the tire against the mold sidewall that causes the treat pattern.

Bias belted tires are constructed similarly to bias tires except belts are placed between the plies and tread. The belts act to coordinate the movement of the plies and treads, and thus resist tread squirm and puncture. Fiberglass belts are used in bias belted tires.

All radial tires are belted. The cord fabric is added with the cord running radially, that is, from bead to bead. The combination of belting and addition of radial plies

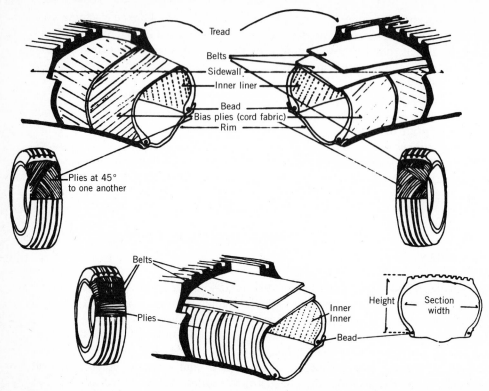

Figure 10.6. Cut-away views of modern passenger tires: bias (top left), bias belted (top right), and radial (bottom).

Table 10.1   Physical Properties of Typical Elastomers

		Pure Gum Vulcanizates		Carbon Black-Reinforced Vulcanizates	
	T_g (°F)	Tensile Strength (kg/cm^{-2})	Elongation (%)	Tensile Strength (kg/cm^{-2})	Elongation (%)
Natural rubber (NR)	−100	210	700	315	600
Styrene–butadiene rubber (NBR)	−60	28	800	265	550
Acrylonitrile–butadiene rubber (NBR)		42	600	210	550
Polyacrylates (ACM)				175	400
Thiokol (T)		21	300	85	400
Neoprene (CR)		245	800	245	700
Butyl rubber (IIR)	−81	210	1000	210	400
Polyisoprene (IR), polybutadiene (PB)	−140	210	700	315	600
Polyurethane elastomers (AU)		350	600	420	500

results in less flex and squirm and longer wear than either the bias or bias belted tires. The belts are mainly steel, aramid, fiberglass, or rayon. These types of passenger tires are illustrated in Figure 10.6. The physical properties of typical elastomers are shown in Table 10.1.

10.9   THE BOUNCE

We know why balls bounce, don't we? Is the bounce of a ball related to energy or probability factors? The answer to the first question, of course, is "Yes." The bounce depends on both factors. This dependency can be illustrated through the following demonstration, which uses a metal ball bearing and a "superball." When dropped onto a hard surface the balls hit the surface and rebound, but why?

The rebound of the ball bearing is largely due to the deformation of metal bonds upon striking the hard surface. This collision pushes the metal atoms into a higher energy situation. The metal atoms then move back to the original, lower energy sites, resulting in a push against the surface and the "bounce." For the ball bearing there is little unoccupied or free volume so the applied force is used to disrupt the primary (metal) bonded iron atoms with little change in the overall order of the iron atoms. Energy is the principal driving force here.

Polymers are most tightly packed when they are arranged in an ordered fashion such as a folded clothesline or thread on a spool. The "superball" is a semitough rubber that is solidified so that the polymer chains are arranged in a highly disorganized, random manner. When the "superball" hits the surface, the decreased space is largely accommodated by a reorganization of the polymer chains into a more ordered, less probable configuration. When the polymer chains return to their original, highly disorganized state, the "push" to occupy the original, predeformation volume is translated into a push against the surface, resulting in the "bounce." In this situation, probability is the major driving force in the bounce. Thus, an apparently similar phenomenon, the "bounce," results from two different free energy contributors in the metal and rubber balls.

GLOSSARY

accelerator: Vulcanization catalyst.

Aluminum chloride: $AlCl_3$.

Aramid: An aromatic nylon.

Boric acid: H_3BO_3.

Buna-N: Copolymer of butadiene and acrylonitrile.

Buna-S: Original name for SBR.

Butadiene: $C{=}C{-}C{=}C$.

Butene: $C{-}C{=}C{-}C$.

Butyl rubber: Copolymer of isobutylene and isoprene (IIR).

Caoutchouc: Natural rubber.

Chloroprene: 2-chlorobutadiene

$$\begin{array}{c} Cl \\ | \\ ({-}C{=}C{-}C{=}C{-}) \end{array}$$

Cis arrangement: Configuration in which substituents or chain extensions are on the same side of the ethylene double bond.

1,2 configuration:

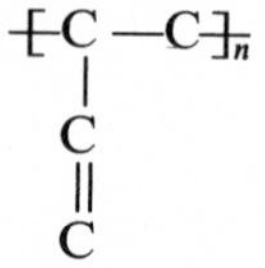

$$\begin{array}{c} +\!\!\!\operatorname{C-C}\!\!\!+_n \\ | \\ \operatorname{C} \\ || \\ \operatorname{C} \end{array}$$

CR: Neoprene.

Cross-link: Primary bonds joining polymer chains.

Elastomer: A general term for natural and synthetic rubbers.

Empirical formula: Simplest formula.

EPDM: Vulcanizable elastomeric copolymer of ethylene and propylene.

Ethylene–propylene copolymers (EP):

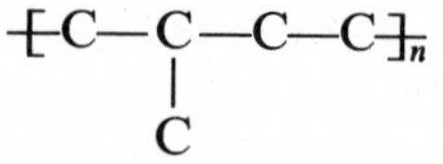

$$\begin{array}{c} +\!\!\!\operatorname{C-C-C-C}\!\!\!+_n \\ | \\ \operatorname{C} \end{array}$$

Formic acid: HCOOH.

Goodyear, Charles: Discoveries of vulcanization (cross-linking of rubber).

GRS: Government rubber styrene; name used for SBR during World War II.

Hevea braziliensis: Natural rubber.

IIR: Butyl rubber.

Intermolecular force: Attraction between atoms on different polymer chains.

IR: Polyisoprene.

Isobutylene:

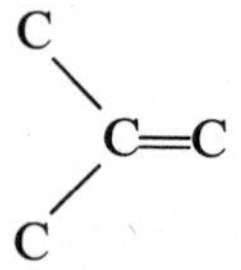

$$\begin{array}{c} \operatorname{C} \\ \backslash \\ \quad \operatorname{C\!=\!C} \\ / \\ \operatorname{C} \end{array}$$

Isoprene:

$$\begin{array}{c} \operatorname{C} \\ | \\ (\operatorname{C\!=\!C-C\!=\!C}) \end{array}$$

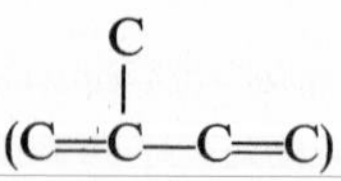

Latex: An aqueous emulsion.

Memory: The process whereby stretched elastomers return to their original dimensions when tension is released.

Methylisoprene:

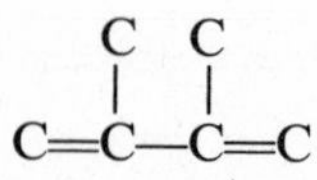

$$\begin{array}{c} \operatorname{C} \quad \operatorname{C} \\ | \quad\; | \\ \operatorname{C\!=\!C-C\!=\!C} \end{array}$$

n: Number of repeating units in a polymer (DP).

Natsyn: A trade name for polyisoprene.

NBR: Copolymer of butadiene and acrylonitrile.

Neoprene: Polychlorobutadiene.

NR: Natural rubber.

Ozone: O_3.

Polyphosphazene: Inorganic elastomer with the repeating unit $-N{=}P(R_2)-$.

Primary bond: Covalent bond of carbon–carbon atoms.

Pyrolysis: Process of thermal degradation.

Random copolymer: A macromolecule containing randomly arranged repeating units of two different monomers.

SBR: Styrene–butadiene elastomers.

Silane: $Si(CH_3)_4$.

Silicon: Si.

Silicone: Incorrect name for polysiloxanes.

$$\begin{array}{ccc} R & & R \\ | & & | \\ {\leftarrow}Si{-}O{-}Si{-}O{\rightarrow}_n \\ | & & | \\ R & & R \end{array}$$

Siloxane: Compounds containing one or more Si—O unit such as

$$-\overset{|}{\underset{|}{Si}}{-}O{\leftarrow}\overset{|}{\underset{|}{Si}}{-}O{\rightarrow}_n$$

Skeletal formula: Structural formula showing carbon–carbon bonds and omitting the hydrogen atoms such as used in this glossary.

Sodium: Na.

SR: Synthetic rubber.

Trans arrangement: Configuration in which substituents or chain extensions are on opposite sides of the ethylene double bond.

Ule, ulei: Names used by Aztecs for the rubber tree and rubber, respectively.

Vinylacetylene: $C{=}C{-}C{\equiv}C$.

Ziegler–Natta catalyst: Catalyst system that produces linear polyethylene and stereoregular vinyl polymers, usually $TiCl_3$ and $Al(CH_3)_3$.

REVIEW QUESTIONS

1. What are the advantages of using the guayule shrub as a source of natural rubber?

2. What does rubber latex have in common with milk?

3. What substance did Charles Goodyear use as a cross-linking agent in his vulcanization process?

4. What is stronger: the sulfur bonds that cross-link the polymer chains or the intermolecular forces between polymer chains in elastomers?

5. A stretched rubber band returns to its original dimensions when the tension is released because of what characteristic quality of elastomers?

6. What is the difference between Buna-S and SBR?

7. Why is butyl rubber more resistant to ozone degradation than natural rubber?

8. What is the difference between DP and *n*?

9. Which of the following configurations is cis?

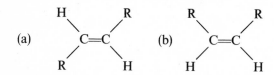

10. What is the difference in the structure of isoprene and butadiene?

11. Which repeating unit provides more elasticity in SBR?

12. What compounds are present in the Ziegler–Natta catalyst?

13. Why is a silicone elastomer more heat resistant than *Hevea braziliensis*?

14. Why is ethylene–propylene copolymer (EPDM) used in place of *Hevea* rubber in white sidewalls of tires?

15. What is the advantage of using an accelerator in the vulcanization of *Hevea* rubber?

16. Define an accelerator.

BIBLIOGRAPHY

Alliger, J., and Sjothum, I. V. (1964). *Vulcanization of Elastomers.* New York: Reinhold.

Baum, V. (1933). *The Weeping Wood.* Long Island, NY: Doubleday, Doran.

Coran, A. Y. (1978). *Science and Technology of Rubber.* New York: Academic Press.

Davis, C. C., and Blake, J. T. (1937). *The Chemistry and Technology of Rubber.* New York: Reinhold.

Kennedy, J. P., and Thornqvist, E. G. M. (1968). *Polymer Chemistry and Synthetic Elastomers.* New York: Wiley–Interscience.

Mark, J. E., and Lal, J. (1982). *Elastomers and Rubber Elasticity,* ACS Symposium Series No. 193. Washington, DC: American Chemical Society.

McGregor, R. B. (1954). *Silicones and Their Uses.* New York: McGraw–Hill.

Morton, M. (1987). *Rubber Technology.* New York: Reinhold.

Roberts, A. D. (1988). *Natural Rubber Science and Technology.* Oxford, England: Oxford University Press.

Tornqvist, E. G. M. (1986). Polyolefin elastomers (Chap. 9). In R. B. Seymour and T. Cheng (Eds.), *History of Polyolefins.* Dordrecht, The Netherlands: Reidel.

Whitby, G. S. (1954). *Synthetic Rubber.* New York: Wiley.

ANSWERS TO REVIEW QUESTIONS

1. Guayule can be grown in the arid areas of northern Mexico and southwestern United States. It can be harvested mechanically and no overseas shipment is involved.

2. They are both emulsions of polymers, that is, *Hevea* rubber and casein.

3. Sulfur (S).

4. The sulfur bonds. These are the primary covalent bonds, which are at least 25 times stronger than the weak intermolecular forces (London or dispersion forces, i.e., secondary bonds).

5. Memory.

6. They are the same, that is, copolymers of butadiene and styrene.

7. Butyl rubber has fewer double bonds. Ozone attacks the ethylene double bonds.

8. They are identical. Each is equal to the number of repeating units in a macromolecule.

9. (b) Both substituents are on the same side of the plane of the ethylenic double bond.

10. Isoprene is 2-methylbutadiene.

11. Butadiene.

12. $TiCl_3$ and $Al(C_2H_5)_3$.

13. The siloxane bonds are stronger than carbon–carbon bonds.

14. EPDM has fewer carbon–carbon double bonds and hence is more resistant to ozone, which causes cracking of *Hevea* rubber.

15. Decreases curing time.

16. Accelerators are catalysts that speed up the vulcanization process.

CHAPTER XI

Paints, Coatings, Sealants, and Adhesives

11.1 HISTORY OF PAINTS

Early humans used crude paintings as a means of communication. The mineral-based paintings of a bison in the Altimara cave in Spain and of a Chinese horse at Laucaux, in France, are at least 15,000 years old. Aboriginal mineral-based paintings, such as the Obiri Rock sketches at Arnhem Land in northern Australia, are at least 5000 years old.

Lacquer, which includes a polymer as its principal component, originated in China at the time of the Chou Dynasty, sometime before the Christian era. The lacquers in

176

China and Japan were based on the sap of specific trees, whereas the lacquers in India and Burma were based on shellac, which is a resinous material exuded by insects. The early painters also used exudates from trees, such as copals, and these resins are still used today. A lacquer is a solution that forms a film by evaporation of the solvent. The Egyptians employed pitch and balsam resins as sealants for ships.

Because of high costs and lack of knowledge, improvements in the lacquer and sealant art were slow. However, pigments, such as white lead ($2PbCO_3 \cdot Pb(OH)_2$), zinc oxide (ZnO), litharge (PbO), red lead (Pb_3O_4), and carbon black (C), and naturally occurring polyunsaturated vegetable oils were produced prior to the Industrial Revolution.

11.2 PAINT

Oleoresinous coatings or paint have been produced from flax seed (linseed oil, *Lininum usitatissium*) and finely divided pigments since the fourteenth century using recipes supplied by the monk Theophilus. Other vegetable oils used in the paint industry are soybean, safflower, tung, oiticica (*Licania rigida*), and menhaden oils. All of these oils contain monounsaturated oleic acid ($C_{17}H_{33}COOH$) and diunsaturated linoleic acid ($C_{17}H_{31}COOH$). Linseed and soybean oil also contain triunsaturated linolenic acid ($C_{17}H_{29}COOH$). The polymerizable unsaturated oil, which is called a binder, polymerizes by cross-linking in the presence of oxygen, and this reaction is catalyzed (accelerated) in the presence of soluble organic acid salts of heavy metals, such as lead or cobalt.

The polymerization (hardening or drying) of these unsaturated oils is a chain reaction, which is similar to the initiation, propagation, and termination that occurs in the addition polymerization of vinyl monomers. In the initiation reaction, the unsaturated oil adds oxygen to produce a hydroperoxide, which decomposes to produce a free radical. These free radicals are responsible for the propagation, which also involves cross-linking. The binder is the film former. The liquid, which includes the binder, is called the vehicle or medium. The unpigmented paint is called a varnish. The solvent that dissolves the binder is sometimes called a "thinner." Although these oil-based paints are still in use today, they have been displaced, to some extent, by lacquers based on man-made resins and waterborne coatings.

The first widely used commercial lacquer was based on pigmented cellulose nitrate. This man-made resin, as well as natural resins such as kauri, was dissolved in ester solvents, such as butyl acetate, and used as an automotive finish (Duco) and textile finish (Pyroxylin).

Although the resins or resin-forming compounds used in paints may be cured or cross-linked after application, they must be linear and flexible when applied to the surface or substrate. The glass transition temperature (T_g) is used to define the temperature above which a polymer is flexible because of the segmental motion of the polymer chains. All coatings must be applied at temperatures above their characteristic glass transition temperatures.

The product mix for coatings is changing from organic solvent-borne to waterborne or high-solids coatings, and the market for industrial and residential coatings continues to grow. Over 1 billion (10^9) gallons of coatings with a value of almost $10 billion ($10^{10}$) are produced annually in the United States by over 1200 paint manufacturers. However, the 10 leading manufacturers produce over 35% of all paint sold in the United States.

The principal steps in the production of coatings are mixing, grinding, and thinning. In most cases, the pigment is mixed with the vehicle to form a heavy paste, which is then ground in a ball mill or by a high-speed impeller. Appropriate solvents (thinners) are then added to these dispersions. The mixing and grinding step for water borne coatings is similar to that used for solution coatings but the liquid dispersion for waterborne coatings is water. The dispersion of the pigments and other additives are then mixed with an aqueous dispersion of the resin. The components of the five major types of coatings (paint) are shown in Figure 11.1.

11.3 PAINT RESINS

The first synthetic paint resin was introduced by Leo Baekeland early in the twentieth century. This soluble resin was produced by heating phenol (C_6H_5OH) and formaldehyde ($H_2C{=}O$) in the presence of rosin. The latter is one of the constituents of the exudate from pine trees.

The most widely used paint resin, which is called an alkyd, was introduced in 1925 by R. Kienle. It was obtained by the reaction of an alcohol, such as glycerol, and an acid, such as phthalic acid. Unsaturated acids, such as linoleic acid, are also incorporated in the alkyd reactants so that this oil-modified resinous product is unsaturated. Hence, oil-modified alkyds will cure or dry in air much like the oil paints. Another ester type of paint called glyptal was introduced by W. Smith in 1901. Glyptal is produced by the condensation of glycerol and phthalic anhydride.

Phthalic anhydride   Glycerol

Glyptal

Many other synthetic polymers, such as chlorinated rubber, polyvinyl chloride, polystyrene, melamine–formaldehyde, silicone, and epoxy resins, are also used as

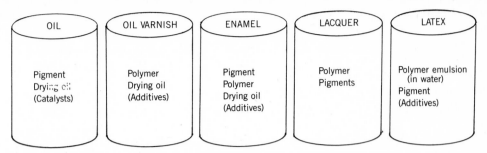

Figure 11.1. Components of the five major types of coatings (paint).

paint resins. In the past, these coatings have been applied as solutions of the resins in volatile solvents. Since the solvents usually are evaporated into the atmosphere, they contribute to atmospheric pollution. Accordingly, alternate methods of application, such as powdered resins and aqueous emulsions of resins, are preferred today in place of the solvent-based application of coatings. The solvent content of coating solutions is also being reduced. These higher-solids coatings, waterborne coatings, powder coatings, and two compound coating systems now account for over 60% of the total paint market.

Melamine–formaldehyde resins (MF)

Poly(vinyl chloride) (PVC)

Polystyrene (PS)

11.4   WATER-BASED PAINTS

Primitive humans used aqueous suspensions of colored clays to decorate their cave dwellings. Tempera paint was also a water-based coating in which eggs were used as the binder. Whitewash, also called whiting, consisted of a dispersion of calcium hydroxide $(Ca(OH)_2)$ in water. Since this was an inferior and temporary coating, the term whitewash is now used to describe a cover-up of vices or crimes.

Starch is water soluble and has been used for centuries as a coating. Over 50,000 tons of starch and chemically modified starch are now used annually in the United States as coatings, primarily for textile sizes. The first commercial water-based paint, which consisted of an ammonia (NH_3) -stabilized solution of casein and dispersed pigments, was introduced in the 1930s.

Emulsions of poly(vinyl acetate) $(CH_2CHOOCCH_3)_n$ containing aqueous dispersions of pigments were used as paint substitutes in Germany during World War II. These so-called water paints were introduced in the United States in 1948 and are now in wide use. Pigmented emulsions of polymethyl methacrylate, $+CH_2C(CH_3)COOCH_3)_n$, and of copolymers of styrene $(H_2C{=}CH{-}C_6H_5)$ are also used as water-based paints. Unlike the previously discussed SBR synthetic rubber, the principal constituent in this copolymer is styrene instead of butadiene. Coalescent agents that contain hydrophilic (water-loving) and lyophilic (resin-loving) groups are usually added to aqueous resin emulsions to assure the formation of continuous films.

Waterborne coatings based on resins with water-soluble groups are also available. For example, an alkyd resin with a large number of water-soluble hydroxyl (OH) groups may be produced from the reaction of phthalic anhydride and an excess of pentaerythritol $((HOCH_2)_4C)$ or by the condensation of ethylene glycol with an excess of phthalic anhydride or the anhydride of trimellitic acid $(HOOC(C_6H_3)C_2O_3)$. Aqueous suspensions of these resins with residual water-soluble groups may be used as waterborne coatings.

11.5   PIGMENTS

Pigments used by the paint industry include iron blue (Prussian blue), which is produced by the precipitation of a soluble ferrocyanide salt, such as yellow prussiate $(K_4Fe(CN)_6 \cdot 3H_2O)$, with iron sulfate $(FeSO_4)$. The ferroferricyanide precipitate is oxidized by air to form a ferriferricyanide.

Chrome yellow $(PbCrO_4)$, ultramarine blue (lapis lazuli), sodium aluminum silicate, white lithopone (barium sulfate, $(BaSO_4,$ zinc sulfide $(ZnS))$, chromic oxide (green cinnabar, Cr_2O_3), white titanium dioxide (TiO_2), and many organic pigments are also used as colorants by the coatings industry. Orr's zinc white was introduced in 1874 by J. Orr. This mixture of barium sulfate and zinc sulfide is now known as lithopone.

It is of interest to note that titanium ore, called ilmenite, was proposed as a black pigment by J. Ryland in 1865. Jebson and Farup extracted white titanium dioxide from this ore in 1880, and this pigment was made commercially in 1927. Titanium dioxide is the most widely used white pigment today. The annual volume of inorganic and organic pigments consumed by the American paint industry is valued at $600 million and $125 million, respectively.

11.6   APPLICATION TECHNIQUES FOR COATINGS

The classic tempera and oil-based paint were applied by brushing and this technique continues to be used by artists and, to a lesser degree, by house painters. This labor-intensive process has been replaced in many instances by less labor-intensive methods, such as dipping, flow coating, curtain coating, roll coating, spraying, powder coating, and electrodeposition.

Dipping is a simple immersion coating process that can be automated. In flow coating, the part to be coated is sprayed or showered with an excess of the coating and then allowed to drain before drying or curing. In curtain coating, the part to be coated is passed through a "curtain" of the coating material. This process is repeated if both sides of the part are to be protected by the coating.

Roll coating is used by "do-it-yourself" applicators and in industrial coating processes. In this process, the coating is transferred from a roller to a flat surface. Air spraying of coatings is fast but inefficient, with much of the coating being wasted by overspray. Nevertheless, this technique is widely used in industry. This system is similar to that used when one applies a coating from an aerosol paint can. Less overspray is encountered when one uses hydrostatic spraying techniques.

A minimum amount of overspray is encountered when the spray is electrically charged and the surface to be coated has the opposite charge. Electrostatic spraying is being used industrially to coat metal parts and nonmetallic objects, such as golf balls.

11.7   END USES FOR COATINGS

The use of tin plate on food cans has been replaced, to some extent, by the use of clear lacquers. A newer application of coatings is for the protection of onshore and offshore installations against marine atmosphere, the protection of steel and concrete against radiation in nuclear reactor environments, and protection of objects in outer space. Space exploration would have been impossible without thermal control surface coatings. The silicone rubber thermal control coatings on the polyester film used in the Skylab mission reflected 75% of the solar energy.

Intumescent paints are also used to protect burning wood. These coatings contain borax ($Na_2B_4O_7$), boric acid (H_3BO_3), sodium silicate, aluminum sulfate ($Al_2(SO_4)_3 \cdot 18H_2O$), or sodium carbonate ($Na_2CO_3 \cdot 1OH_2O$), which release water of hydration, sodium bicarbonate ($NaHCO_3$), which releases carbon dioxide (CO_2), and diammonium phosphate (($NH_4)_2HPO_4$) and melamine–formaldehyde resin, which contribute to intumescence or foam formation.

11.8   SOLVENT SELECTION

Before the advent of waterborne and powder coatings, paint technicians spent considerable time trying to discover appropriate solvents for resins used in coatings. It is of interest to note that although cellulose nitrate is insoluble in ethanol (C_2H_5OH) and ethyl ether (($C_2H_5)_2O$), Menard produced collodion by dissolving this polymer in an equimolar mixture of these two solvents in the 1950s. Other solvent systems have also been developed empirically, but more scientific guidelines are now available.

Paint chemists obtained useful solvency data by determining the temperature (aniline point) at which equal volumes of aniline ($C_6H_5NH_2$) and an unknown solvent became turbid when cooled. Paint chemists also obtained kauri–butanol values by determining the volume of unknown solvent that would produce turbidity when added to a solution of kauri copal resin in n-butanol ($H(CH_2)_4OH$).

These and other empirical tests have been largely replaced by solubility parameters developed by J. Hildebrand in the 1920s. These values can be calculated from the square root of the cohesive energy density ($(CED)^{1/2}$). CED is a measure of the intermolecular forces present in 1 mol of liquid. The solubility parameter values can be used to select solvents or mixtures of solvents for polymers.

11.9   SEALANTS

Sealants, such as classic putty, are paintlike products that are formulated for the filling of cracks and voids. The classic putty that was used as a sealant for window glass was based on linseed oil. It hardened in the presence of a drier by reacting with oxygen from the atmosphere.

Modern sealants are based on butyl rubber, acrylic polymers, polyurethane, polyolefin sulfides (Thiokol), neoprene, silicones, and chlorosulfonated polyethylene. These sealants are used for sealing fabricated building units and fuel tanks. Thiokol is used as a highway and bridge sealant and was the sealant used in the construction of the World Trade Building in New York City.

$$\text{-}\!\!\left(\text{CH}_2\text{CH}_2\text{—SSSS}\right)_{\!n}$$

Polyethylene tetrasulfide (Thiokol)

Hot melt butyl rubber sealants are used in automotive windshields and as automotive sealants. Both solvent-borne and waterborne sealants are available. Thiokol sealant is used as the binder in solid rocket propellants. This low-molecular-weight liquid (LP2) is obtained by reduction of polyethylene sulfide (Thiokol) to produce a prepolymer with thio terminal groups (SH). The prepolymer (LP2) is oxidized in situ to produce a high-molecular-weight polymer. This process is similar to that used in the cold waving of hair.

$$\left[\text{-CH}_2\overset{\displaystyle \overset{\text{CH}_3}{|}}{\underset{\displaystyle \underset{\text{CH}_3}{|}}{\text{C}}}\text{CH}_2\text{CH}=\overset{\displaystyle \overset{}{\underset{\displaystyle \underset{\text{CH}_3}{|}}{\text{C}}}\text{CH}_2\text{-}\right]_n$$

Butyl rubber

11.10   HISTORY OF ADHESIVES

In contrast to coatings, which must adhere to one surface only, adhesives are used to join two surfaces together. Resinous adhesives were used by the Egyptians at least 6000 years ago for bonding ceramic vessels. Other adhesives, such as casein (from milk), starch and sugar (from plants), and glues (from animals and fish), were first used about 3500 years ago.

Combinations of egg white and lime $(Ca(OH)_2)$ as well as sodium silicate (Na_2SiO_3) were used in the first century, and a glue works or factory was built in Holland in 1690. Animal glue is produced by dissolving the calcium phosphate $(Ca_3(PO_4)_2)$ and calcium carbonate $(CaCO_3)$ in bones by heating with hydrochloric acid (HCl). The residue, ossein, plus collagen from animal skins, is treated with lime, extracted with hot water, and concentrated by evaporation to form an adhesive.

Adhesives are also made from fish skins, dextrins (degraded starch), and gum arabic. An aqueous solution of the latter is called mucilage. Animal glue continues to be used for gummed tapes, labels, and match heads. Casein is still used as an adhesive in wallboard, and starch and dextrin are still used for making corrugated board, but the use of these natural adhesives is decreasing, accounting for less than 10% of the entire adhesives market. Over 5 million tons of adhesives at a cost of over $3 billion are used annually in the United States.

11.11   ADHESION

An adhesive is an agent that binds together two or more surfaces. The surfaces adhered may be as smooth as steel or as rough as masonry blocks; these surfaces may require one or many coatings. Secondary bonds between the adhesive and adhered surfaces are required for good adhesion. Primary bonds may be formed by the addition of cross-linking agents. Hydrogen and polar bonds may also bond two surfaces together. Polar groups are present in many adhesives, such as cyano- and acrylic-based glues.

Polyacrylonitrile containing
a cyano group

Poly(methyl acrylate)
containing an ester group

11.12   TYPES OF ADHESIVES

In solvent-based adhesives, the polymer is dissolved in an appropriate solvent. Solidification occurs after the evaporation of the solvent. A good bond is formed if the solvent attacks or actually dissolves some of the plastic (adherend).

Adhesives may be solvent-based, latex-based, pressure-sensitive, or reactive adhesives. Solvent-based adhesives, such as model airplane glue, depend on the evaporation of the solvent for the formation of a bond (solvent weld) between the polymer (adherend) and the surface to be adhered. This type of adhesive is used to join poly(vinyl chloride) pipe in a process called solvent welded pipe (SWP).

Latex-based adhesives should be used at temperatures above the glass transition temperature of the adhesive resin. This type of adhesive is widely used for bonding pile to the backing of carpets.

Pressure-sensitive adhesives must also be applied as a highly viscous solution at a temperature above the glass transition temperature of the polymer. The application of pressure causes the adhesive to flow to the surface to be adhered, for example, adhesive tape.

Holt melt adhesives, which may be used in electric "glue guns," are applied as molten polymers. Plywood is produced by the impregnation of thin sheets of wood by a reactive resin adhesive that cures after it has been applied. Phenolic, urea, melamine, and epoxy resins are used as the reactive adhesives.

11.13   RESINOUS ADHESIVES

Unsaturated polyester resins and polyurethanes are used for automobile body repair and for bonding polyester cord to the rubber in tires. Both polyester and epoxy resins are used to bond fibrous glass and aramid fibers in reinforced plastic composites.

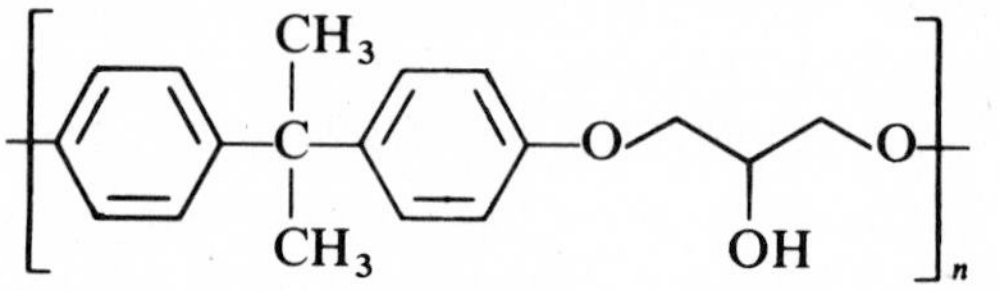

Epoxy resin

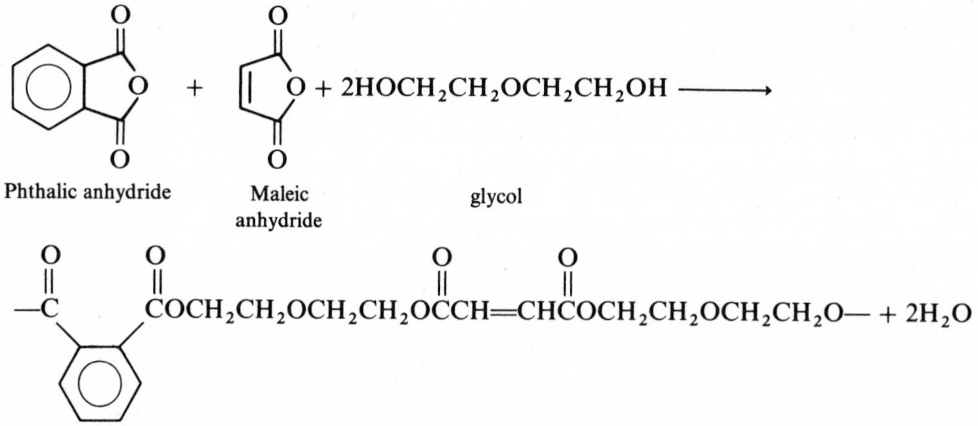

Phthalic anhydride   Maleic anhydride   glycol

Unsaturated polyester

A solution of natural rubber in naphtha was used by MacIntosh to produce a waterproof cloth laminate in the nineteenth century and comparable systems continue to be used today. Blends of neoprene and phenolic resins are used as contact adhesives, in which the adhesive is applied to both surfaces, which are then pressed together.

"White glue," which is used as a general-purpose adhesive, consists of a polyvinyl acetate emulsion. Copolymers of ethylene and vinyl acetate are used as hot melt adhesives. Anaerobic adhesives, which cure when air is excluded, consist of mixtures of dimethacrylates and hydroperoxide. "Super-glue" or "Krazy-glue" contains butyl-α-cyanoacrylate, which polymerizes spontaneously in the presence of moist air or on dry glass surfaces. This superior adhesive is used in surgery and for mechanical assemblies.

$$CH_2=C \begin{matrix} CN \\ \\ COO-C_4H_9 \end{matrix}$$

Butyl-α-cyanoacrylate

$$CH_2=\overset{CN}{C}COOCH_3$$

Methyl-α-cyanoacrylate

GLOSSARY

Acrylic polymer: Poly(methyl methacrylate) or poly(ethyl acrylate).

Adhesion: The degree of attachment between a film and another surface.

Adhesive: An agent that binds two surfaces together.

Alkyd: Resin produced by the condensation of glycerol and phthalic acid.

Anaerobic: Free of oxygen.

Aniline point: Temperature at which a 50–50 mixture of aniline and unknown solvent becomes turbid when cooled.

Aqueous: Watery.

Aramid fiber: Aromatic nylon fibers.

Baekeland, Leo: Inventor of commercial phenolic resins.

Balsam: An aromatic exudate from trees or shrubs, such as Canadian balsam.

Binder: Film-forming constituent of a coating system.

Borax: $Na_2B_4O_7$.

Boric acid: H_3BO_3.

Butyl rubber: A copolymer of isobutylene ($H_2C=C(CH_3)_2$) and isoprene ($H_2C=C(CH_3)CH=CH_2$).

Casein: Milk protein.

Chlorinated rubber: Product of the reaction of chlorine and natural rubber.

Chrome yellow: $PbCrO_4$.

Chromium oxide: Cr_2O_3, a green pigment.

Coalescent agent: Substance added to emulsions to ensure the formation of continuous films.

Coating, curtain: The deposition of a curtain of paint on a flat surface followed by draining.

Cohesive energy density (CED): Energy of intermolecular forces between molecules.

Collagen: Gelatinlike protein.

Collodion: A solution of cellulose nitrate in a mixture of ethanol and ethyl ether.

Copal: A natural resin obtained from tropical trees.

Copolymer: A polymer with more than one type of repeating units in its chain.

Dextrin: Degraded starch.

Diammonium phosphate: $(NH_4)_2HPO_4$.

Drier: A soluble salt of a heavy metal and an organic acid (a catalyst for the polymerization of unsaturated oils).

Drying: Polymerization of unsaturated oil in the presence of oxygen.

Electrodeposition: Deposition of a coating from a waterborne system on an object of opposite charge.

Emulsion: A permanent aqueous suspension of a polymer.

Enamel: A term used for ceramic coatings and also for polymer solution coatings.

Ester gum: Glycerol ester of abietic acid (rosin).

Ethylene glycol: $HO(CH_2)_2OH$.

Formaldehyde: H_2CO.

Glass transition temperature: The temperature at which a glassy polymer becomes flexible when the temperature is increased.

Glue: Adhesive usually derived from animals or fish.

Glycerol: $HOCH_2CH(OH)CH_2OH$.

Glyptal: Resin produced by the condensation of glycerol and phthalic anhydride.

Gum arabic: Salts of arabic acid obtained from mimosa plants (acacia).

Hildebrand, J.: Developer of the solubility parameter concept.

Hydroperoxide: $ROOH$.

Ilmenite: Titanium ore.

Intumescent paint: One that forms a protective tarry sponge when burned.

Iron blue: Prussian blue; ferriferricyanide.

Kauri: A copal resin.

Kauri-butanol value: Volume of unknown solvent that causes turbidity when added to a solution of kauri copal resin in 1-butanol.

Lacquer: A solution of a film-forming resin (binder).

Linolenic acid: A diunsaturated acid ($C_{17}H_{31}COOH$).

Linseed oil: Oil from flax seed.

Lithopone: $BaSO_4$ and ZnS.

MacIntosh: A laminate of natural rubber and cloth.

Menhaden oil: An unsaturated (drying) oil obtained from menhaden (Moss bunker) fish.

Mucilage: An aqueous solution of gum arabic.

Neoprene: Polychloroprene,

$$\left[\begin{array}{c} \overset{\displaystyle H}{\underset{\displaystyle H}{\overset{|}{\underset{|}{C}}}} - \overset{\displaystyle Cl}{\overset{|}{C}} = \overset{\displaystyle H}{\overset{|}{C}} - \overset{\displaystyle H}{\underset{\displaystyle H}{\overset{|}{\underset{|}{C}}}} \end{array}\right]_n$$

Oiticica oil: An unsaturated (drying) oil obtained from the Brazilian oiticica tree (*Licania rigida*).

Oleic acid: A monounsaturated acid ($C_{17}H_{33}COOH$).

Oleoresinous: A material based on unsaturated vegetable oils and a drier.

Paint: A liquid system consisting of a solid (pigment) and a liquid (vehicle).

Pentaerythritol: $(HOCH_2)_4C$.

Phenol: C_6H_5OH.

Phthalic acid: $C_6H_4(COOH)_2$.

Pigment: A colorant.

Pitch: A bituminous substance based on asphalt, wood tar, or coal tar.

Polyurethane: The reaction product of a diol (HOROH) and a diisocyanate (OCNRNCO).

Pressure-sensitive adhesive: A viscous solution that flows under pressure to produce an adhered system, for example, adhesive tape.

Putty: A sealant or caulking material based on a mixture of filler, drier, and linseed oil.

Reinforced plastic composite: Polyester-bonded fiberglass composite.

Rosin: Pine resin.

Safflower oil: An unsaturated (drying) oil obtained from safflower seed (*Carthamus*).

Sealant: A crack filler.

Shellac: A resinous material secreted by insects that feed on the lac tree.

Silicone: An inorganic polymer with the repeating unit —Si(R_2)—O—.

Sodium bicarbonate: $NaHCO_3$.

Sodium carbonate: Na_2CO_3.

Sodium silicate: Na_2SiO_3.

Solubility parameter: A measure of solvency; these values generally increase with the polarity of the solvent.

Substrate: A surface to be coated.

Super-glue: An adhesive based on butyl-α-cyanoacrylate.

SWP: Solvent welded pipe.

Teflon: Polytetrafluoroethylene.

Tempera: A paint based on egg binder.

Thinner: Paint solvent.

Thiokol: The first American synthetic rubber produced by the condensation of ethylene dichloride and sodium polysulfide.

Titanium dioxide: TiO_2, a white pigment.

Trimellitic acid: $C_6H_3(COOH)_3$.

Tung oil: China wood oil (*Aleutites cordata*), an unsaturated (drying) oil.

Ultramarine blue: Lapis lazuli (sodium aluminum silicate).

Unsaturated oil: Oil with carbon–carbon double bonds.

Varnish: An unpigmented oil-based paint.

White glue: Polyvinyl acetate.

Whitewash: Aqueous dispersion of calcium hydroxide ($Ca(OH)_2$).

REVIEW QUESTIONS

1. Which will produce a tack-free film first; a layer of lacquer or a layer of classic paint?

2. Why must a classic paint be applied in thin layers?

3. Which is more highly unsaturated: linseed oil or mineral oil?

4. What color is lithopone?

5. What polymer is present in collodion?

6. Describe the relationship between chain length and viscosity.

7. What is the function of a paint drier?

8. Is a polymer more ductile or more brittle when it is cooled below the glass transition temperature?

9. Why is whitewash not permanent?

10. Why must a coalescent agent be present in emulsion coatings?

11. What are the names of two polyester coatings based on phthalic acid or anhydride and glycerol or ethylene glycol?

12. Why does putty harden?

13. What is mucilage?

14. What is ester gum?

15. Why is butyl rubber called a copolymer?

16. Why must butyl-α-cyanoacrylate be kept in a moisture-free container?

17. Will an anaerobic adhesive polymerize in the presence of moist air?

18. What polar groups are present in "Super-glue"?

BIBLIOGRAPHY

Banon, A. (1974). *Paint and Coatings Handbook*. Farmington, MI: Structure Publishing Co.

Brewer, G. E. (1973). *Electrodeposition of Coatings*, Advances in Chemistry Series. Washington, DC: American Chemical Society.

Cagle, C. V. (1975). *Handbook of Adhesive Bonding*. New York: McGraw–Hill.

Herman, B. S. (1975). *Adhesives*. Park Ridge, NJ: Noyes Data Corp.

Krumbhaar, W. (1973). *The Chemistry of Synthetic Surface Coatings*. New York. Reinhold.

Lambourne, R. R. (1987). *Paint and Surface Coatings*. New York: Halsted.

Lee, H. (1975). *Adhesive Science and Technology*. New York: Plenum.

Martens, C. R. (1981). *Waterborne Coatings*. New York: Van Nostrand–Reinhold.

Miles, F. D. (1955). *Cellulose Nitrate*. London: Oliver & Boyd.

Morgans, W. N. (1982/1984). *Outlines of Paint Technology* (Vols. I and II). High Wycombe, Bucks., England: Charles Griffin & Co.

Nylen, P., and Sunderland, E. (1965). *Modern Surface Coatings*. New York: Interscience.

Ott, E., Spurling, H. M., and Grafflin, M. W. (1955). *Cellulose and Cellulose Derivatives*. New York: Interscience.

Payne, H. F. (1954). *Organic Coating Technology*. New York: Wiley.

Remmer, E., and Samsonov, G. V. (1973). *Protective Coatings on Metal*. New York: Plenum.

Seymour, R. B. (1960). *Hot Organic Compounds*. New York: Reinhold.

Seymour, R. B. (Ed.). (1979). *Plastic Mortars, Sealants and Caulking Compounds*, ACS Symposium Series. Washington, DC: American Chemical Society.

Seymour, R. B. (1982). *Plastics vs. Corrosives*. New York: Wiley.

Seymour, R. B. (1987). Coatings, colorants and paints. In *Encyclopedia of Physical Science and Technology*. Orlando, FL: Academic Press.

Seymour, R. B. (1990). *Organic Coatings Handbook*. New York: Elsevier.

Seymour, R. B. (Ed.). (1990). *Organic Coatings: Their Origin and Development*. New York: Elsevier.

Seymour, R. B., and Carraher, C. E. (1987). *Polymer Chemistry: An Introduction* (2nd ed.). New York: Dekker.

Stewart, J. R., and Seymour, R. B. (1948). *National Paint Dictionary*. Washington, DC: Steward Research Laboratories.

ANSWERS TO REVIEW QUESTIONS

1. The lacquer. It hardens by evaporation of the solvent.

2. Because it hardens by the reaction of oxygen from the air. The oxygen cannot diffuse readily through thick layers of paint.

3. Linseed oil contains linolenic acid.

4. White.

5. Cellulose nitrate.

6. Viscosity increases as chain length increases.

7. It catalyzes the polymerization (hardening or drying) of the unsaturated oil in the paint.

8. More brittle. It is glasslike below T_g.

9. There is no resinous binder present.

10. The polymer will not form a continuous film below its T_g in the absence of a coalescent agent.

11. Alkyd and glyptal.

12. It polymerizes (dries) when exposed to air.

13. A solution of gum arabic in water.

14. A glyceryl ester of rosin.

15. The repeating units of both isobutylene and isoprene are present in the backbone.

16. It polymerizes in the presence of moisture.

17. No.

18. Cyano and ester groups.

Thermoplastics

12.1   INTRODUCTION

All polymers are classified as either thermoplastics, that is, linear or branched polymers that can be reversibly softened by heating and solidified by cooling, or thermosets, that is, cross-linked polymers that cannot be softened by heating without degradation. The word plastic is derived from the Greek word *plastikos*, meaning able to be molded. Both thermoplastics and thermoset prepolymers can be molded into desirable shapes.

It is of interest to note that the human desire to produce shaped articles was satisfied in the early cultures by chipping stone, chiseling wood, casting bronze, and shaping warmed tortoise shell and horn. Artisans who shape ivory are still called horners. Since these products could be molded by heat, the advent of ebonite (hard rubber) and celluloid in the nineteenth century provided a new outlet for shaping.

Shellac, gutta-percha, balata, casein, and bitumens are naturally occurring thermoplastics. Derivatives of natural rubber and cellulose, that is, cyclized rubber, cellulose nitrate, and cellulose acetate, are also thermoplastics. However, the first synthetic moldable plastic was a thermoset, that is, the reaction product of phenol and formaldehyde, which was produced commercially by Leo Baekeland in the early 1900s. The thermosets, which were the principal plastics prior to World War II and now account for less than 10% of all moldable plastics, are described in Chapter 13.

Each year the United States consumes about 56,000 million pounds of plastic and synthetic resins, or about 225 pounds for every citizen. The use of lightweight plastics has helped increase gas mileage in automobiles (Table 12.1), and this trend will increase with plastic car bodies being more widely used in the near future.

It is important to recognize that of the total yearly U.S. oil and gas consumption, 60% is used as stationery fuels in home heating and fuel to run power plants, 33% is used as transportation fuels, and 7% is used for the manufacture of petrochemicals, including fertilizer, rubber, paints, fibers, solvents, and medicines. Only about 2.5% is employed for polymer applications, yet this 2.5% makes possible the production of many useful products.

12.2   HIGH-DENSITY POLYETHYLENE

The alkanes, such as paraffin wax, high-density polyethylene (HDPE), and other polyolefins have the empirical formula $H+CH_2)_nH$. Nevertheless, the degree of polymerization (DP) of paraffin wax and lower-molecular-weight alkanes is too low to permit entanglement of the polymer chains. Another highly branched polyolefin, called elasterite, occurs naturally in the fossil *Fungus subterraneus* but was never used as a commercial plastic.

The first synthetic polyethylene was produced by von Peckman in the 1890s by the catalytic decomposition of diazomethane (CH_2N_2). W. Carothers, the coinventor of

Table 12.1   Major Plastics Applications in Automobiles

Area	Application	Material (usual)
Interior	Crash pad	Urethane, ABS, PVC
	Headrest pad	Urethane, PVC
	Trim, glove box	Polypropylene, PVC, ABS
	Seat	Urethane
	Upholstery, carpet	PVC, nylon
Exterior	Fender apron	Polypropylene
	Front end	Unsaturated polyester
	Wheel covers	ABS, polyphenylene oxide
	Fender extension	Unsaturated polyester, nylon
	Grille	ABS, polyphenylene oxide
	Lamp housing (rear)	Polypropylene
	Styled roof	PVC
	Bumper sight shield	EPDM rubber, urethane
	Window louvers	Poly(butylene terephthalate)
Under the hood	Ducts	Polypropylene
	Battery case	Polypropylene
	Fan shroud	Polypropylene
	Heater and air conditioning	Unsaturated polyester
	Electrical housing and wiring	Phenolic, PVC, silicone
	Electronic ignition components	Poly(butylene terephthalate)

nylon, produced a low-molecular-weight linear polyethylene in the early 1930s by the coupling of decamethylene dibromide ($Br(CH_2)_{10}Br$) in the presence of sodium metal. C. Marvel also produced HDPE in the early 1930s by the polymerization of ethylene in the presence of lithiumalkyl (LiR) and an arsonium compound. This polymer was investigated by duPont, but that company failed to recognize the potential use of HDPE at that time.

It is of interest to note that this noncommercial polymer, which occurs naturally, was also synthesized by three distinctly different techniques prior to World War II. HDPE is now a commercial plastic with over 4 million tons being produced annually in the United States.

The first commercial HDPE was produced independently by J. Hogan and R. Banks in the United States and later by Ziegler in Germany in the early 1950s. Ziegler's synthesis was related to that used by Marvel. K. Ziegler and coworkers used aluminumtriethyl ($Al(C_2H_5)_3$) and titanium trichloride ($TiCl_3$) for their polymerization catalyst in what they called the "aufbau" or building-up reaction. Ziegler was awarded the Nobel prize in 1963.

Hogan and Banks used chromic oxide (CrO_3), supported on silica (SiO_2), as their catalyst system for making HDPE. The polymer obtained by the German and American chemists was a linear crystalline polymer. The regularity in the HDPE chain favored the formation of crystals, and this crystalline structure contributed to the higher specific gravity of HDPE (0.96). In contrast, the specific gravity of the highly amorphous branched polyethylene (LDPE) is about 0.91. A space-filling model and simulated portions of HDPE and LDPE chains are shown in Figures 12.1 and 12.2. Physical properties of commercial polyethylene are listed in Table 12.2.

The terms high-density and low-density polyethylene are derived from their densities. Density is a measure of the weight of material contained within a given volume. Thus the air at sea level has a density about 1.2×10^{-3} grams per cubic centimeter (g/cc) of volume. Wood has a density of about 0.1 to 1.4 g/cc. Most plastics and organic liquids have densities of about 0.7 to 1.0 g/cc, whereas water has a density of 1.0 g/cc at ordinary temperatures. Denser materials include the metals, such as mercury and tungsten.

Density is related to how tightly material can be packed. Thus linear polyethylene (HDPE) can be tightly packed since linear chains can be efficiently folded as noted in Figure 12.2. Conversely, branched polyethylene (LDPE) packs less firmly as a result of the presence of the branches, which prohibit close, regular folding. Thus, more ethylene units can be packed within a specific volume for linear polyethylene than for branched polyethylene, resulting in a higher weight per volume and consequently higher density for linear polyethylene.

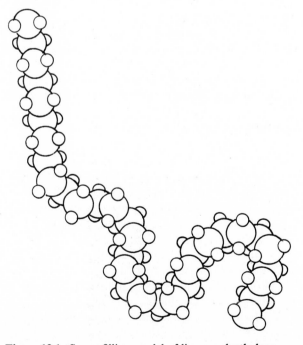

Figure 12.1. Space-filling model of linear polyethylene.

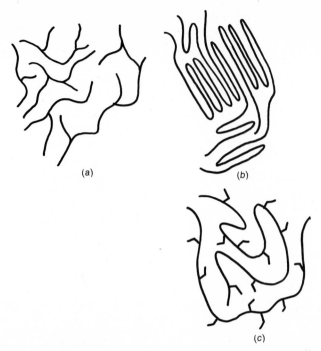

Figure 12.2. Representations of different forms of polyethylene to illustrate branching and nonbranching aspects. (*a*) Low-density polyethylene (LDPE). (*b*) High-density polyethylene (HDPE). (*c*) Linear low-density polyethylene copolymer with 1-butene (LLDPE). The 1-butene discourages crystallization.

Specific gravity is simply the ratio of the density of the material compared to the density of water. Employing the units of g/cc, the density and specific gravity of a material are essentially the same since the density of water is 1.0 g/cc and a number divided by one is the number itself.

12.3   LOW-DENSITY POLYETHYLENE

The development of low-density polyethylene (LDPE) was based on less than 1 g of a residue that was accidentally produced by E. Fawcett and R. Gibson in 1933 in their unsuccessful attempt to condense ethylene and benzaldehyde at 340°F and at extremely high pressure (45,000 psi). However, they did produce a trace of polyethylene.

Larger amounts of LDPE were obtained when a trace of oxygen was used as an initiator. The first full-scale LDPE plant went "on stream" on the day of the outbreak of World War II. LDPE, which is a highly branched polymer, was used advantageously as an insulator for coaxial cable in radio detecting and ranging (radar). Over 4 million tons of LDPE are produced annually in the United States.

Unlike the diazomethane decomposition, alkylene dichloride coupling, and coordination-catalyzed polymerization used to produce HDPE, LDPE is produced by a free radical-initiated chain reaction in which the original oxygen initiator has

Table 12.2 Properties of Typical Polyethylenes

Property	HDPE	LDPE	LLDPE	PP	PP (40% talc)
Melting point (T_m, °F)	275	200	122	170	165
Glass transition temp. (T_g, °F)	—	20	—	—	—
Processing temp. (°F)	450	400	450	450	450
Molding pressure (10^3 psi)[a]	15	10	10	15	15
Mold shrinkage (10^{-3} in./in.)	25	30	20	20	12
Heat deflection temp. under flexural load of 264 psi (°F)	190	110	—	130	175
Maximum resistance to continuous heat (°F)	175	110	130	125	160
Coefficient of linear expansion (10^{-6} in./in., °F)	40	150	125	40	30
Compressive strength (10^3 psi)	30	—	—	65	75
Impact strength Izod (ft-lb/in. of notch)[b]	2	No break	No break	1.0	0.5
Tensile strength (10^3 psi)	35	30	33	50	45
Flexural strength (10^3 psi)	30	—	—	60	80
% elongation	200	300	400	400	5
Tensile modulus (10^3 psi)	155	35	45	50	45
Flexural modulus (10^3 psi)	150	30	50	200	500
Shore hardness	D70	D50	D55	R90	R100
Specific gravity	0.96	0.91	0.93	0.91	1.25
% water absorption	0.01	0.01	0.01	0.01	0.02
Dielectric constant					
Dielectric strength (V/mil)	500	750	700	500	500
Resistance to chemicals at 750°F[c]					
Nonoxidizing acids (20% H_2SO_4)	S	S	S	S	S
Oxidizing acids (10% HNO_3)	Q	Q	Q	Q	Q
Aqueous salt solutions (NaCl)	S	S	S	S	S
Polar solvents (C_2H_5OH)	S	S	S	S	S
Nonpolar solvents (C_6H_6)	Q	Q	Q	Q	Q
Water	S	S	S	S	S
Aqueous alkaline solutions (NaOH)	S	S	S	S	S

[a] psi/0.145 = kPa (kilopascals).
[b] ft-lb/in. of notch/0.0187 = cm · N/cm of notch.
[c] S = Satisfactory, Q = questionable, and U = unsatisfactory.

been replaced by organic peroxides. In the following equation the free radical (R·) adds to ethylene (H_2C=CH_2) to produce a new free radical (RCH_2—$CH_2\cdot$), which adds to other ethylene molecules in a series of propagating steps.

$$R-CH_2-CH_2^{\cdot} + n\,CH_2{=}CH_2 \rightarrow R{+}CH_2-CH_2{+}_n CH_2-CH_2^{\cdot}$$

As shown by the data in Table 12.2, LDPE has a lower modulus (is more flexible) and has a lower melting point than HDPE.

12.4   ULTRAHIGH-MOLECULAR-WEIGHT POLYETHYLENE

A minimum or threshold molecular weight (about 100 $\overline{DP}$) is required for entanglement of HDPE. Since high-molecular-weight polymers are difficult to process, polymers with molecular weights slightly above the threshold molecular weight are usually produced commercially. However, the toughness of HDPE and other polymers increases with molecular weight. Hence, ultrahigh-molecular-weight polyethylene (UHMWPE) ($\overline{DP}$ = 1 million) is produced commercially for use where unusual toughness is essential, such as in trash cans and liners for coal freighters.

12.5   CROSS-LINKED POLYETHYLENE

LDPE cross-links when exposed to high-energy radiation. The cross-linked product, which is insoluble in solvents even at elevated temperatures, is used as heat-shrinkable tubing. The stretched, cross-linked product has "elastic memory" and returns to its original dimensions when heated.

12.6   LINEAR LOW-DENSITY POLYETHYLENE

Commercial copolymers in which both ethylene and 1-butene ($H_2C=CH-CH_2CH_3$) are present as repeating units in the polymer chain are linear, but because of the bulky pendant (C_2H_5) groups they occupy greater volume and have a lower specific gravity than HDPE. Linear low-density polyethylene (LLDPE) may be produced at low pressure in the gaseous phase or in solution. Higher homologues such as 1-octene ($H_2C=CH(CH_2)_5CH_3$) may also be used as the comonomers in LLDPE. New coordination catalysts, which are related to those used for making HDPE, are also used in the production of LLDPE.

LDPE is characterized by good flexibility and hence can be used as a film and in squeeze bottles. HDPE is stiffer and more heat resistant and is used as rigid pipe. LLDPE is stronger than LDPE and can be used as thinner films for making bags, for example.

12.7   OTHER COPOLYMERS OF ETHYLENE

In addition to copolymers of 1-olefins, such as LLDPE, there are several other commercial copolymers of ethylene. The copolymer of ethylene and vinyl acetate is an amorphous copolymer that may be cast as a clear film or used as a melt coating. The copolymer of ethylene and methacrylic acid ($CH_2=C(CH_3)COOH$) is also a moldable thermoplastic. This copolymer when partially neutralized to form monovalent and divalent metal-containing materials is called ionomerar. These ionomer salts have a stable cross-linked structure at ordinary temperatures but can be injection molded. These tough copolymers are used as golf ball covers in place of balata.

$$\left[-CH_2CH_2-\right]_n \left[-CH_2\underset{\underset{COO^-}{|}}{\overset{\overset{CH_3}{|}}{C}}-\right]_n$$

Ethylene–methacrylic acid copolymers (ionomers)

Both HDPE and polypropylene are high-melting crystalline polymers. However, the random copolymer of these two comonomers is an amorphous, low-melting elastomer. It is customary to add a cross-linking monomer, such as dicyclopentadiene, to the comonomers to produce a vulcanizable elastomer (EPDM). EPDM is used as the white sidewalls of tires and as single-ply roofing material.

$$\left[-CH_2CH_2-\right]_n \left[-CH_2\underset{\underset{CH_3}{|}}{CH}-\right]_n$$

Ethylene–propylene copolymer

These ethylene–propylene copolymers are also employed in other automotive applications such as radiator and heater hoses, seals, mats, weather strips, bumpers, and body parts. Nonautomotive applications include coated fabrics, gaskets and seals, hoses and wire, and cable insulators.

The block copolymer of ethylene and propylene, which contains long sequences of ethylene and propylene repeating units, is a clear, moldable copolymer and is used in place of HDPE in many applications. Its specific gravity is similar to that of LDPE.

12.8   POLYPROPYLENE

Nobel laureate K. Ziegler patented HDPE but failed to include polypropylene (PP) in his patent application. However, many other chemists used the Ziegler catalyst $(TiCl_3 \cdot Al(C_2H_5)_3)$ to produce PP in the early 1950s. Nobel laureate G. Natta of Montedison, W. Baxter of duPont, and E. Vanderburg of Hercules filed for patents for the production of PP using the Ziegler catalyst. J. Hogan and R. Banks of Phillips and A. Zletz of Amoco filed for patents using supported metal oxide catalysts. In 1973 the U.S. Patent Office granted a patent for PP to Natta, but reversed its decision in favor of Hogan and Banks in 1983.

As shown by the data in Table 12.2, PP has a higher melting point than HDPE and its properties are improved by adding 40% talc as a filler. In addition to being used for injection-molded articles and extruded pipe, it is also used as a fiber for indoor–outdoor carpet. Over 3.5 million tons of isotactic PP are produced annually in the United States.

$$\left[-CH_2\underset{\underset{CH_3}{|}}{CH}-\right]_n$$

Polypropylene (PP)

It is of interest to note that the cationic polymerization of propylene was investigated by Butlerov in 1873. Some 80 years later, Fontana used aluminum bromide ($AlBr_3$) and hydrobromic acid (HBr) to produce PP at $-96°F$. This amorphous PP had a glass transition temperature of $-22°F$.

Natta's wife referred to the amorphous PP as atactic PP. The more regular crystalline structures in which the pendant methyl groups were all on one side of the polymer chain or had an alternating arrangement were called isotactic and syndiotactic PP, respectively. The structure of isotactic PP is shown in Figures 12.3 to 12.5.

12.9   OTHER POLYOLEFINS

Butlerov produced amorphous, low-melting polyisobutylene in 1873 by the cationic polymerization of isobutylene by using boron trifluoride (BF_3).

$$\left[CH_2 - \underset{\underset{CH_3}{|}}{\overset{\overset{CH_3}{|}}{C}} \right]_n$$

Polyisobutylene (PIB)

This polymer is used as a chewing gum base, and as a caulking material, but when cold it flows much like unvulcanized rubber. This deficiency was overcome by Sparks and Thomas, who produced butyl rubber by copolymerizing isobutylene with small amounts (10%) of isoprene ($H_2C{=}C(CH_3)CH{=}CH_2$). Butyl rubber is resistant to permeation by gases, and this property is enhanced by chlorination.

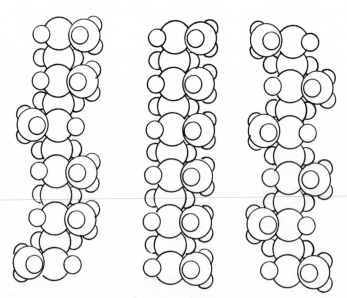

Figure 12.3. Atactic (left), isotactic (middle), and syndiotactic (right) polypropylene. (Some of our youth tell us it is really a model of poly(teddy bears).)

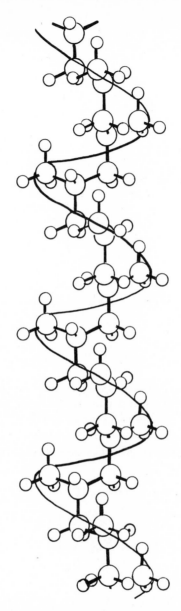

Figure 12.4. Stick-and-ball model of the helical conformation of isotactic polypropylene.

Polybutene-1, $\text{+CH}_2\text{—CH(C}_2\text{H}_5\text{)}\text{+}$, is produced by the Ziegler-catalyzed poly-merization of butene-1. The gas barrier properties of polybutene-1 are inferior to those of butyl rubber.

Polymethylpentene (TPX) is produced by the Ziegler polymerization of 4-methylpentene ($\text{H}_2\text{C}=\text{CH—CH}_2\text{—CH(CH}_3\text{)}_2$. Because of the bulky pendant group, TPX has a relatively high volume and low specific gravity (0.83). This high-melting (465°F) transparent polymer is used for laboratory ware.

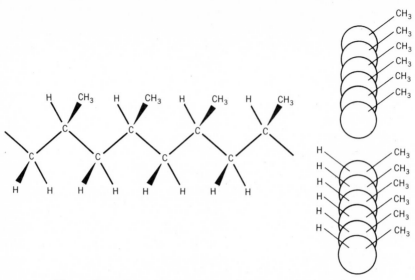

Figure 12.5. Illustrations representing isotactic polypropylene.

12.10   POLYSTYRENE

Crude styrene monomer ($CH_2{=}CH(C_6H_5)$) was obtained by Neuman, who pyrolyzed (thermally decomposed) a storax balsam. A similar product was also produced by the pyrolysis of amber. This monomer is now obtained by the dehydrogenation of ethylbenzene ($C_6H_5C_2H_5$). Polystyrene (PS) is obtained by the free radical polymerization of styrene monomer. A ball-and-stick model of PS is shown in Figure 12.6.

$$\left[CH_2-CH\right]_n$$

Polystyrene (PS)

Polystyrene has been available commercially since the 1930s. This amorphous, brittle, clear polymer may be injection molded, extruded, or expanded to produce plastic foam. Much of the commercial PS is toughened by blending with elastomers to produce high-impact polystyrene (HIPS). PS is produced at an annual rate of 2.5 million tons in the United States. Its properties are listed in Table 12.3.

12.11   STYRENE COPOLYMERS

In addition to the SBR elastomer described in Chapter 10, a less rubbery copolymer with a lower percentage of butadiene is used as a tough plastic. Styrene–acrylonitrile copolymers, SANs, have relatively high heat deflection temperatures. Because of their

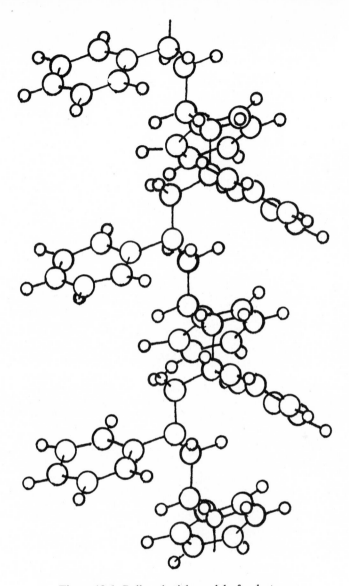

Figure 12.6. Ball-and-stick model of polystyrene.

thermal stability, SANs are employed in the production of "dishwasher-safe" houseware, such as blender bowls, humidifier parts, detergent dispensers, and refrigerator vegetable and meat drawers. Also, fiberglass-reinforced automotive battery cases and dashboard components are molded from SAN.

Blends of SAN and butadiene–acrylonitrile rubber (NBR) have superior impact resistance. Sheets of this acrylonitrile–butadiene–styrene (ABS) terpolymer are thermoformed for the production of suitcases, crates, and appliance housings. The weather resistance and clarity of ABS is improved by replacing the acrylonitrile by methyl methacrylate. The properties of SAN and ABS are listed in Table 12.3.

Table 12.3   Properties of Typical Styrene Polymers

Property	PS	HIPS	SAN	ABS
Melting point (T_m, °F)				
Glass transition temp. (T_g, °F)	100	95	120	115
Processing temp. (°F)	350	400	350	350
Molding pressure (10^3 psi)[a]	15	15	15	20
Mold shrinkage (10^{-3} in./in.)	5	5	4	6
Heat deflection temp. under flexural load of 264 psi (°F)	175	185	210	200
Maximum resistance to continuous heat (°F)	175	175	200	180
Coefficient of linear expansion (10^{-6} in./in., °F)	40	—	40	35
Compressive strength (10^3 psi)	12	—	14	7
Impact strength Izod (ft-lb/in. of notch)[b]	0.4	2.5	0.5	2
Tensile strength (10^3 psi)	12	30	10	5
Flexural strength (10^3 psi)	15	50	12	10
% elongation	2	40	2	20
Tensile modulus (10^3 psi)	400	250	550	350
Flexural modulus (10^3 psi)	425	200	500	400
Rockwell hardness	M65	R65	R83	R110
Specific gravity	1.04	1.04	1.07	1.2
% water absorption	0.02	0.02	0.2	0.4
Dielectric constant	2.5	3.0	2.5	3.0
Dielectric strength (V/mil)	550	—	425	400
Resistance to chemicals at 75°F[c]				
Nonoxidizing acids (20% H_2SO_4)	S	S	S	S
Oxidizing acids (10% HNO_3)	Q	Q	Q	Q
Aqueous salt solutions (NaCl)	S	S	S	S
Polar solvents (C_2H_5OH)	S	S	S	S
Nonpolar solvents (C_6H_6)	U	U	U	U
Water	S	S	S	S
Aqueous alkaline solutions (NaOH)	S	S	S	S

[a]psi/0.145 = kPa (kilopascals).
[b]ft-lb/in. of notch/0.0187 = cm · N/cm of notch.
[c]S = Satisfactory, Q = questionable, and U = unsatisfactory.

ABS terpolymers are actually a family of polymers that can be used as foams, plastics and elastomers. Acrylonitrile repeating units contribute good strength, heat stability, and chemical resistance. Styrene units contribute rigidity, processability, and good gloss, whereas butadiene repeating units contribute impact strength, toughness, and unsaturation that can form cross-links. As shown in Figure 12.7, ABS plastics and rubbers are used in automotive grills, trim, instrument panels, and bumpers and as

Figure 12.7. Blow-molded Cycolac (Borg-Warner Chemicals, Division of General Electric). ABS bumpers were chosen because of processibility, ability to reproduce mold detail, and adequate impact strength.

appliance housings and cabinets.

Acrylonitrile–butadiene–styrene terpolymer (ABS)

Styrene–butadiene rubber (SBR)

$$\left[\begin{array}{c} CH_2CH \\ | \\ CN \end{array}\right]_m \left[\begin{array}{c} CH_2CH \\ \end{array}\right]_n$$

Styrene–acrylonitrile copolymer (SAN)

$$\left[\begin{array}{c} CH_2CH \\ | \\ CN \end{array}\right]_n \left[\begin{array}{c} CH_2CH=CHCH_2 \\ \end{array}\right]_m$$

Nitrile rubber (NBR)

12.12   POLY(VINYL CHLORIDE)

Unlike polyethylene and polystyrene, poly(vinyl chloride) (PVC) has no natural counterpart, but like other widely used (general-purpose) plastics, PVC has been known for many years and is produced by free radical polymerization at an annual rate of 4 million tons in the United States. PVC, $\left(CH_2CHCl\right)_n$, and its monomer, vinyl chloride ($H_2C=CHCl$), were described by Regnault and Baumann in the nineteenth century, but this polymer could not be readily fabricated by methods available to the industry prior to the 1930s.

$$\left[\begin{array}{c} CH_2CH \\ | \\ Cl \end{array}\right]_n$$

Poly(vinyl chloride) (PVC)

$$\left[(CH_2-CH)_m(CH_2-CH)_n\right]$$
$$\begin{array}{cc} | & | \\ Cl & O \\ & | \\ & C=O \\ & | \\ & CH_3 \end{array}$$

Poly(vinyl chloride-*co*-vinyl acetate)

The difficulty of fabrication was associated with its tendency to decompose at temperatures used for molding and extrusion. W. Semon reduced the processing temperature and increased the flexibility of PVC by adding large amounts of a high-boiling liquid called a plasticizer. Doolittle and Powell reduced the processing temperature by copolymerizing vinyl chloride with vinyl acetate ($H_2C=CH(OOCCH_3)$).

During World War II, German chemists added heat stabilizers, which made it possible to mold unplasticized PVC. About one-half of the present PVC production is rigid PVC. The brittleness of rigid PVC can be overcome by blending with impact modifiers, such as chlorinated polyethylene. The heat resistance of PVC is improved by chlorination (PVDC). The properties of rigid and plasticized PVC are summarized in Table 12.4.

The toxic tricresyl phosphate plasticizer used by Semon has been replaced by diethylhexyl phthalate (DOP). However, extremely large quantities of this plasticizer have been claimed to be toxic when ingested by laboratory animals.

PVC and the copolymer with vinyl acetate are used as protective coatings, cable sheathing, pipe, and film. The plasticized PVC may be used as a plastisol, in which finely divided PVC is suspended in a liquid plasticizer, such as dioctyl phthalate ($C_6H_4(CO_2(CH_2)_7CH_3)_2$), and fused by heating at 300°F in a mold.

Tricresyl phosphate

Table 12.4 Properties of Typical Vinyl Polymers

Property	Rigid PVC	Plasticized PVC
Melting point (T_m, °F)		
Glass transition temp. (T_g, °F)	85	85
Processing temp. (°F)	325	365
Molding pressure (10^3 psi)[a]	25	20
Mold shrinkage (10^{-3} in./in.)	4	20
Heat deflection temp. under flexural load of 264 psi (°F)	150	—
Maximum resistance to continuous heat (°F)	140	125
Coefficient of linear expansion (10^{-6} in./in., °F)	35	65
Compressive strength (10^3 psi)	8	1
Impact strength Izod (ft-lb/in. of notch)[b]	0.5	2
Tensile strength (10^3 psi)	6	2
Flexural strength (10^3 psi)	10	—
% elongation	60	300
Tensile modulus (10^3 psi)	450	—
Flexural modulus (10^3 psi)	400	—
Shore hardness	275	475
Specific gravity	1.4	1.2
% water absorption	0.1	0.2
Dielectric constant	2.5	3.0
Dielectric strength (V/mil)	400	350
Resistance to chemicals at 75°F[c]		
Nonoxidizing acids (20% H_2SO_4)	S	S
Oxidizing acids (10% HNO_3)	S	Q
Aqueous salt solutions (NaCl)	S	S
Polar solvents (C_2H_5OH)	S	S
Nonpolar solvents (C_6H_6)	S	Q
Water	S	S
Aqueous alkaline solutions (NaOH)	S	S

[a]psi/0.145 = kPa(kilopascals).
[b]ft-lb/in. of notch/0.0187 = cm · N/cm of notch.
[c]S = Satisfactory, Q = questionable, and U = unsatisfactory.

12.13   VINYL CHLORIDE COPOLYMERS

In addition to its copolymer with vinyl acetate (Vinylite), vinyl chloride is also copolymerized with vinylidene chloride (H_2C=CCl_2) (Saran, Pliovic). Relatively ductile copolymers of vinyl chloride and alkyl vinyl ethers are produced by cationic polymerization using boron trifluoride (BF_3) as an initiator at $-70°F$.

$$+CH_2CCl_2+_n$$
Poly(vinylidene chloride)

12.14   FLUOROCARBON POLYMERS

In spite of their similarity in structure to PVC and PVDC, the fluorine counterparts were not discovered until the 1930s, and even then their discovery by R. Plunkett was accidental. He was investigating tetrafluoroethylene as an aerosol and discovered solid polytetrafluoroethylene (PTFE) in the storage tank. The polymer (Teflon) is now produced by the free radical polymerization of tetrafluoroethylene. PTFE is a temperature-resistant polymer with excellent nonsticking properties (lubricity).

PTFE was an essential construction material for the separation of the hexafluorides of uranium isotopes during World War II. Because of its resistance to solvents and corrosives, it is used as gaskets and as a coating in cooking ware. The properties of PTFE and polychlorotrifluoroethylene (PCTFE) are summarized in Table 12.5.

$$+CF_2CF_2+_n \qquad +CF_2CFCl+_n$$
Polytetrafluoroethylene (PTFE)   Polychlorotrifluoroethylene (PCTFE)

The polymers of monochlorotrifluoroethylene ($CClF$=CF_2) (PCTFE), vinylidene fluoride (CH_2=CF_2), PVDF, and vinyl fluoride (CH_2=CHF) have lower resistance to heat and corrosives and have reduced lubricity in accordance with the reduced fluorine content. The flexibility of polyfluorocarbons is increased by copolymerizing with ethylene. The copolymer of vinylidene fluoride and hexafluoropropylene is a heat-resistant elastomer.

12.15   ACRYLIC POLYMERS

Acrylic acid (H_2C=$CHCOOH$) was synthesized in 1843, and ethyl methacrylate (H_2C=$C(CH_3)COOC_2H_5$) was synthesized and polymerized in 1865 and 1877, respectively. Otto Rohm produced acrylic plastics in the early 1900s and a lacquer-based on acrylic polymer was marketed by Rohm and Haas in 1927 in Germany and in 1931 in the United States. One of the first uses of acrylic polymers was as an interlining for automobile windshields, but poly(methyl methacrylate) sheet (Plexiglas, Lucite) soon became the principal use of acrylic plastics.

Poly(methyl methacrylate) (PMMA), $+CH_2$–$CH(CH_3)COOCH_3+$, has a light transmittancy of about 92% and has good resistance to weathering. It is widely used in

Table 12.5 Properties of Typical Polyfluorocarbons

Property	PTFE	PCTFE
Melting point (T_m, °F)	325	220
Processing temp. (°F)	—	500
Molding pressure (10^3 psi)[a]	—	5
Mold shrinkage (10^{-3} in./in.)	—	10
Heat deflection temp. under flexural load of 264 psi (°F)		
Maximum resistance to continuous heat (°F)	300	250
Coefficient of linear expansion (10^{-6} in./in., °F)	35	25
Compressive strength (10^3 psi)	2	6
Impact strength Izod (ft-lb/in. of notch)[b]	3	3
Tensile strength (10^3 psi)	4	5
Flexural strength (10^3 psi)	—	10
% elongation	200	150
Tensile modulus (10^3 psi)	65	200
Flexural modulus (10^3 psi)	80	200
Rockwell hardness	D60 (Shore)	R85
Specific gravity	2.2	2.1
% water absorption	0	0
Dielectric constant		
Dielectric strength (V/mil)	500	550
Resistance to chemicals at 750°F[c]		
Nonoxidizing acids (20% H_2SO_4)	S	S
Oxodizing acids (10% HNO_3)	S	S
Aqueous salt solutions (NaCl)	S	S
Polar solvents (C_2H_5OH)	S	S
Nonpolar solvents (C_6H_6)	S	S
Water	S	S
Aqueous alkaline solutions (NaOH)	S	S

[a]psi/0.145 = kPa (kilopascals).
[b]ft-lb/in. of notch/0.0187 = cm·N/cm of notch.
[c]S = Satisfactory, Q = questionable, and U = unsatisfactory.

thermoformed signs, aircraft windshields, and bathtubs. The properties of PMMA are summarized in Table 12.6.

Poly(methyl acrylate)

Poly(methyl methacrylate) (PMMA)

Poly(methyl methacrylate) is used as an automobile lacquer and polyacrylonitrile, $+CH_2-CHCN+_n$, is used as a fiber. Poly(ethyl acrylate), $+CH_2-CHCOOC_2H_5+_n$,

Table 12.6   Properties of Typical Poly(methyl Methacrylate)

Property	
Glass transition temp. (T_g, °F)	100
Processing temp. (°F)	350
Molding pressure (10^3 psi)[a]	15
Mold shrinkage (10^{-3} in./in.)	3
Heat deflection temp. under flexural load of 264 psi (°F)	185
Maximum resistance to continuous heat (°F)	180
Coefficient of linear expansion (10^{-6} in./in., °F)	40
Compressive strength (10^3 psi)	15
Impact strength Izod (ft-lb/in. of notch)[b]	0.5
Tensile strength (10^3 psi)	10
Flexural strength (10^3 psi)	15
% elongation	5
Tensile modulus (10^3 psi)	400
Flexural modulus (10^3 psi)	400
Rockwell hardness	M80
Specific gravity	1.2
% water absorption	0.2
Dielectric constant	3.0
Dielectric strength (V/mil)	450
Resistance to chemicals at 75°F[c]	
Nonoxidizing acids (20% H_2SO_4)	S
Oxidizing acids (10% HNO_3)	Q
Aqueous salt solutions (NaCl)	S
Polar solvents (C_2H_5OH)	S
Nonpolar solvents (C_6H_6)	Q
Water	S
Aqueous alkaline solutions (NaOH)	Q

[a]psi/0.145 = kPa (kilopascals).
[b]ft-lb/in. of notch/0.0187 = cm · N/cm of notch.
[c]S = Satisfactory, Q = questionable, and U = unsatisfatisfactory.

is more flexible and has a lower softening temperature than PMMA. Poly(hydroxy-ethyl methacrylate), $-(CH_2-C(CH_3)COOC_2H_4OH)-_n$, is used for contact lenses and poly(butyl methacrylate) is used as an additive in lubricating oils.

12.16   POLY(VINYL ACETATE)

Vinyl acetate and its polymer were described by Klatte in 1912, and the polymer (PVAc) was produced commercially under the trade name of Elvacet and Gelva in 1920. Because of its low softening point, PVAc is not used as a moldable plastic but is used as an adhesive and in waterborne coatings.

Over 200,000 tons of PVAc are produced annually in the United States. Some of this polymer is hydrolyzed to produce a water-soluble polymer (poly(vinyl alcohol),

PVA), and some of the PVA is reacted with butyraldehyde to produce poly(vinyl butyral) (PVB). Poly(vinyl butyral) is used as an inner layer of safety windshield glass.

$$\left[CH_2CH \atop \begin{array}{c} | \\ OCOCH_3 \end{array} \right]_n$$

Poly(vinyl acetate) (PVAc)

$$\left[\begin{array}{ccc} & CH_2 & \\ & / \quad \backslash & \\ CH_2\!-\!CH & & CH \\ | & & | \\ O & & O \\ \backslash & & / \\ & CH & \\ & | & \\ & (CH_2)_2CH_3 & \end{array} \right]_n$$

Poly(vinyl butyral) (PVB)

$$\left[CH_2CH \atop \begin{array}{c} | \\ OH \end{array} \right]_n$$

Poly(vinyl alcohol) (PVA)

12.17 POLY(VINYL ETHERS)

Vinyl ethers, such as vinyl isobutyl ether, are readily polymerized by Lewis acids, such as boron trifluoride, to produce polymers that have excellent adhesive properties. The copolymer of vinyl isobutyl ether and maleic anhydride (Gantrez) is used as a water-soluble component of floor waxes.

$$\begin{array}{c} \{CH_2\!-\!CH\} \\ | \\ O \\ | \\ CH_2 \\ | \\ CH(CH_3)_2 \end{array}$$

Poly(vinyl isobutyl ether)

12.18 CELLULOSICS

Cellulose, which is a polymer made up of D-glucose repeating units, occurs widely, but because of strong intermolecular hydrogen bonds between the oxygen atoms in one molecule and the hydrogen atoms in another molecule, it cannot be molded by standard procedures. However, this high-molecular-weight linear polymer can be processed and fabricated when it is converted to derivatives. Thus, A. Parkes in England (1862) and J. and I. Hyatt in the United States (1869) were able to soften cellulose nitrate and shape it into useful articles. Parkes used cottonseed oil, and the Hyatts used camphor to soften (plasticize) cellulose nitrate (erroneously called nitrocellulose). These pioneer plastics were called Parkesine and Celluloid.

From a social and economical viewpoint, it is of interest to note that Leominster, Massachusetts, became the center of the fabrication of celluloid products because of the development of plastic fabrication machinery, and the National Plastics Museum is located in that city. Hyatt's firm (Merchant's Manufacturing Co.) in Newark, New Jersey, became the nation's largest producer of Celluloid. This firm was purchased by E. I. duPont de Nemours & Co. in order to start that firm's plastics operations.

Other large U.S. firms, such as Celanese, Eastman, Hercules, and Monsanto, also entered the plastics business via Celluloid. Hyatt's incentive for producing Celluloid was a $10,000 award offered by a producer of billiard balls (Pheland and Collendar) for a substitute for ivory. Although Celluloid was widely used for shirt fronts, collars, combs, and brush handles in the nineteenth century, its use today is limited.

Because of its explosive nature, Celluloid could not be extruded or injection molded. However, cellulose diacetate was produced by the partial saponification of cellulose triacetate by G. Miles in 1905. This product, which is flammable but not explosive, continues to be used as a molding resin and for the production of films and fibers.

A mixed ester of cellulose called cellulose acetate butyrate, which was developed in 1935, is a tough transparent plastic that is widely used for molding steering wheels, ballpoint pens, and typewriter keys. Ethers of cellulose, such as ethylcellulose, have been molded and extruded to produce molded articles and sheets. Ethylcellulose melts have also been used for tool handles and waterproof packaging.

12.19   PLASTICS PROCESSING

A. Introduction

Both natural and synthetic polymers must be processed before use. The seeds must be separated from cotton in the ginning process, pigments and driers must be added to oleoresinous paints, and the latex of *Hevea* rubber or gutta-percha must be coagulated to obtain the solid elastomer plastic. Synthetic polymers must also be compounded and fabricated into useful shapes. Plastics are converted into their final shapes by utilizing a variety of techniques and machinery.

B. Casting

One of the simplest and least expensive methods for the production of plastic articles is casting. In this process, which is illustrated in Figure 12.8, a prepolymer, such as a catalyzed epoxy resin, is placed in a mold and allowed to harden, preferably with additional heat. This technique may also be used with urethane reactants (RIM), phenolic resins, unsaturated polyesters, PVC plastisols, and acrylic resins.

With the exception of plastisols, most of these processes are exothermic and thus the articles should be small or the mold must be cooled. Plastisols, which consist of a

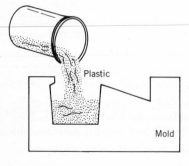

Plastic

Mold

Figure 12.8. Illustration of the casting method of molding plastics.

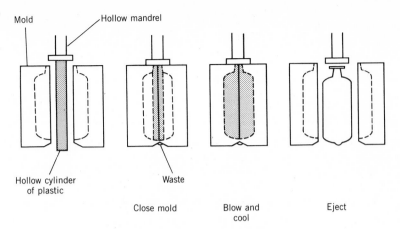

Figure 12.9. Blow molding technique.

dispersion of a finely divided polymer, usually PVC, in a liquid plasticizer, must be heated to at least 300°F to fuse the plasticizer–polymer mixture. Polymers, like ethylcellulose and ethylene–vinyl acetate copolymers, which can be melted without decomposition, can be cast as hot melts. Solutions of polymers can be cast as films.

Polymer concrete is produced by a casting process. Simulated marble consists of a filled-peroxy-catalyzed unsaturated polyester prepolymer that polymerizes *in situ*. Comparable mortars consisting of filled-catalyzed phenolic, epoxy, or polyester resins are used for joining brick and tile. Casting is used in manufacturing both thermosetting and thermoplastic resins for making eyeglass lenses, plastic jewelry, and cutlery handles.

C. Blow Molding

Blow molding and plug-assisted vacuum thermoforming are employed to make hollow items such as bottles and many hollow, thin-walled toys and bowls. For blow molding, a plastic parison is placed in the mold and air is applied through the opening of the cylinder-shaped plastic, blowing the plastic toward the mold walls (Figure 12.9). In the plug-assisted molding sequence the plastic resin is present as a sheet.

D. Injection Molding

In injection molding, a large volume of thermoplastics is injection molded to produce a variety of articles at a rapid rate. As shown in Figure 12.10, the polymer pellets may be heated, softened, and formed or forced (injected) by a ram into a closed, cooled mold. The split mold is opened and closed after the molded article is ejected and the cycle is then repeated. As shown in Figure 12.11, a reciprocating preplasticating screw that moves forward to eject the softened polymer may be used in place of the ram.

In contrast to slow compressive molding, injection molding is rapid. Complex parts may be produced in a few seconds in multicavity molds. Containers, gears, honeycombs, and trash cans are produced by the injection molding of selected thermoplastics (Figures 12.12 to 12.16).

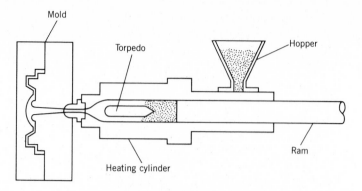

Figure 12.10. Injection molding technique.

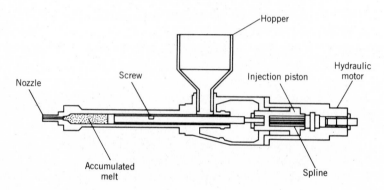

Figure 12.11. Modified injection molding technique.

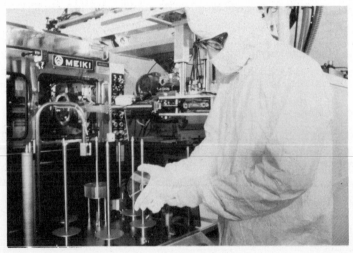

Figure 12.12. A Mobay laboratory technician inspects a compact disc that was molded on the Meiki injection molding system shown in the background.

212

Figure 12.13. A Sailor robot automatically removes a compact disc from the mold.

Figure 12.14. A Mobay lab technician places combinations of preformed glass reinforcement for a bumper beam in the RIM tool.

E. Laminating

In laminating, sheets of metal foil, paper, other plastic, or cloth are treated with a plastic resin. They are then run through rollers that squeeze the sheets together and heat them as shown in Figure 12.17. Paneling and electronic circuits are examples of products produced through this process, which is similar to making sandwiches.

Calendering is similar to laminating except rollers spread melted resin over the sheets to be covered, providing a protective coating as in the case of playing cards and treated wallpapers (Figure 12.18). This is similar to spreading jelly on a slice of bread, with the jelly being the resin.

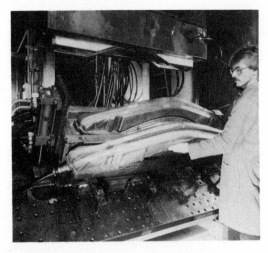

Figure 12.15. A finished bumper beam is removed from the mold less than a minute after the polyurethane mixture is injected into the tool.

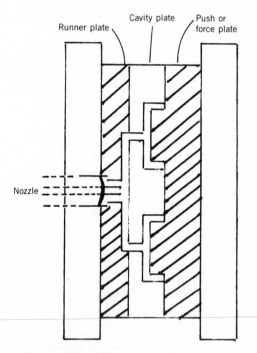

Figure 12.16. Injection mold in closed position.

F. Compression Molding

There are a wide variety of molding techniques. Simple molding entails squeezing plastic between two halves of a mold. It is similar to making waffles, where the batter is the plastic and the waffle iron the mold.

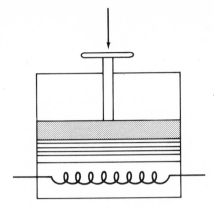

Figure 12.17. Assembly illustrating the laminating process.

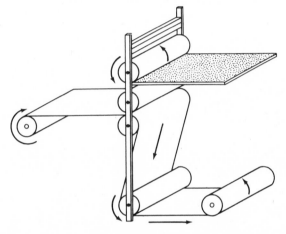

Figure 12.18. Assembly illustrating the calendering process.

One of the simplest molding techniques is compression molding, which is illustrated in Figure 12.20. In this molding process, a heated hydraulic press is used to soften plastic pellets and shape the plastic in a mold. When thermosets are used, the prepolymer is completely polymerized in the closed hot mold and is then ejected when the mold is opened. When thermoplastics are molded by compression molding, the mold cavity must be cooled before ejecting the plastic article.

The labor-intensive compression molding process may be upgraded by preheating a preform of the molding powder in a transfer pot and forcing this softened prepolymer into hot multicavity molds. Transfer molding is illustrated in Figure 12.21.

G. Rotational Molding

One of the more versatile molding techniques is rotational molding in which a hollow mold containing a resin powder or a liquid plastisol is heated and rotated simultaneously on two perpendicular axes. The mold is then cooled and the hollow object, such as a pipe fitting, is removed.

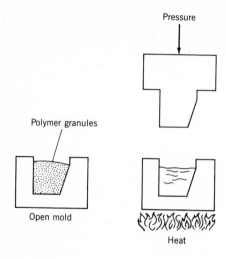

Figure 12.19. Compression molding technique.

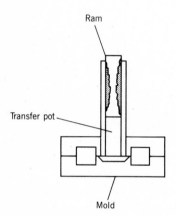

Figure 12.20. Transfer molding.

H. Calendering

One of the most commonly used techniques for making thermoplastic or elastomeric sheet is calendering. As shown in Figure 12.18, the polymer is transported through heated rollers, like those in a rubber mill, to a series of heated wringer-type rollers, which press the polymer into a continuous sheet of uniform thickness. The calendering process is used to produce sheeting for upholstery and for thermoforming.

I. Extrusion

The extrusion process is similar to squeezing toothpaste from its tube. As shown in Figure 12.21, pipe, rods, or profiles may be produced by the extrusion process. In this process, thermoplastic pellets are fed from a hopper to a rotating screw. The polymer is transported through heated, compacting, and softening zones and then forced through a die and cooled after it leaves the die. The extrusion process has been used to coat metal wire and to form coextruded sheet for packaging. Over 1 million tons of

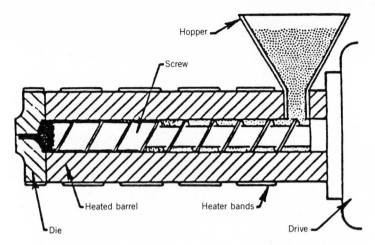

Figure 12.21. Details of screw and extruder zones.

extruded pipe are produced annually in the United States. Figure 12.22 illustrates film formation employing the extrusion process, and Figures 12.23 and 12.24 show an automated extrusion production line.

J. Thermoforming

Thermoplastic sheet, produced by extrusion through a slit die, calendering, or hot pressing of several calendered sheets, is readily thermoformed by draping over a mold and using a plunger or vacuum to force the sheet into the shape of the mold. As

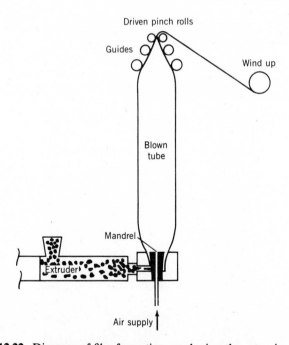

Figure 12.22. Diagram of film formation employing the extrusion process.

Figure 12.23. The entire extrusion line in Mobay's laboratories is controlled by state-of-the-art microprocessor technology, which was designed to optimize processing parameters and reduce start-up times. A specialist is shown here working at the central terminal.

Figure 12.24. An overview of the new Mobay multipurpose extrusion line.

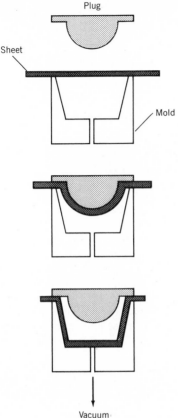

Figure 12.25. Steps in plug-assisted thermoforming.

illustrated in Figure 12.25, refrigerator liners or suitcases may be produced by vacuum sheet thermoforming.

K. Reinforced Plastics

Fiberglass-reinforced plastics (FRP) are fabricated by casting procedures using mixtures of the casting resins and glass or graphite fibers. In the simplest hand lay-up technique, a catalyzed resin, such as an unsaturated polyester resin, is placed on a male form. This gel coat formulation is followed by a sequential buildup of layers of catalyzed resin-impregnated glass mat. The composite is removed after it hardens. The curing step may be accelerated by heating. This technique may be modified by spraying a mixture of chopped fibers and the catalyzed prepolymer onto the form.

In a more sophisticated approach, a continuous resin-impregnated filament is wound around a rotating mandrel and cured as shown in Figure 12.26. In another modification, a bundle of resin-impregnated filaments is drawn through a heated die. Fishing rods and pipe are produced by this pultrusion technique.

L. Conclusion

A review of the many polymers and blends available and the many fabrication techniques that can be used to produce finished articles should demonstrate the

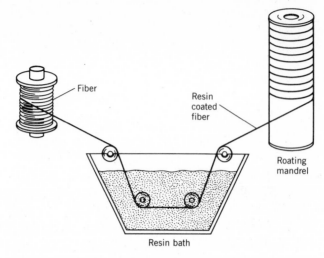

Figure 12.26. Illustration of the filament winding technique.

versatility of polymers. Because of this versatility, the polymer industry has grown at an unprecedented rate and will be the world's largest and most important industry before the beginning of the twenty-first century.

GLOSSARY

Accelerator: Catalyst for the vulcanization of rubber.

Banbury mixer: An intensive mixer.

Blow molding: The production of hollow articles, such as bottles, by air blowing a short section of pipe within a two-piece mold.

Blown film: Film produced by air blowing a warm pipe-like extrudate after it leaves the die.

Calendering: The formation of a sheet by passing a thermoplastic through a series of heated rolls.

Compounder: A technician who mixes polymers and additives.

Compression molding: The curing of a heated polymer under external pressure in a mold.

Elastomer: General term for rubbery materials.

Epoxy resin: The reaction product of bisphenol A and epichlorohydrin. The molecule has terminal epoxy groups

$$
\begin{array}{cc}
\text{H} & \text{H} \\
| & | \\
-\text{C} & -\text{C}- \\
\diagdown & \diagup \\
& \text{O} \\
\end{array}
$$

and multiple hydroxyl groups (OH).

Extrusion: The production of pipe and profiles by the continuous forcing of a heated, softened polymer through a die and cooling the extrudate.

Fiber: A threadlike structure with a length to diameter ratio of at least 100 to 1.

Filament winding: The winding and curing of a resin-impregnated filament around a rotating mandrel.

Gel coat: An unfilled and unreinforced polymer usually formed as the first (outer layer) of a fiber-reinforced composite.

GRP: Glass-reinforced plastic.

Gutta-percha: A nonelastic naturally occurring *trans*-polyisoprene.

***Hevea* rubber:** The most widely used natural rubber.

Injection molding: A rapid process for the production of molded articles by injecting a heat-softened thermoplastic into a closed, cooled split mold, ejecting the solid molded part, and repeating this process.

Latex: A stable suspension of a polymer in water.

Melamine–formaldehyde prepolymers: A low-molecular-weight product of the condensation of melamine and formaldehyde.

Oleoresinous paint: Coating based on polymerizable unsaturated oils.

Parison: A short section of thermoplastic pipe or tubing.

Plastisol: A dispersion of a finely divided polymer (usually PVC) in a liquid plasticizer.

Polyester: The product of the condensation of a diol $(R(OH_2))$ and a dicarboxylic acid $(R(COOH)_2)$.

Polymer concrete: A solid formed by mixing an aggregate containing an initiator and a polymer that polymerizes *in situ*.

Pultrusion: The curing of a resin-impregnated bundle of filaments by drawing through a heated die.

Quaterary ammonium compound: R_4N^+, Cl^- or R_4N^+, OH^-.

Rayon: Regenerated cellulose fiber.

RIM: Reaction injection molding, in which polymerization takes place in the mold.

Rotational molding: The production of hollow articles by heating finely divided polymer particles in a rotating mold.

Simulated marble: A filled polyester produced by the room-temperature polymerization of a filled unsaturated polyester prepolymer.

Stabilizer: An additive used to retard degradation of polymers.

Thermoforming: The forming of trays and other three-dimensional articles by heating and pressing a sheet of thermoplastic over a mold.

Thermoplastic elastomer: An elastic polymer that does not require cross-linking for dimensional stability.

Thermoset: A cured polymer that cannot be softened by heat without decomposition.

Thinner: Solvent.

Transfer molding: An improved compression molding process in which a thermoset preform is preheated before being transformed to the hot mold.

Vehicle: The binder (resin) and solvent in a paint.

Vulcanization: Cross-linking, usually with sulfur.

REVIEW QUESTIONS

1. Are most commercial plastics thermoplastics or thermosets?

2. Is a cross-link a covalent bond or a hydrogen bond?

3. Which of the following natural polymers is not a hydrocarbon: gutta-percha, balata, casein?

4. Which is more highly crystalline: HDPE or LDPE?

5. Which is an alkane: polyethylene or paraffin?

6. What is the DP of HDPE with a molecular weight of 14,000?

7. Why is ABS used for making suitcases?

8. A PVC plastisol is a thermoplastic but is sometimes called a thermoset by nonpolymer technicians. Why?

9. How would you dissolve poly(acrylic acid)?

10. How do poly(methyl methacrylate) and poly(ethyl acrylate) differ in their empirical formulas?

11. How do polymethyl methacrylate and polyethyl acrylate differ in softening point?

12. Why is polyhydroxyethyl methacrylate used for contact lenses?

13. What is the formula for the methyl radical?

14. Which has the larger specific volume: HDPE or LDPE?

15. Which would yield the larger area of film of similar thickness for a given weight: LPDE or HDPE?

16. Which has the larger bulky pendant group: polypropylene or TPX?

17. Which can be readily cross-linked: polyisobutylene or butyl rubber?

18. Which will have the higher melting point: atactic or isotactic PP?

19. Which would be more likely to crystallize: atactic PP or a block copolymer of ethylene and propylene?

20. Which would be more readily soluble in water: PVA or PVC?

21. In the early 1900s, most plastics were celluloid. Why is this no longer true?

BIBLIOGRAPHY

Birley, A. W., and Scott, M. J. (1982). *Plastic Materials*. New York: Chapman & Hall.

Boundy, R. H., and Boyer, R. F. (1952). *Styrene, Its Polymers, Copolymers and Derivatives*. New York: Reinhold.

Horn, H. B. (1960). *Acrylic Resins*. New York: Reinhold.

Juran, R. (1987). *Modern Plastics Encyclopedia* (Vol. 64, No. 10A). New York: McGraw–Hill.

Kaufman, M. H. (1969). *Polyvinyl Chloride*. New York: Gordon & Breach.

Kresser, T. O. J. (1960). *Polypropylene*. New York: Reinhold.

Miles, C. D., and Briston, J. H. (1979). *Polymer Technology*. New York: Chemical Publishing Co.

Rudner, M. A. (1958). *Fluorocarbons*. New York: Reinhold.

Sarretnik, M. A. (1972). *Plastisols and Organisols*. New York: Van Nostrand.

Seymour, R. B. (1975). *Modern Plastics Technology*. Reston, VA: Reston.

Seymour, R. B. (1982). *Plastics vs. Corrosives*. New York: Wiley–Interscience.

Seymour, R. B. (Ed.). (1982). *History of Polymer Science and Technology*. New York: Dekker.

Seymour, R. B., and Mark, H. F. (Eds.). (1988). *Applications of Polymers*. New York: Plenum.

Stannett, V. F. (1950). *Cellulose Acetate Plastics*. London: Temple Press.

Teach, W. C., and Kessling, G. C. (1960). *Polystyrene*. New York: Reinhold.

Ulrich, H. (1982). *Introduction to Industrial Polymers*. Munich: Hanson Publishers.

Yarsley, V. E., Flavell, W., Adamson, P. S., and Perkins, N. G. (1964). London: ILIFFE Publishers.

Young, R. J. (1981). *Introduction to Polymers*. New York: Chapman & Hall.

ANSWERS TO REVIEW QUESTIONS

1. Thermoplastic.

2. Covalent bond.

3. Casein is a polyamide (protein).

4. HDPE.

5. Both have the empirical formula $H(CH_2)_nH$.

6. $14,000/28 = 500$.

7. ABS is a tough plastic.

8. It undergoes an irreversible physical change from a liquid to a solid when heated.

9. In an alkaline solution such as aqueous sodium hydroxide.

10. They are identical ($C_5H_8O_2$).

11. The softening point of poly(ethyl acrylate) is much lower than that of PMMA.

12. The hydrophilic hydroxy group absorbs water and keeps the polymer soft.

13. $\cdot CH_3$.

14. LDPE, the specific volume is the reciprocal of the density.

15. LDPE.

16. TPX (polymethylpentene).

17. Butyl rubber has a few double bonds that can be cross-linked; polyisobutylene has no double bonds.

18. Isotactic PP is a solid; atactic PP is a soft gummy substance.

19. The block copolymer providing the sequences were linear and ordered.

20. PVA because of the hydroxyl pendant groups present.

21. Many other less costly, more readily moldable, and less flammable thermoplastics are available commercially.

CHAPTER **XIII** _____

Thermosets

13.1 INTRODUCTION

Thermoplastics and thermosets form the two major groups of plastics. They share many common processing sequences (see Section 12.9) and are plastic in having properties offering flexible dimensional stability, that is, they can be bent, to some extent, yet are also rigid. "Plastic" cups, rulers, bottle caps, et cetera are representative examples of plastics. Thermoplastics can be further divided according to general-purpose plastics and engineering plastics. The topics of general-purpose plastics and engineering plastics are covered in Chapters 12 and 14. Thermosets are discussed in this chapter.

The commercialization of phenol–formaldehyde plastics preceded the large-scale production of most of the commercial synthetic thermoplastics. Vulcanized rubber, which was introduced by Charles Goodyear in 1838, was a low-density cross-linked elastomer. Hard rubber, which was invented by Nelson Goodyear, was a high-density cross-linked plastic. The cross-linking agent in both of these thermosets was sulfur.

224

Glyptal coatings, developed by W. Smith at the beginning of the twentieth century, were produced by the condensation of glycerol ($HOCH_2CH(OH)CH_2OH$) and phthalic anhydride ($C_6H_4C_2O_3$). Since the secondary hydroxyl group in glycerol was less reactive than the terminal primary hydroxyl groups, the prepolymer was a linear prepolymer with one unreacted hydroxyl group in each repeating unit. This thermoplastic prepolymer was converted to a thermoset by heating after it was applied to the surface of metals.

Oleoresinous paints were also applied as thermoplastic prepolymers. These prepolymers were cross-linked by an addition polymerization reaction with oxygen in the presence of driers. The term thermoset applies to all cross-linked polymers, regardless of whether the cross-links were formed by heating, irradiation, or chemical reaction.

13.2 PHENOLIC RESINS

Products that at the time were called "goos, gunks, and messes" were produced by some of the world's most renowned chemists in the late 1890s. Unfortunately, these chemists were unfamiliar with the importance of functionality. Later chemists, such as Leo Baekeland, recognized that the combination of reactants with difunctional groups such as diols ($R(OH)_2$) and dicarboxylic acids ($R(COOH)_2$) produced linear polymers. They also knew that a trifunctional reactant, such as phenol (C_6H_5OH), when reacted with a difunctional reactant, such as formaldehyde, ($H_2C=O$), produced an infusible cross-linked polymer. The three reactive sites in phenol are in the 2, 4, and 6 positions for reaction with formaldehyde.

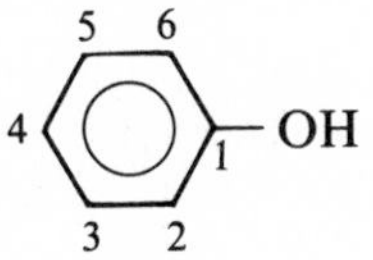

The problem of uncontrolled cross-linking was solved by Baekeland, who used an insufficient amount of the difunctional reactant (formaldehyde) with the trifunctional phenol, in the presence of an acid, to produce a linear prepolymer. This prepolymer contained reactive centers that could react with additional formaldehyde from hexamethylenetetramine in a controlled secondary reaction.

Thus, Baekeland was able to place a mixture of the prepolymer and hexamethylenetetramine in a mold and obtain a thermoset article by heating the mixture. The concept recognized by Baekeland may be demonstrated by the following reactions:

The thermoplastic prepolymer obtained from the reaction of phenol with an insufficient amount of formaldehyde is called a novolak resin. Most phenolic molding powders contain novolak PF, hexamethylenetetramine, pigment, and a filler, such as wood flour. Wood flour is a finely divided, fibrous wood filler obtained by the attrition grinding of debarked wood.

A linear product called a resole resin is also obtained by reacting equivalent amounts of the reactants in the presence of an alkali. This product is a viscous liquid

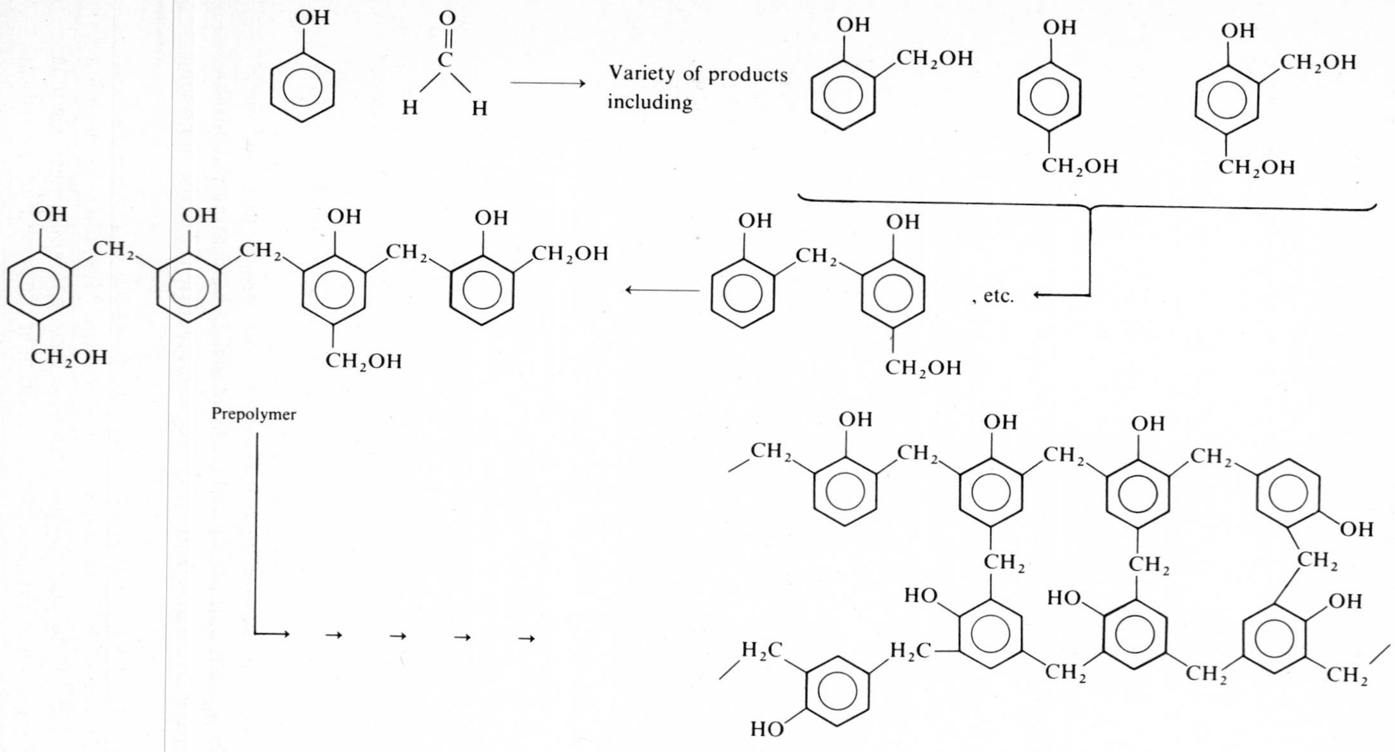

Variety of products including

, etc.

Prepolymer

Complex 3-D matrix

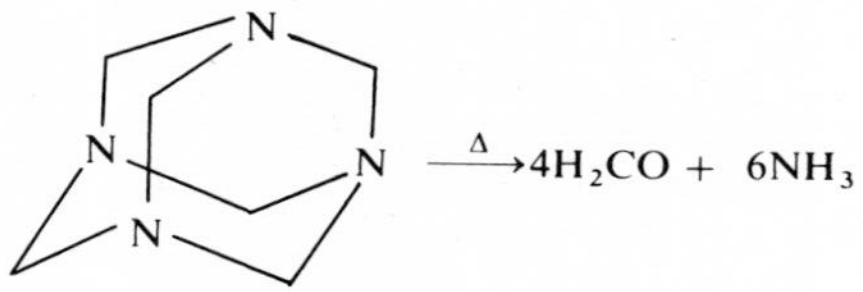

$$\xrightarrow{\Delta} 4H_2CO + 6NH_3$$

Hexamethylenetetramine   Formaldehyde   Ammonia

while it is cooled. This liquid prepolymer is converted to a solid when heated or when an acid is added.

Linear PF resins are also obtained by the condensation of formaldehyde with a para-substituted phenol, such as p-phenylphenol, which has a functionality of 2. The properties of phenolic resins are shown in Table 13.1.

Over 1.5 million tons of phenolic polymers are produced annually in the United States, using the same formulations developed by Baekeland in the early 1900s. Molded wood flour-filled phenolic resins are used for electrical insulators. Almost 600,000 tons of phenolic resins are also used annually in the United States as

Table 13.1   Properties of Typical Phenolic and Amino Plastics

Property	PF, Wood Flour Filled	UF, Cellulose Filled	MF, Cellulose Filled
Processing temp. (°F)	350	300	350
Molding pressure (10^3 psi)[a]	15	15	15
Mold shrinkage (10^{-3} in./in.)	6	10	10
Heat deflection temp. under flexural load of 264 psi (°F)	350	275	300
Maximum resistance to continuous heat (°F)	300	260	250
Coefficient of linear expansion (10^{-6} in./in., °F)	20	15	15
Compressive strength (10^3 psi)	25	35	28
Impact strength Izod (ft-lb/in. of notch)[b]	0.4	0.3	0.3
Tensile strength (10^3 psi)	7	10	8
Flexural strength (10^3 psi)	10	12	9
% elongation	0.5	1	3
Tensile modulus (10^3 psi)	1200	1200	1200
Flexural modulus (10^3 psi)	1000	1400	1100
Rockwell hardness	M105	E90	E95
Specific gravity	1.4	1.5	1.6
% water absorption	0.7	0.5	0.5
Dielectric constant	6	6	5
Dielectric strength (V/mil)	325	350	275
Resistance to chemicals at 75°F[c]			
Nonoxidizing acids (20% H_2SO_4)	S	Q	Q
Oxidizing acids (10% HNO_3)	S	U	U
Aqueous salt solutions (NaCl)	S	S	S
Polar solvents (C_2H_5OH)	S	S	S
Nonpolar solvents (C_6H_6)	S	Q	Q
Water	S	S	S
Aqueous alkaline solutions (NaOH)	U	Q	Q

[a]psi/0.145 = kPa (kilopascals).
[b]ft-lb/in. of notch/0.0187 = cm · N/cm of notch.
[c]S = Satisfactory, Q = questionable, and U = unsatisfactory.

adhesives for plywood. Phenofoam, which is produced by adding a gaseous propellant to PF, has the lowest flammability of all commercial plastic foams.

OH

p-Phenylphenol

   PF resins are used as molding resins (about 35%), laminating resins (about 10%), bonding resins, coatings and adhesives (about 35%), and in ion-exchange resins. The major adhesive use is in the manufacture of plywood. Phenolic resins are widely used in the coatings industry in varnishes. They are employed as bonding resins in the production of abrasive (grinding) wheels, sandpaper, and brake linings. Decorative laminates formed from impregnating wood and paper for countertops, printed circuits, and wall coverings account for about 10% of its uses. The impregnated material is dried in an oven, then hot pressed and molded into the desired shape. Fillers are typically used in molding applications to improve impact properties and to reduce cost. Common fillers are fiberglass, nylon and other fibers, cloth, and cellulosic materials, such as cotton and wood flour. Because of its nonconductive nature, PF is used in molding TV and radio cabinets, appliance parts, and automotive parts.

   The ability of PF resins to withstand high temperatures for a brief time has led to their use in the construction of missile nose cones. Metals vaporize, but PF resins decompose, leaving a carbon (similar to graphite) protective coating.

13.3   UREA RESINS

Resins are produced by the reaction of tetrafunctional urea

$$(H_2N-\overset{\displaystyle O}{\overset{\displaystyle \|}{C}}-NH_2)$$

and formaldehyde and were described by Holzer in 1884 and John and Pollak in 1918. In spite of Baekeland's early work with thermosets, these polymers were not commercialized until the mid-1920s.

   Urea–formaldehyde resin (UF) is produced by the condensation of urea with an insufficient amount of formaldehyde under alkaline conditions (Figure 13.1). The liquid resin is mixed with fillers, such as wood flour, α-cellulose, and an acid. The mixture is dried and densified on a rubber mill, cooled, and granulated. Almost 700,000 tons of UF are produced annually in the United States.

$$H_2NCONH_2 \xrightarrow{\text{CH}_2\text{O}} HOCH_2NHCONH_2 \xrightarrow{\text{CH}_2\text{O}} HOCH_2NHCONHCH_2OH$$

$$\rightarrow \text{Urea–formaldehyde resin (UF)}$$

Figure 13.1. Formation of urea–formaldehyde resins.

Urea resins are much lighter in color than the dark phenolic resins. They are used as adhesives for particleboard and laminated paper and foams. α-Cellulose-filled urea resins are used as molding resins.

Unless stabilizers are present, UF adhesives and foams may release some formaldehyde, which is considered to be toxic. The properties of UF and melamine formaldehyde (MF) are summarized in Table 13.1. UF and MF are called amino plastics.

Urea resins offer less heat and moisture resistance and are softer than the melamine resins but they are also generally less expensive (about 50¢/lb compared to about 75¢/lb for PF resins). Almost all urea molded products are cellulose filled. Because of their good solvent and grease resistance, surface hardness, mar resistance, and easy colorability, they are widely employed for bonding wood in furniture and plywood. They are also added to cotton and rayon textiles to impart crease resistance, shrinkage control, water repellency, fire retardance, and stiffness. Urea-based enamels are used for coating kitchen appliances, including dishwashers and refrigerators. These enamels are called baked enamels and contain urea resins as well as alkyd resins. About 60% of the urea–formaldehyde resins are now employed in the production of particleboard.

13.4 MELAMINE RESINS

Melamine, which is shown by the following structure, has six reactive sites (two on each amine group). It may be condensed with formaldehyde to produce light-colored, heat-resistant plastics (Figure 13.2).

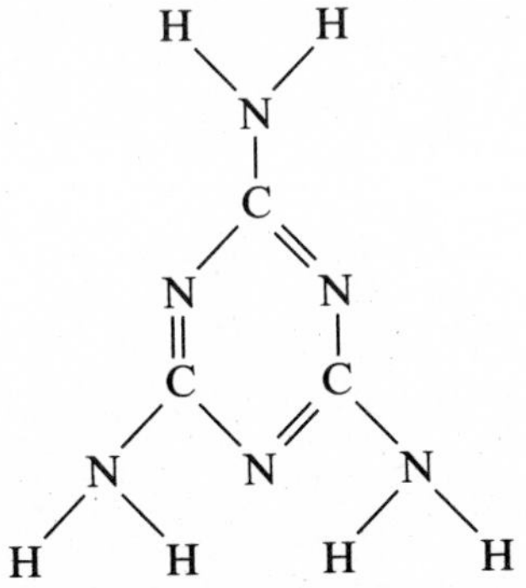

Melamine

Melamine was characterized by Liebig in 1843, but in spite of the information available on phenolic and urea resins, melamine resins (MF) were not produced commercially until the mid-1900s. MF resins are used for the production of decorative

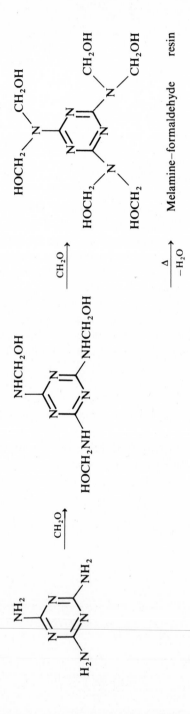

Figure 13.2. Condensation of melamine with formaldehyde to form thermoset plastics.

230

laminates. α-Cellulose-filled MF resins are used for molding dinnerware (Melmac). The properties of a cellulose-filled MF molded plastic are shown in Table 13.1. The annual production of MF in the United States is about 90,000 tons.

A distinct advantage of urea and melamine resins over phenolic PF resins is the fact that they are clear and colorless, thus aiding in ease of coloration. Melamine resins have poorer impact strength and moisture and heat resistance compared to PFs but they are harder. Melamine resins are predominantly cellulose filled although fiberglass and cotton fabrics are also employed. The production of dinnerware (such as plates, cups, and serving bowls) using cellulose-filled UF resins is the single largest use for these resins. Melamine resins are also used for the production of laminates, such as tops for counters, cabinets, and tables. Typically a core of phenolic-impregnated paper is overlaid with melamine-impregnated sheets to produce Formica. Melamine formulations are also employed as automotive finishes.

13.5 ALKYDS–POLYESTER RESINS

The first polyester plastic was a cross-linked resin produced by Berzelius in 1847 by the condensation of difunctional tartaric acid and trifunctional glycerol. Commercial resinous coatings, called Glyptal resins, were produced by W. Smith in 1902 by the condensation of difunctional phthalic acid and trifunctional glycerol. The linear polyester prepolymer obtained at moderate temperatures by the reaction with the two primary hydroxyl groups was converted to a cross-linked plastic by a reaction with the residual (secondary) hydroxyl group after the reaction temperature was increased (Figure 13.3).

In the 1920s, R. Kienle coined the term alkyd for these polyester resins, which he obtained by the condensation of alcohols and acids. Many alkyd resins are produced by the condensation of difunctional reactants, alcohol and carboxylic acid, in the presence of unsaturated acids, such as oleic acid. The double bonds in these alkyds serve as sites for cross-linking in the presence of oxygen. Commercial alkyd molding powders were produced in 1948.

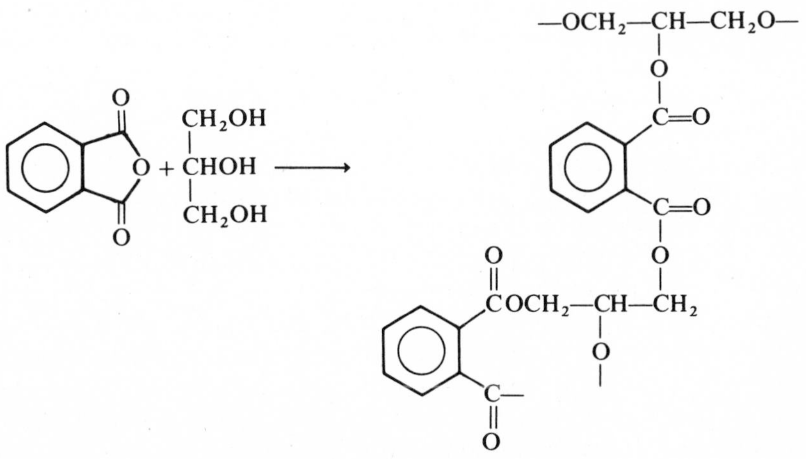

Figure 13.3. Glyptal resin formation.

Tartaric acid

Glycerol

Phthalic acid

Oleic acid (derived from vegetable fats and oils,
such as olive oil and soybean oil)

The most widely used polyester plastics are fiberglass-reinforced polyester plastics (FRP), which were introduced by Ellis in 1940. These FRP materials are based on unsaturated alkyd-type polyester prepolymer (polyethylene maleate), which is dissolved in styrene monomer. The prepolymer solution is cross-linked by the addition of peroxide-type initiators, such as benzoyl peroxide.

Mixtures of chopped fiberglass and polyester prepolymers and fiberglass mat impregnated with polyester prepolymer are called bulk molding compounds (BMC) and sheet molding compounds (SMC), respectively.

The so-called vinyl ester resins (Derakane) are also used for the production of FRP. These vinyl esters are formed by the reaction of bisphenol A, acrylic acid, and ethylene oxide. Other polyesters used in FRP are produced from esters of bisphenol A and fumaric acid, isophthalic acid (m-phthalic acid), ethylene glycol, and maleic anhydride. The properties of a typical fiberglass-reinforced polyester laminate are shown in Table 13.2. Over 1 million tons of FRP are used annually in the United States in boats, panels, and automotive components.

Alkyd resins are primarily used in organic coating applications. They are used in some lacquers along with natural resins such as shellac and in varnish-type coatings as drying oils or resins. Prepolymers (partially polymerized but still thermoplastics) are used as molding resins in the production of fiberglass-reinforced laminates.

About 80% of unsaturated polyesters (excluding alkyd resins) are used to produce reinforced products, including electrical, marine, and transportation applications (Figure 13.4). Speedboat and motorboat hulls are generally produced by the SMC process, in which a mixture of unsaturated polyester resin, fibers (often fiberglass), and fillers is held between sheets of polyethylene film until it thickens to a leathery sheet. These sheets are molded under pressure to give fiber-reinforced plastic hulls. Shower stalls and industrial tubs are also made by the SMC process.

Table 13.2 Properties of Typical Reinforced Polyesters

Property	Alkyd Mineral Filled	BMC Polyester	SMC Polyester
Processing temp. (°F)	300	300	300
Molding pressure (10^3 psi)[a]	15	1	1
Mold shrinkage (10^{-3} in./in.)	2	4	2
Heat deflection temp. under flexural load of 264 psi (°F)	425	375	425
Maximum resistance to continuous heat (°F)	400	350	400
Coefficient of linear expansion (10^{-6} in./in., °F)	15	10	15
Compressive strength (10^3 psi)	25	20	20
Impact strength Izod (ft-lb/in. of notch)[b]	0.5	10	12
Tensile strength (10^3 psi)	6	10	12
Flexural strength (10^3 psi)	2	2.0	2.5
% elongation	1	4	3
Tensile modulus (10^3 psi)	1500	2000	2000
Flexural modulus (10^3 psi)	2000	2000	2000
Rockwell hardness	E98	60 (Barcol)	60 (Barcol)
Specific gravity	2	1.9	2
% water absorption	0.5	0.5	0.5
Dielectric constant	5	4	4
Dielectric strength (V/mil)	400	400	400
Resistance to chemicals at 75°F[c]			
Nonoxidizing acids (20% H_2SO_4)	S	Q	Q
Oxidizing acids (10% HNO_3)	U	U	U
Aqueous salt solutions (NaCl)	S	S	S
Polar solvents (C_2H_5OH)	S	Q	Q
Nonpolar solvents (C_6H_6)	U	U	U
Water	S	S	S
Aqueous alkaline solutions (NaOH)	Q	U	U

[a] psi/0.145 = kPa (kilopascals).
[b] ft-lb/in. of notch/0.0187 = cm · N/cm of notch.
[c] S = Satisfactory, Q = questionable, and U = unsatisfactory.

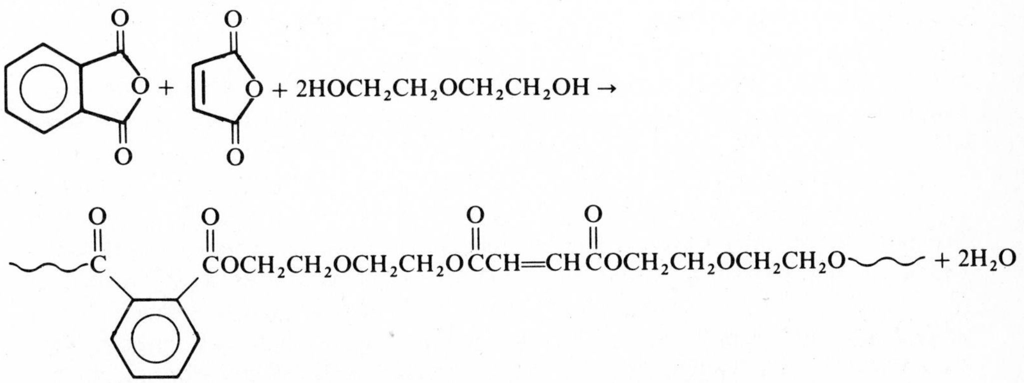

Figure 13.4. Unsaturated polyester resin formation.

13.6   EPOXY RESINS

The word epoxy is derived from the Greek prefix *epi*, meaning between, and the English suffix of oxygen. Epoxy resins, which are obtained by the condensation of bisphenol A and epichlorhydrin, were described by Linderman in the last part of the nineteenth century and patented by P. Schlack in 1934. These linear prepolymers contain hydroxyl (OH) and epoxy

groups, which may be cross-linked by reaction with cyclic anhydrides, such as maleic anhydride, at elevated temperatures or with polyamines $(R(NH_2)_n)$ at moderate temperatures.

Epoxy resins

Bisphenol A   Epichlorhydrin   Maleic anhydride

Epoxy resins (EP) may be flexibilized by using fatty acid diamines (Versamid) as the cross-linking or curing agent. The properties of typical epoxy resins are shown in Table 13.3.

Over 225,000 tons of epoxy resins are used annually in the United States as adhesives, coatings, and encapsulating compositions. Fiberglass-reinforced epoxy resins are used as aircraft components. About 110,000 tons of EP are used annually as coatings and 45,000 tons are used for GRP (FRP).

The wide range of applications of EP is a result of the versatility of the system. By varying ratios of materials present in the prepolymers, large differences in curing rate and hardness are possible. There are two major types of surface coatings. The first is room (ambient) temperature cured. These are cross-linked using polyamides and polyamines. The second type of coating is heat cured. These are cured using formaldehyde resin, anhydrides, and polycarboxylic acids. Phenol–formaldehyde resin–epoxy products include drum and tank linings, wire coatings, impregnation

Table 13.3   Properties of Typical Epoxy and Silicone Resins

Property	Glass Fiber-Reinforced EP	Mineral-Filled Silicone
Processing temp. (°F)	300	300
Molding pressure (10^3 psi)[a]	2	3
Mold shrinkage (10^{-3} in./in.)	1	1
Heat deflection temp. under flexural load of 264 psi (°F)	350	500
Maximum resistance to continuous heat (°F)	325	500
Coefficient of linear expansion (10^{-6} in./in., °F)	20	15
Compressive strength (10^3 psi)	25	12
Impact strength Izod (ft-lb/in. of notch)[b]	2.5	4
Tensile strength (10^3 psi)	12	6
Flexural strength (10^3 psi)	20	12
% elongation	4	5
Tensile modulus (10^3 psi)	3000	—
Flexural modulus (10^3 psi)	3000	1500
Rockwell hardness	M105	M85
Specific gravity	1.8	1.9
% water absorption	0.1	0.2
Dielectric constant	4	3
Dielectric strength (V/mil)	300	300
Resistance to chemicals at 75°F[c]		
Nonoxidizing acids (20% H_2SO_4)	S	Q
Oxidizing acids (10% HNO_3)	U	U
Aqueous salt solutions (NaCl)	S	S
Polar solvents (C_2H_5OH)	S	S
Nonpolar solvents (C_6H_6)	S	Q
Water	S	S
Aqueous alkaline solutions (NaOH)	S	S

[a]psi/0.145 = kPa (kilopascals).
[b]ft-lb/in. of notch/0.0187 = cm · N/cm of notch.
[c]S = Satisfactory, Q = questionable, and U = unsatisfactory.

varnishes, and food and beverage can coatings. Epoxy coatings can be used as powder coatings, thus eliminating the use of solvents.

Printed circuit boards are made from fiberglass laminates and epoxy resins. Epoxy resins are used as binders for floor surfaces that are subject to heavy traffic, in patching concrete, and in casting, encapsulation, and potting electrical equipment. The common adhesive sold in hardware stores is generally a two-component liquid or paste that cures to give an epoxy resin.

13.7   SILICONES

Polysiloxanes $-\!(\text{Si}\!-\!\text{O}\!-\!\text{Si})\!-$ were investigated by F. Kipping prior to World War I. Since he believed that these compounds were ketones, he called them silicones. Unfortunately, he did not think these inorganic polymers were useful and did not attempt to make them on a large scale. Nevertheless, these silicones were commercialized by G.E. and Dow-Corning Corporation during the early 1940s.

The first silicones were produced by the hydrolysis of chloromethylsilanes. The dichlorodimethyl silane ($Cl_2Si(CH_3)_2$) produces linear polymers that cross-link when some trichloromethyl silane ($Cl_3Si(CH_3)$) is present. The original silicones were produced by the reaction of magnesium with methyl chloride (CH_3Cl) in the Grignard reaction. They are now produced by heating methyl chloride with silicon in the presence of a copper catalyst. Either chloroalkylsilanes or methoxyalkysilanes will polymerize in the presence of water to produce silicones.

Silicones are available as low-molecular-weight fluids, molding resins, and elastomers. As shown in Table 13.3, these polymers have excellent resistance to solvents and retain their properties at elevated temperatures.

13.8 POLYURETHANES

Organic isocyanates (RNCO) were synthesized by Wurtz in 1819. These active materials, particularly phenyl isocyanate (C_6H_5NCO), were used in qualitative organic chemistry to characterize alcohols by a reaction that produced urethane ($C_6H_5NHCOOR$). In the mid-1930s, Otto Bayer used difunctional reactants to produce polyurethanes (PUR). Fibers, plastics, coatings, adhesives, and foams were produced from this versatile reaction. The adhesive properties were accidentally discovered when a PUR molding stuck tenaciously to the metal mold. The foam, which was called "imitation Swiss cheese" by Bayer's critics, was discovered when an organic acid was present in the reaction. Traces of water in the reactants will also produce carbon dioxide (CO_2), which causes foaming to take place.

The most widely used diisocyanate is tolylene diisocyanate (TDI, H_3C—$C_6H_3(NCO)_2$). Hydroxyl-terminated low-molecular-weight polyesters and polyethers are used as the diols $HO(RO)_nH$. The extent of cross-linking is controlled by the amount of triol, such as glycerol, present in the reactants.

Because of its versatility, PUR is used in a wide variety of applications ranging from foundation garments to bowling pin coatings to upholstery to automobile tires. Over 750,000 tons of flexible PUR foam, over 450,000 tons of rigid PUR foam, and almost 50,000 tons of PUR elastomers are used commercially each year in the United States. One of the fastest growing applications of PUR is in reaction injection molding (RIM) in which the reactants (diisocyanate and diol) are mixed and the polymer is formed rapidly under very little pressure in a mold. The properties of PURs are summarized in Table 13.4.

13.9 PLASTIC COMPOSITES

Although all mixtures of polymers and additives are composites, the term composite is used primarily for reinforced plastics. Asbestos-filled phenolics and α-cellulose-filled ureas and melamines are also plastic composites, but the emphasis is on those composites containing fibers with greater aspect ratios (l/d).

It was fortuitous that commercial fiberglass and commercial unsaturated polyester resins were introduced almost simultaneously in the late 1930s by Slater and Thomas

Table 13.4 Properties of Typical Polyurethanes

Property	RIM (PUR)	PUR, 50% Mineral Filled
Processing temp. (°F)	25	25
Molding pressure (10^3 psi)[a]	—	—
Mold shrinkage (10^{-3} in./in.)	2.0	2.0
Heat deflection temp. under flexural load of 264 psi (°F)		
Maximum resistance to continuous heat (°F)		
Coefficient of linear expansion (10^{-6} in./in., °F)	60	40
Compressive strength (10^3 psi)	20	—
Impact strength Izod (ft-lb/in. of notch)[b]	25	5
Tensile strength (10^3 psi)	10	5
Flexural strength (10^3 psi)	20	5
% elongation	50	10
Tensile modulus (10^3 psi)	25	5
Flexural modulus (10^3 psi)	100	30
Rockwell hardness	D90 (Shore)	R40
Specific gravity	1.05	1.7
% water absorption	0.2	0.4
Dielectric constant	6	6
Dielectric strength (V/mil)	400	600
Resistance to chemicals at 75°F[c]		
Nonoxidizing acids (20% H_2SO_4)	Q	Q
Oxidizing acids (10% HNO_3)	U	U
Aqueous salt solutions (NaCl)	S	S
Polar solvents (C_2H_5OH)	U	U
Nonpolar solvents (C_6H_6)	Q	Q
Water	S	S
Aqueous alkaline solutions (NaOH)	Q	Q

[a]psi/0.145 = kPa (kilopascals).
[b]ft-lb/in. of notch/0.0187 = cm·N/cm of notch.
[c]S = Satisfactory, Q = questionable, and U = unsatisfactory.

and by Foster and Ellis, respectively. Fiberglass has limited use by itself, and unsaturated polyesters are too brittle for commercial use as plastics, but the combination (FRP) has unusually good properties and is now produced in the United States at an annual rate of over 1 million tons.

Since FRP is used for making boat hulls, car bodies, fishing rods, and golf club shafts, it is commonplace and erroneously referred to as fiberglass. Of course, the composite would be useless without both the resinous continuous phase and the discontinuous reinforcing phase. Although improved unsaturated polyesters dominate the market, other prepolymers, based on silicones, vinyl esters, and epoxy resins, are also used for the production of plastic composites.

The original FRP composites were made by impregnating fiberglass mat with a catalyzed (initiated) prepolymer. This hand lay-up technique has been supplemented by a spray-up technique in which chopped fibers and prepolymers are applied by a special spray gun. More uniform FRP composites are produced by bulk molding and sheet molding. The production of FRP articles may be automated by filament winding and pultrusion.

Asbestos and cellulose fibers have been used, to a limited extent, as reinforcements for thermosetting resins. Because of its toxicity, asbestos is being displaced by other fibers, and the use of cellulose fibers is limited because of their high water absorption and poor resistance to elevated temperatures.

Aromatic polyamides (aramids) and boron fibers are also used as reinforcements for thermosets. Boron filaments are produced by the chemical vapor decomposition (CVD) process in which boron trichloride is heated with hydrogen to produce hydrogen chloride and boron. The latter is deposited uniformly on a tungsten or graphite filament.

The second most widely used reinforcing filament is graphite. This strong fiber may be produced by the pyrolysis of polyacrylonitrile filaments (PAN) or by the thermal treatment of pitch. Graphite-reinforced resinous composites have outstanding strength given their light weight. Because of the high cost of graphite fibers, they are sometimes mixed with glass fibers. Composites based on these hybrid fibers have properties that are superior to FRP.

The early research emphasis was on reinforced thermosets and some plastic technologists believed that the use of reinforcements in thermoplastics was not advantageous. However, molding compounds of chopped fiberglass and most thermoplastics are now available commercially. The properties of high-performance plastics have been upgraded considerably by the addition of glass or graphite fibers. Over 100 million tons of reinforced thermoplastics (RTP) are now used annually in the United States.

GLOSSARY

α-Cellulose: High-molecular-weight cellulose, insoluble in 17.5% NaOH.
Alkyd: A generic name for unsaturated polyester resins.
Amino plastic: Urea and melamine plastics.
Baekeland, Leo: Inventor of phenolic resins.
Bakelite: Trade name for PF.
Bayer, Otto: Inventor of polyurethanes.
Benzoyl peroxide: $C_6H_5COOOOCC_6H_5$.
Cross-link: Intermolecular primary valence bond.
Diamine: A compound with two amino (NH_2) groups.
Drier: A paint catalyst, usually the heavy metal salt of an organic acid.
Epoxy group:

$$-\underset{}{\overset{H}{C}}-\underset{}{\overset{H}{C}}-$$
$$O$$

Epoxy resin: A resin obtained by the reaction of bisphenol A and epichlorohydrin. The terminal groups in epoxy resins are epoxy groups.

Ethylene oxide:

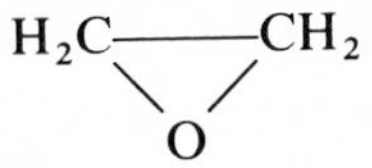

$$\text{H}_2\text{C}\underset{\displaystyle \text{O}}{\diagdown\diagup}\text{CH}_2$$

Fiberglass: Fibers obtained by the melt spinning of glass.

Formaldehyde: H_2CO.

FRP: Fiberglass-reinforced plastics.

Fumaric acid: The trans isomer of maleic acid,

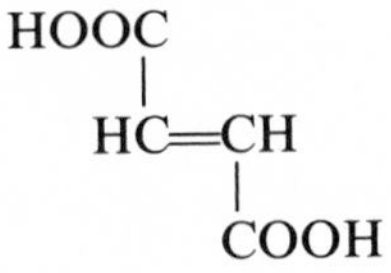

$$\begin{array}{c}\text{HOOC}\\ |\\ \text{HC}=\text{CH}\\ |\\ \text{COOH}\end{array}$$

Glycerol: $(H_2COH)CHOH$.

Glyptal: A resin obtained by the reaction of glycerol and phthalic anhydride.

Goodyear, Charles: Inventor of vulcanized rubber.

Goodyear, Nelson: Inventor of hard rubber.

Grignard reagent: RMgX, where R = an alkyl or aryl group.

Hexamethylenetetramine: The reaction product of ammonia (NH_3) and formaldehyde.

Isocyanate group: NCO.

Isophthalic acid: *m*-Phthalic acid.

Laminate: A composite consisting of resin bonded to reinforcing sheets.

Linear polymer: A polymer with a continuous chain (thermoplastic).

Melamine resin (MF): The reaction product of melamine and formaldehyde.

Meta group: A group in position 3 or 5 on a substituted benzene.

Methacrylic acid: $H_2C=C(CH_3)COOH$.

Novolak: A thermoplastic prepolymer produced by the reaction of an insufficient amount of formaldehyde and phenol under acidic conditions.

Oleic acid: A monounsaturated acid, $C_{17}H_{33}COOH$.

Oleoresinous paint: A paint based on curable unsaturated oils.

Ortho group: A group in position 2 or 6 in a substituted benzene.

***P*-Phenyl phenol:**

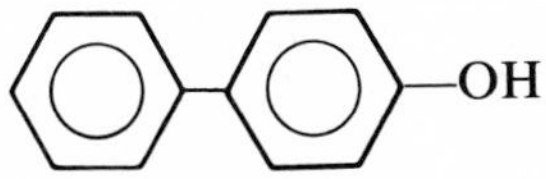

Phenol: C_6H_5OH.

Phenol–formaldehyde resin (PF): The reaction product of phenol and formaldehyde produced under controlled conditions.

Phenolic resin (PF): The reaction product of phenol and formaldehyde.

Plywood: A composite consisting of thin sheets of wood bonded together by an adhesive, such as a phenolic resin.

Polyester: A product obtained by the reaction of a dihydric alcohol and a dicarboxylic acid.

Polyurethane (PUR): The reaction product of a diisocyanate ($R(NCO)_2$) and a diol ($R(OH)_2$).

Prepolymer: A low-molecular-weight polymer that can be converted to a useful higher-molecular-weight polymer by heat or by the addition of a catalyst.

Primary hydroxyl group: A hydroxy group bonded to a carbon atom that is joined to two hydrogen atoms,

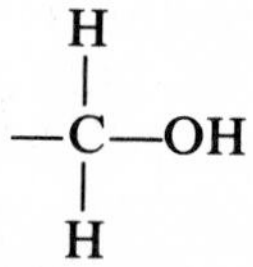

$$\begin{array}{c} H \\ | \\ -C-OH \\ | \\ H \end{array}$$

Silane: SiH_4, the simplest silicon hydride.

Silicone: A polysiloxane,

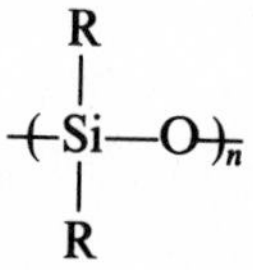

$$\begin{array}{c} R \\ | \\ \left(\!\!- Si-O \!-\!\right)_{\!n} \\ | \\ R \end{array}$$

Styrene: $C_6H_5CH{=}CH_2$.

Tartaric acid: A dihydroxy, dicarboxylic four-carbon compound.

TDI: Tolylene diisocyanate,

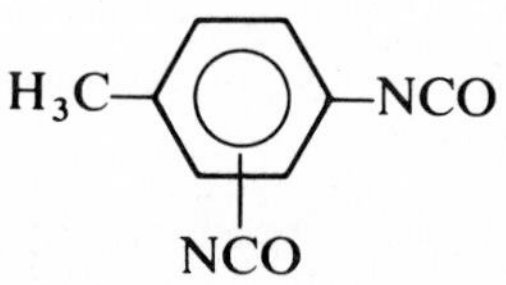

$$H_3C-\!\!\bigcirc\!\!-NCO$$
$$NCO$$

Thermoplastic: A linear or branched fusible polymer.

Thermoset: A cross-linked polymer.

Urea:

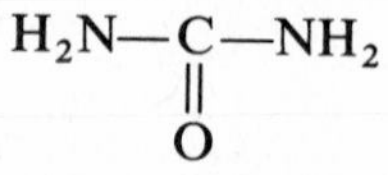

$$\begin{array}{c} H_2N-C-NH_2 \\ || \\ O \end{array}$$

Urea resin (UF): The reaction product of urea and formaldehyde.

Vinyl ester resin: A product obtained by the reaction of bisphenol A, ethylene oxide, and methacrylic acid.

Wood flour: Finely divided wood fibers obtained by attrition grinding.

REVIEW QUESTIONS

1. Is the reaction product of *p*-phenylphenol and formaldehyde a thermoset?

2. Which is more reactive in ester formation: a primary or a secondary alcohol?

3. What is the functionality of glycerol?

4. What is the functionality of ethanol?

5. What is the functionality of *m*-phenylphenol with formaldehyde?

6. What is the function of hexamethylenetetramine in a novolak molding compound?

7. Which will produce a thermoset when heated: a novolak or a resole resin?

8. What is the difference between wood flour and sawdust?

9. What is the functionality of urea?

10. How many carbon–carbon double bonds are in a molecule of oleic acid?

11. What is the functionality of melamine?

12. What is the repeating group in a polyurethane?

13. What is the formula for bisphenol A?

14. What is the structural difference between fumaric acid and maleic acid?

15. Silicon dioxide,

$$\begin{array}{ccc} & | & | \\ +Si & —O—Si—O+ \\ & | & | \\ & & O \\ & & | \\ & & —Si— \\ & & | \end{array}$$

sand is abrasive but silicones are lubricants. Why?

BIBLIOGRAPHY

Boenig, H. V. (1964). *Unsaturated Polyesters: Structure and Properties.* Amsterdam: Elsevier.

Bruins, P. R. (1976). *Unsaturated Polyester Technology.* New York: Gordon & Breach.

Carswell, T. S. (1957). *Phenoplasts.* New York: Interscience.

Dombrow, B. A. (1957). *Polyurethanes.* New York: Reinhold.

Dostal, C. A. (Ed.). (1987). *Composites.* Metals Park, OH: AMC International.

Doyle, E. N. (1971). *The Development and Use of Polyurethane Products.* New York: McGraw–Hill.

Elias, H. G. (1987). *Megamolecules.* Berlin: Springer-Verlag.

Lawrence, J. R. (1960). *Polyester Resins.* New York: Reinhold.

Martens, C. R. (1961). *Alkyd Resins.* New York: Reinhold.

Martin, R. W. (1958). *The Chemistry of Phenolic Resins.* New York: Wiley.

McGregor, R. R. (1959). *Silicones and Their Use.* New York: Reinhold.

Noll, N. (1968). *Chemistry and Technology of Silicones.* New York: Academic Press.

Potter, N. G. (1970). *Epoxide Resins.* London: Plastics and Rubber Institute.

Rochow, R. G. (1951). *An Introduction to the Chemistry of the Silicones.* Chapman & Hall.

Saunders, J. H., and Frisch, K. C. (1962, 1964). *Polyurethanes—Chemistry and Technology,* Vols I and II. New York: Wiley.

Seymour, R. B. (Ed.). (1982). *History of Polymer Science and Technology.* New York: Dekker.

Seymour, R. B., and Carraher, C. E. (1987). *Polymer Chemistry: An Introduction* (2nd ed.). New York: Dekker.

Seymour, R. B., and Deanin, R. D. (Eds.). (1987). *Polymeric Composites: Their Origin and Development.* Utrecht, Holland: VNU Science Press.

Seymour, R. B., and Mark, H. F. (1988). *Applications of Polymers.* New York: Plenum.

Skeist, I. (1960). *Epoxy Resins.* New York: Reinhold.

Vale, C. P., and Taylor, W. G. R. (1964). *Aminoplastics.* London: ILIFFE.

Weeton, J. W., Peters, D. M., and Thomas, K. L. (1987). *Engineer's Guide to Composite Materials.* Metals Park, OH: ASM, Inc.

Whelan, A., and Hudson, J. A. (1975). *Developments in Thermosetting Plastics.* London: Applied Science Publishers.

Whitehouse, A. A. K., Pritchette, G. K., and Barnett, G. (1967). *Phenolic Resins.* London: Plastics and Rubber Institute.

ANSWERS TO REVIEW QUESTIONS

1. No, *p*-phenylphenol is bifunctional.

2. Primary alcohol.

3. 3.

4. 1.

5. 3, the two ortho and para groups are reactive.

6. It supplies formaldehyde when heated.

7. A resole resin.

8. Wood flour has a fibrous structure.

9. 4.

10. 1.

11. 6.

12. RNHCOOR.

13. $HOC_6H_4C(CH_3)_2C_6H_4OH$.

14. Maleic acid is a cis isomer and fumaric acid is a trans isomer.

15. The groups in silicones in contact with another surface are oily alkyl groups and not abrasive siloxane groups (—Si—O—Si—O) which are highly crosslinked and rigid.

CHAPTER XIV _____

Engineering Plastics

14.1   INTRODUCTION

The term "engineering plastic" is used to describe a material that can be easily and readily machined, milled, drilled, or otherwise have its shape modified while remaining in the solid state—much like metal. Such plastics must be rigid and tough over a temperature range of 32 to 212°F, but they must deform when the load or shock is too great, yielding and deforming rather than simply cracking or breaking in two. This property is called impact resistance, that is, the ability to withstand shock without undergoing brittle failure.

Reinforced thermosets are often employed in engineering applications but such thermosets, which were covered in Chapter 13, are not usually considered as engineering plastics by all plastics engineers. The term "high-performance" or "engineering plastic" simply means that certain plastics perform exceptionally well under adverse conditions under which general-purpose thermoplastics perform poorly.

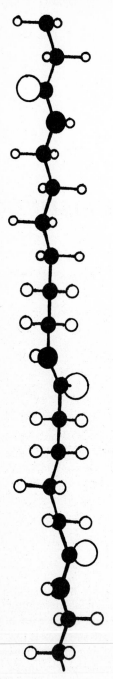

Figure 14.1. Ball-and-stick model of nylon 6,6.

14.2 NYLONS

Although nylon 6,6 (Figure 14.1) and nylon-6 are used primarily for fibers, they are also used as engineering plastics at an annual rate of over 250,000 tons in the United States. The aromatic polyamides (nylons), such as Kevlar, have heat deflection temperatures (HDUL) as high as 480°F.

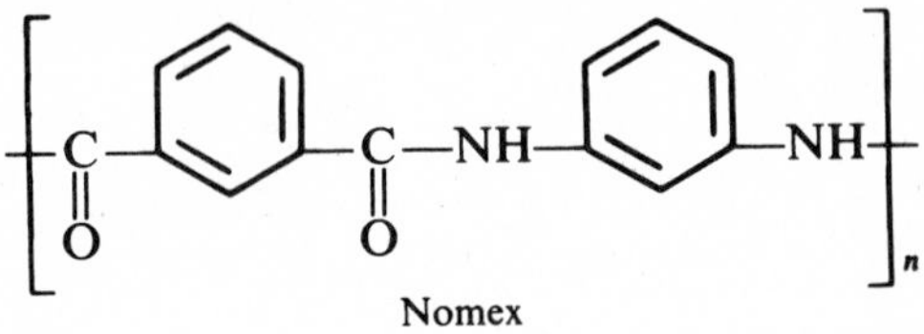

Nomex

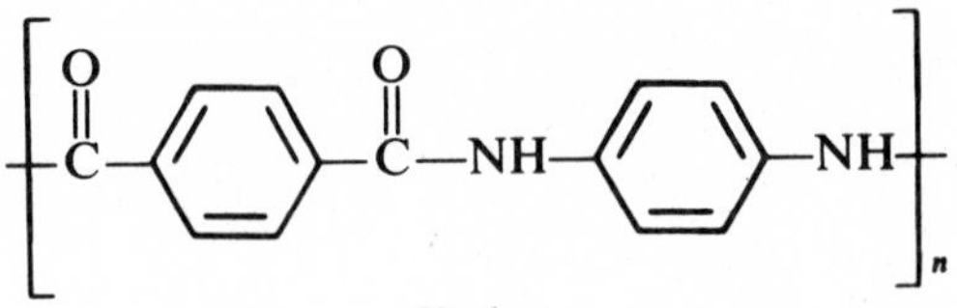

Kevlar

The HDUL is the temperature at which a simple beam of plastic deflects a definite distance under a specified load. Thus, Nomex and Kevlar, which have high HDUL, can be used in place of ceramics and metals in many instances.

Kevlar is employed in tires and "bulletproof" vests because of its strength and lightness. Kevlar has five times the strength of steel. The aromatic polyamides, aramids, are also used for the production of heat-resistant clothing and in smoke stack filtration. Kevlar has a HDUL above 570°F and can withstand temperatures to 930°F without decomposition. These aromatic polyamides are now being used in blends and composites and the polymers themselves can be synthesized and processed to give fibers with a wide range of physical properties.

Kevlar–graphite fibers are employed, along with fibrous glass and other fibers, in the composite materials called fiber-reinforced plastics, or FRPs. Their lightness, high stiffness, and superior strength allow them to replace metal in many applications. Other everyday applications for the Kevlar–graphite fibers are in the production of fishing rods, golf club shafts, skis, ship masts, bicycle frames, and tennis rackets. Significant amounts are now employed in the production of Boeing's 757 and 767 airplanes. These Kevlar–graphite composites are dark in color and can be coated, but since the composites need not be protected from the elements, painting is generally omitted.

Polyamides are also produced by the reaction of aromatic diisocyanates, such as tolylene diisocyanate (TDI), with aromatic diamines, such as phenylenediamine. Another heat-resistant nylon is produced by the reaction injection molding (RIM) process similar to that used to produce large polyurethane (PUR) articles, such as automobile bumpers.

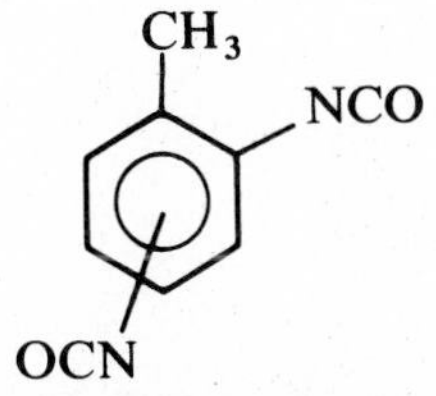

Tolylene diisocyanate

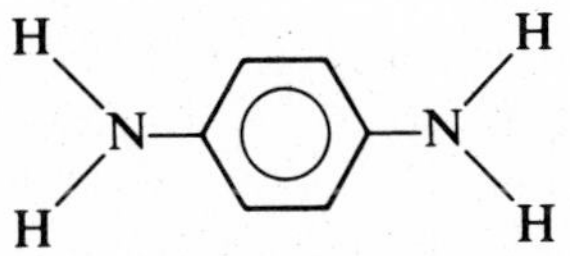

p-Phenylenediamine

Table 14.1   Comparative Data for Nylon 6,6 and Reinforced Nylon 6,6 (Zytel, Vydine)

Property	Unfilled	30% Glass Reinforced
Tensile strength (psi)	8,500	26,000
Elongation (%)	150	4
Compressive strength (psi)	4,900	24,000
Flexural strength (psi)	15,000	41,000
Flexural modulus (psi)	290,000	1,300,000
Notched Izod impact (ft-lb/in. of notch)	1	2.4
Coefficient of expansion (cm/cm, °C)$(10)^{-6}$	50	17.5
Heat deflection temp. (°F)	135	485

In the RIM PUR process, a dihydric alcohol and diisocyanate are injected simultaneously into a mold cavity, where they react to form the polymer in the shape of the mold. Polypropylene glycol and caprolactam are also injected into the molds in the RIM nylon process (Nyrim).

$$\underset{\text{Dialcohol}}{HO-R-OH} + \underset{\text{Diisocyanate}}{OCN-R'-NCO} \longrightarrow \underset{\text{Polyurethane}}{\left(O-R-O-\overset{\overset{O}{\|}}{C}-\overset{\overset{H}{|}}{N}-R'-\overset{\overset{H}{|}}{N}-\overset{\overset{O}{\|}}{C}\right)}$$

Other nylon-related high-performance polymers are based on copolymers of polyamides and imides and blends of nylon 6,6 with nylon-6 or poly(ethylene terephthalate) (PET). All of these commercial high-performance plastics can be fabricated by the injection molding process. Nylon plastics are used for bicycle wheels, lawnmower blades, roller skate wheels, and crankcases. The properties of nylons are shown in Table 14.1.

14.3 POLYACETALS

In 1859 Butlerov recognized that a polymer precipitates in aqueous formaldehyde solutions (formalin), but little was done with this polymeric sediment except to discard it. However, in the 1950s, polymer chemists at duPont produced this type of polymer from pure formaldehyde ($H_2C{=}O$) under controlled conditions and obtained a linear high-molecular-weight polyoxymethylene (POM).

Since POM decomposes when heated, the terminal hydroxyl groups (OH) were esterified by acetic anhydride in a process called capping. The stable capped POM is sold under the trade name of Delrin.

$$\underset{\substack{\text{Acetic} \\ \text{anhydride.}}}{CH_3-\overset{\overset{O}{\|}}{C}-O-\overset{\overset{O}{\|}}{C}-CH_3} + \underset{\substack{\text{Polyoxymethylene (showing the} \\ \text{end group noted by asterisks)}}}{\overset{*}{HO}\,CH_2{\left(CH_2-O\right)_n}CH_2{\overset{*}{O}}\,H}$$

$$\underset{\substack{\text{Capped POM} \\ \text{(Delrin)}}}{CH_3-\overset{\overset{O}{\|}}{C}-O-CH_2{\left[CH_2-O\right]_n}CH_2-O-\overset{\overset{O}{\|}}{C}-CH_3}$$

Table 14.2   Properties of High-Performance Plastics

Property	ASTM Method	Acetal Copolymer	Acetal Homopolymer
Specific gravity (g/cm)	D792	1.410	1.425
Tensile strength at yield (psi)	D638	8,800	10,000
Elongation at break (%)	D638	60	25
Tensile modulus (psi, $\times 10)^5$	D638	4.10	5.20
Flexural strength (psi)	D790	13,000	14,100
Flexural modulus (psi, $\times 10$)	D790	3.75	4.10
Fatigue endurance limit (psi/no. of cycles)	D671	4200 per	5000 per
Compressive stress at 10% dilation (psi)	D695	16,000	18,000
Rockwell hardness (M)	D785	80	94
Notched Izod impact (ft-lb/in. of notch)	D256	1.3	1.4
Tensile impact (ft-lb/in.^{2})	D1822	70	94
Water absorption, 24 h immersion (%)	D570	0.22	0.26
Tabor abrasion, 1000-g load, Cs-17 wheel (mp/1000 cycles)	D1044	14	20

Polymer chemists at Celanese also obtained a stable polyacetal by copolymerizing formaldehyde with homologous compounds, such as ethylene oxide

$$H_2C\text{———}CH_2$$
$$\diagdown \diagup$$
$$O$$

The copolymer (Celcon) is also stable at relatively high temperatures.

These injection moldable high-performance plastics, which have HDUL values of about 300°F, are produced at an annual rate of 65,000 tons. Polyacetals are used for molded door handles, tea kettles, pump impellers, shoe heels, and plumbing fixtures. The properties of these polymers are shown in Table 14.2.

14.4   POLYCARBONATES

Einhorn produced a high-melting, clear polyester of carbonic acid in 1898 by the reaction of phosgene and hydroquinone or resorcinol. Commercial polycarbonates (PC) were produced in the 1950s by the General Electric Company and Bayer Company by the condensation of bisphenol A and phosgene ($COCl_2$). This tough, transparent polymer (Lexan, Merlon) is produced at an annual rate of 130,000 tons.

Polycarbonates are processed by all the standard plastic methods (Section 12.9). They are used in glazing (40%), appliances (15%), signs, returnable bottles, solar collectors, business machines, and electronics. They show good creep resistance, good thermal stability, and a wide range of use temperatures (about −60 to 270°F). Coatings are generally used on PC sheets to improve mar and chemical resistance.

Bisphenol A   Phosgene

Polycarbonate

Blends of PC and ABS (Bayblend) or with poly(butylene terephthalate) (Xenoy) are tough, heat-resistant (HDUL 390°F) plastics. The properties of polycarbonates are shown in Table 14.3.

14.5 POLY(ARYL ESTERS)

The heat resistance of Carothers' poly(alkyl esters) was not high enough to withstand the temperature of the hot ironing process. However, poly(aryl esters) (now simply called polyesters), which were produced by J. R. Whinfield and J. T. Dickson in 1941 by the condensation of ethylene glycol and terephthalic acid, met most of the specifications for a useful synthetic fiber (PET). Nevertheless, because of inferior molding machines and inadequate plastic technology, it was not possible to injection mold these polyesters until recently.

Since the ease of processing and fabricating polyesters is related to the number of methylene groups (CH_2) in the repeating units, polymer chemists in several firms produced poly(butylene terephthalate) (PBT) for use as a moldable engineering plastic. PBT has four methylene groups whereas PET has only two of these

Table 14.3 Comparative Data for Polycarbonate and Reinforced Polycarbonate (Lexan, Merlon)

Property	Unfilled	Reinforced 30% Glass
Tensile strength (psi)	9,500	19,000
Elongation (%)	110	4
Compressive strength (psi)	12,500	18,000
Flexural strength (psi)	13,500	23,000
Flexural modulus (psi)	340,000	11,000,000
Notched Izod impact (ft-lb/in. of notch)	14	2
Coefficient of expansion (cm/cm, °C) $(10)^{-6}$	68	22
Heat deflection temp. (°F)	270	295

Figure 14.2. A 1.5-L bottle molded from Kodapak PET, a polyester manufactured by Eastman Chemical Products, Inc.

Table 14.4   Comparative Data for Poly(ethylene Terephthalate) and Reinforced PET (Rynite)

Property	Unfilled	30% Glass Reinforced
Tensile strength (psi)	9,500	23,000
Elongation (%)	150	3
Compressive strength (psi)	13,000	23,000
Flexural strength (psi)	16,000	33,000
Flexural modulus (psi)	400,000	1,300,000
Notched Izod impact (ft-lb/in. of notch)	0.5	2
Coefficient of expansion (cm/cm, °C) $(10)^{-6}$	65	30
Heat deflection temp. (°F)	100	435

flexibilizing groups in each repeating unit.

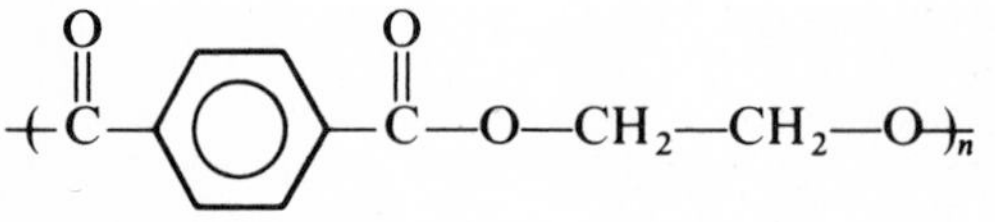

$$+\overset{\overset{\displaystyle O}{\parallel}}{C}-\hspace{-2pt}\left\langle \bigcirc \right\rangle\hspace{-2pt}-\overset{\overset{\displaystyle O}{\parallel}}{C}-O-CH_2-CH_2-O\hspace{-2pt}\Big)_{\overline{n}}$$

PET

$$+\overset{\overset{\displaystyle O}{\parallel}}{C}-\hspace{-2pt}\left\langle \bigcirc \right\rangle\hspace{-2pt}-\overset{\overset{\displaystyle O}{\parallel}}{C}-O-CH_2-CH_2-CH_2-CH_2-O\hspace{-2pt}\Big)_{\overline{n}}$$

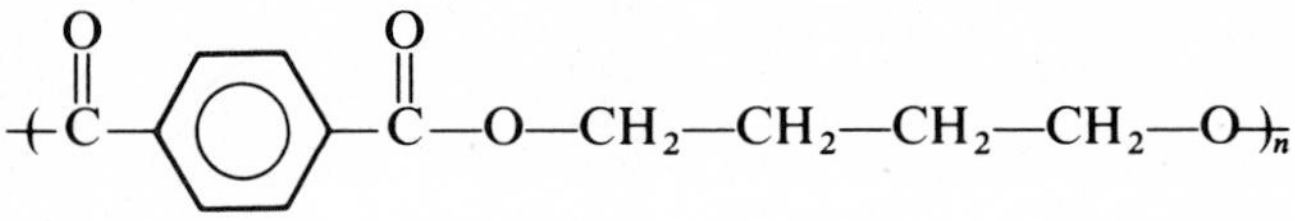

PBT

In addition to its use in soft drink bottles (Figure 14.2), PET is used for molded automobile parts. Over 500,000 tons of polyester engineering plastics are produced annually in the United States. The properties of saturated polyesters are shown in Tables 14.4 and 14.5.

Table 14.5   Comparative Data for Poly(butylene Terephthalate) and Reinforced PBT (Valox, Celanex, Petra)

Property	Unfilled	30% Glass Reinforced
Tensile strength (psi)	8,000	18,000
Elongation (%)	150	3
Compressive strength (psi)	12,000	21,000
Flexural strength (psi)	14,000	27,000
Flexural modulus (psi)	350,000	1,100,000
Notched Izod impact (ft-lb/in. of notch)	1	1.5
Coefficient of expansion (cm/cm, °C) $(10)^{-6}$	75	25
Heat deflection temp. (°F)	150	425

PBT has a melting point around 435°F and is generally processed at about 480°F, whereas PET has a melting point around 520°F. This is a direct consequence of the presence of the addition of two methylene units in PBT, which allows easier fabrication of PBT by injection molding and extrusion procedures and through blow molding. PBT is employed for under-the-hood automotive parts, including fuse cables, pump housings, and electrical connectors, and for selected automotive exterior parts.

A more hydrophobic, stiffer polyester was introduced in 1958 by Eastman–Kodak as Kodel polyester. It contains a cyclohexanedimethanol moiety in place of the simple methylene moiety present in PET and PBT. Copolyesters base on cyclohexanedimethanol and ethylene glycol as the diols are blow-molded into bottles for shampoos and liquid detergents. Other similar copolyesters are processed by extrusion into tough, clear fibers employed to package hardware and other heavy items.

Cyclohexanedimethanol   Dimethyl Terephthalate

Kodel

14.6 POLY(PHENYLENE OXIDE)

Poly(phenylene oxide) (PPO, Noryl), which is produced by the copper chloride-catalyzed oxidative coupling of a disubstituted phenol, was invented by A. Hay in the 1960s. The polymeric product is difficult to mold but the blend of PPO and polystyrene is readily injection molded. This modified PPO is produced at an annual rate of 90,000 tons in the United States. This unusual engineering plastic, which is also called poly(phenylene ether), is used for window frames, beverage glasses, electrical switches, business machines, solar energy collectors, and wheel covers. The properties of PPO are shown in Table 14.6.

2,6-Xylenol   Poly(phenylene oxide)

14.7 POLY(PHENYLENE SULFIDE)

Polyphenylene sulfide (PPS, Ryton) is produced by the condensation of p-dichlorobenzene and sodium sulfide at a rate of more than 35,000 tons annually. This

Table 14.6   Comparative Data for Poly(phenylene Oxide) and Reinforced PPO (Noryl)

Property	Unfilled	30% Glass Reinforced
Tensile strength (psi)	8,000	17,500
Elongation (%)	50	4
Compressive strength (psi)	14,000	18,000
Flexural strength (psi)	13,000	21,500
Flexural modulus (psi)	380,000	1,100,000
Notched Izod impact (ft-lb/in. of notch)	5	2
Coefficient of expansion (cm/cm, °C) $(10)^{-6}$	50	20
Heat deflection temp. (°F)	230	290

high-melting polymer (550°F) is used for quartz halogen lamps, pistons, circuit boards, and appliances. The properties of PPS are shown in Table 14.7.

$$n\,Na_2S_x \;+\; n\,Cl\!-\!\!\bigcirc\!\!-\!Cl \longrightarrow \big(\!\bigcirc\!-\!S_x\big)_{\!n}$$

Sodium sulfide   *p*-Dichlorobenzene   Poly(phenylene sulfide)

14.8   POLY(ARYL SULFONES)

Polysulfones form a family of engineering plastics with outstanding high-temperature stability, with a ceiling use temperature around 320°F (Tables 14.8 and 14.9). Examples are given in Figure 14.3. Since they are relatively high priced ($10/lb), they are employed only in specific applications where high-temperature stability is needed, including as electrical conductors and mechanical parts. Like Kevlar–graphite composites and fibers, polysulfones too are replacing metals. This is particularly true

Table 14.7   Comparative Data for Poly(phenylene Sulfide) and Reinforced PPS (Ryton)

Property	Unfilled	30% Glass Reinforced
Tensile strength (psi)	9,500	19,500
Elongation (%)	1.5	1
Compressive strength (psi)	16,000	21,000
Flexural strength (psi)	14,000	29,000
Flexural modulus (psi)	550,000	1,200,000
Notched Izod impact (ft-lb/in. of notch)	0.5	1.4
Coefficient of expansion (cm/cm, °C) (10)	49	22
Heat deflection temp. (°F)	275	485

Table 14.8 Comparative Data for Polysulfone and Reinforced Polysulfone (Vitrex, Udel)

Property	Unfilled	30% Glass Reinforced
Tensile strength (psi)	—	14,500
Elongation (%)	3	1.5
Compressive strength (psi)	40,000	19,000
Flexural strength (psi)	—	20,000
Flexural modulus (psi)	400,000	1,000,000
Notched Izod impact (ft-lb/in. of notch)	1	1
Coefficient of expansion (cm/cm, °C) (10)	55	25
Heat deflection temp. (°F)	345	350

Table 14.9 Ceiling (top) Use Temperatures for Selected Engineering Thermoplastics

Polymer	Ceiling Use Temperature[a]	
	°C	°F
Polybenzimidazole	400	752
Polyimides	260 (prolonged)	500
	480 (short)	895
Kevlar	500	932
Nomex	360	680
Udel	160 (prolonged)	320
	800 (short)	1472
PPS	150	302
PPO	130	266
PC	130	266
Nylon 6,6	100	212
PBT	90	194
ABS	80	176

[a]Unless noted, the temperatures are for extended use.

for complex-shaped objects since the polysulfones can be injection molded into these shapes without the machining and other procedures that are required for metals. Films and foil of polysulfones are used for flexible printed circuitry.

14.9 POLYIMIDES

Intractable polyimides (PI) produced by the condensation of pyromellitic anhydride and various polyimides have been available for several years under the trade names of Kaptan and Kinel (Table 14.10). Moldable PI is now available. The synthesis of

Trade Name	Polymer Unit	T_g (°F)
Astrel (3M Corp.)	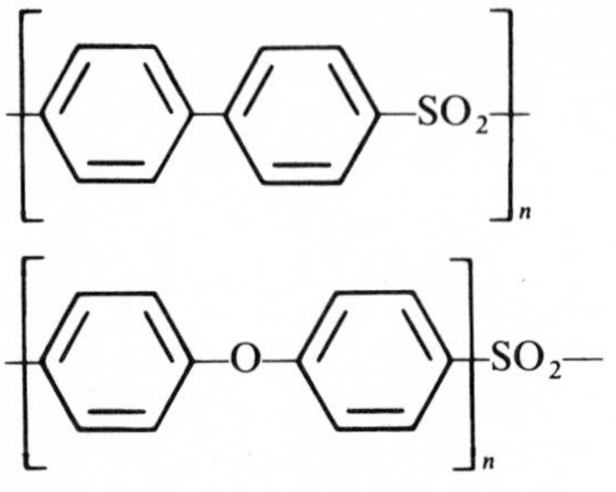	545
Poly(ether sulfone) 720 P (ICI)	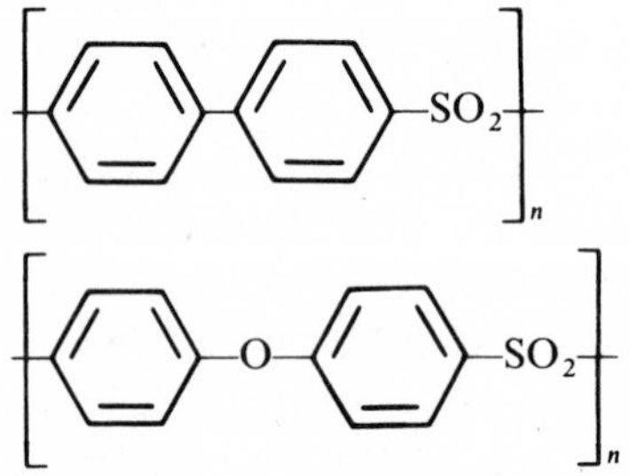	480
Poly(ether sulfone) 200 P (ICI)	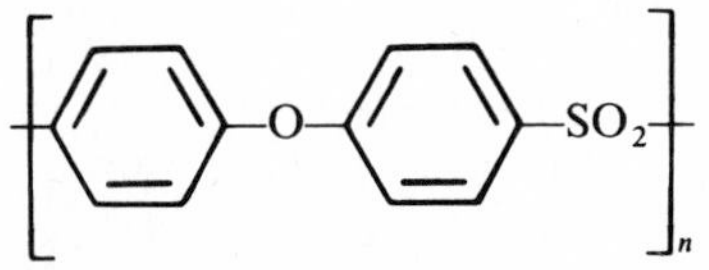	445
Udel (Union Carbide)	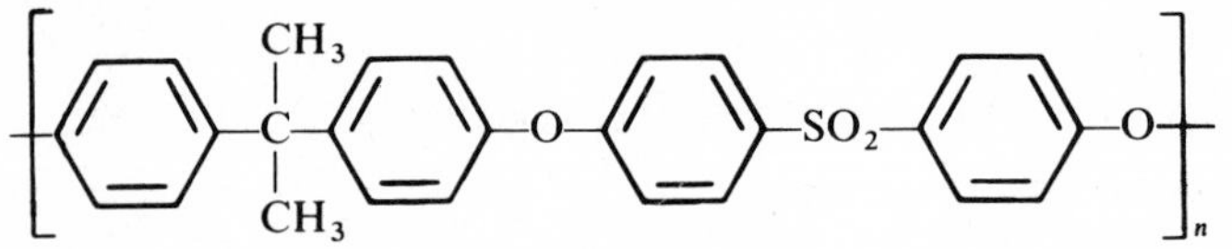	375

Figure 14.3. Commercially available polysulfones.

Table 14.10   Comparative Data for Polyimide and Reinforced PI

Property	Unfilled	30% Graphite Filled
Tensile strength (psi)	13,000	7,500
Elongation (%)	9	3
Compressive strength (psi)	35,000	17,500
Flexural strength (psi)	24,000	14,000
Flexural modulus (psi)	500,000	700,000
Notched Izod impact (ft-lb/in. of notch)	1.5	0.7
Coefficient of expansion (cm/cm, °C) $(10)^{-6}$	50	38
Heat deflection temp. (°F)	600	680

polyimides is shown in the following reaction.

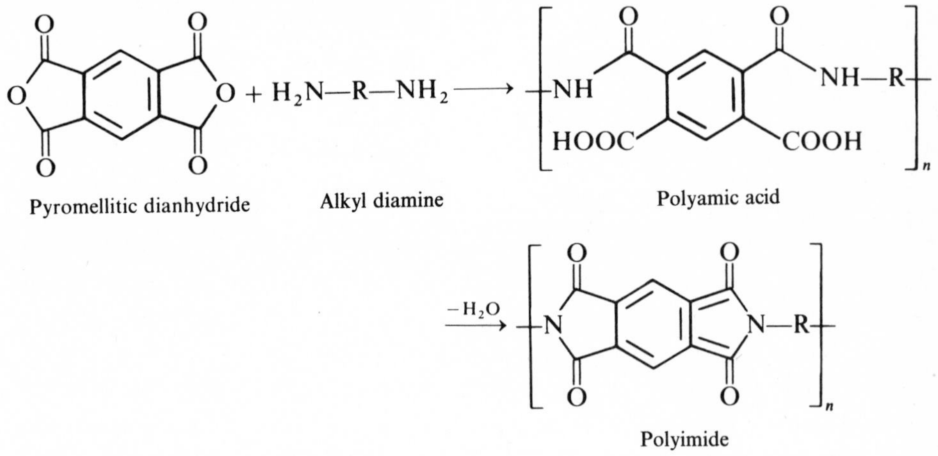
O
O
O + H_2N—R—NH_2
O
O
Pyromellitic dianhydride
Alkyl diamine
NH
NH—R
HOOC
COOH
n
Polyamic acid
$-H_2O$
N
N—R
Polyimide

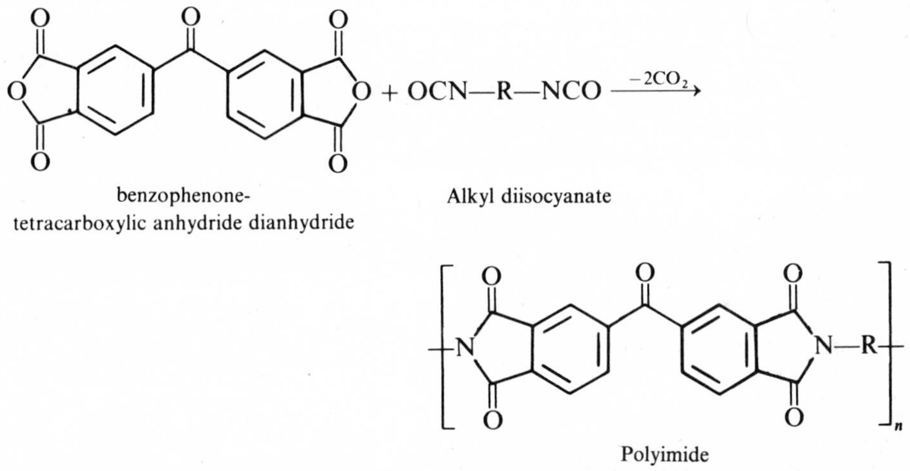
O + OCN—R—NCO
$-2CO_2$
benzophenone-
tetracarboxylic anhydride dianhydride
Alkyl diisocyanate
N
N—R
n
Polyimide

A polyether imide (PEI, Ultem) and a polyamide imide (PAI, Torlon) are available commercially (Table 14.11). The latter has been used to fabricate a Ford prototype engine.

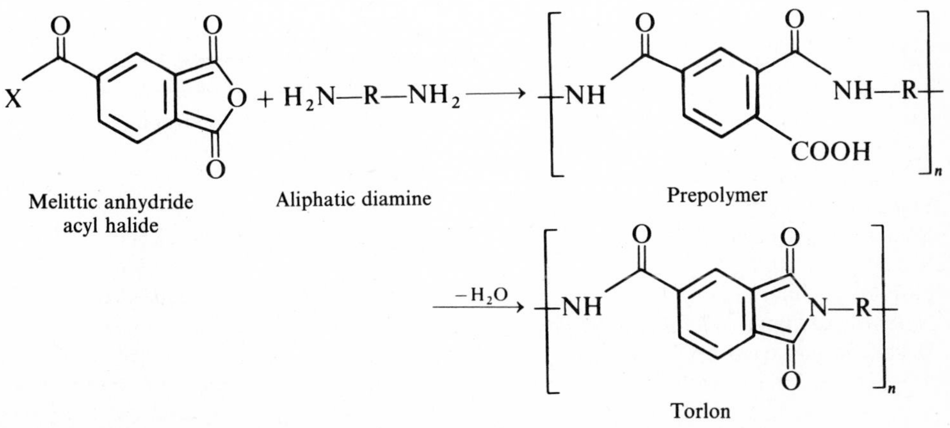
X
O + H_2N—R—NH_2
Melittic anhydride
acyl halide
Aliphatic diamine
NH
NH—R
COOH
n
Prepolymer
$-H_2O$
NH
N—R
Torlon

Table 14.11   Comparative Data for Poly(ether Imide) and Reinforced PEI

Property	Unfilled	30% Glass Filled
Tensile strength (psi)	15,200	24,500
Elongation (%)	7.5	0.2
Compressive strength (psi)	20,300	23,500
Flexural strength (psi)	21,000	33,000
Flexural modulus (psi)	480,000	1,200,000
Notched Izod impact (ft-lb/in. of notch)	1.0	2.0
Head deflection temp. (°F)	392	410

14.10   POLY(ETHER ETHER KETONE)

Poly(ether ether ketone) (PEEK) was invented by J. Rose, who is also the inventor of poly(ether sulfones). Because of the presence of sulfone (SO_2) and carbonyl ($C=O$) stiffening groups, both have excellent resistance to elevated temperatures. The ether groups contribute flexibility and moldability. A 60% glass mat-reinforced PEEK retains 60% of its flexural modulus at 570°F. PEEK is used for nuclear waste containers, jet engine components, and electrical appliances. The properties of PEEK are summarized in Table 14.12.

PEEK

14.11   OTHER ENGINEERING THERMOPLASTICS

Polysiloxanes can be elastomers, thermoplastics, or thermosets depending on the particular structure and intended end use.

Table 14.12   Comparative Data for Poly(ether Ether Ketone) and Reinforced PEEK

Property	Unfilled	30% Graphite Filled
Tensile strength (psi)	10,044	23,500
Elongation (%)	100	3
Compressive strength (psi)	—	22,500
Flexural strength (psi)	—	39,500
Flexural modulus (psi)	—	1,250,000
Notched Izod impact (ft-lb/in. of notch)	1.6	2.7
Heat deflection temp. (°F)	320	594

Polybenzimidazoles, which were developed for aerospace applications, possess good stiffness and strength but are generally difficult to process. They have one of the highest use temperatures (about 750°F) of any organic polymer.

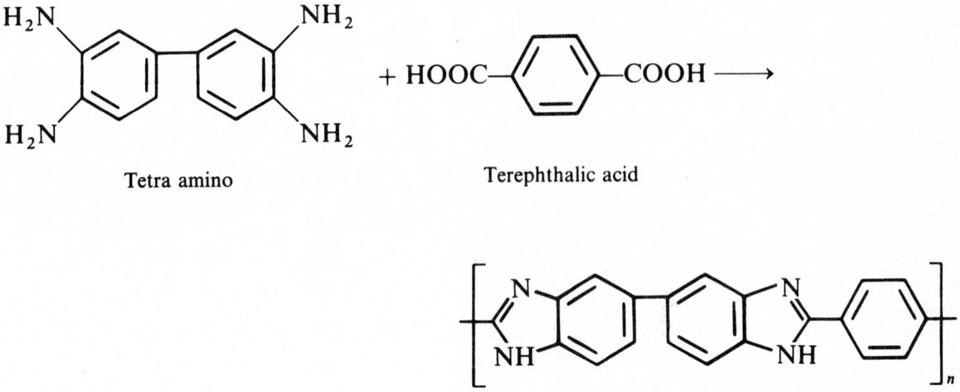

$$H_2N \quad \cdots \quad NH_2 \quad + \quad HOOC- \bigcirc -COOH \longrightarrow$$

Tetra amino   Terephthalic acid

Polybenzimidazole

Polyphosphazenes are in the early stages of development and have many potential and actual uses. They exhibit a broad service range temperature (-85 to $250°F$), have outstanding resistance to fuels, oils, and chemicals, and possess good mechanical properties.

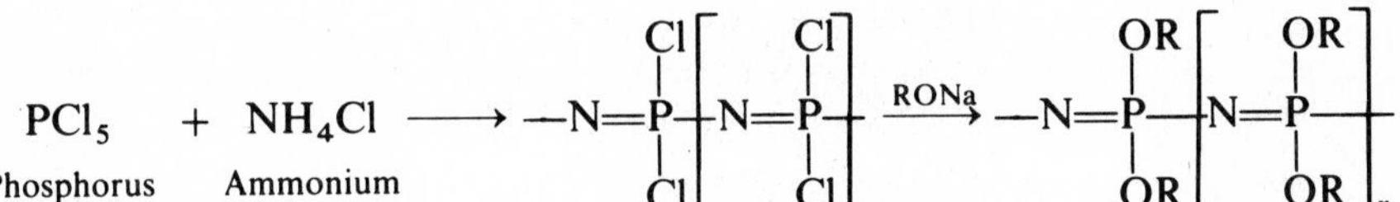

$$PCl_5 \quad + \quad NH_4Cl \longrightarrow$$

Phosphorus   Ammonium
pentachloride   chloride

Polyphosphazene

Polytetrafluoroethylene and other highly halogenated (i.e., containing Group VII substituents like chlorine, fluorine, and bromine) polymers are used as coatings as well as thermoplastics. Polytetrafluoroethylene (PTFE, Teflon) is insoluble in all organic solvents and is processed only by ram extrusion, cutting, machining, and sinter molding techniques.

$$-\!\!\left[CF_2CF_2\right]_{\!\!n}$$

Polytetrafluoroethylene (PTFE)

PTFE has outstanding resistance to chemical attack. In fact, pipes that are employed to convey molten, liquid sodium metal are made from PTFE. Although PTFE is normally stable to metallic sodium, a very violent, exothermic reaction can occur when sodium strips off the fluoride ions on PTFE to form sodium fluoride, which may create a fire and/or explosion. PTFE has a high impact strength but it cold flows (creeps) and has low wear resistance and tensile strength. Because of its excellent "lubricity" (low friction) and outstanding hydrophobic (water-hating) nature, it is widely employed in easy-clean and nonstick cookware, such as frying pans, muffin pans, and cake pans.

PTFE is also used in high-temperature cable insulation and molded electrical applications. Reinforced PTFE is used as seals and bushings in compressor hydraulic applications, pipe lines, and automotive applications, as a specialty tape to ensure closure of pipe fittings, as a seal for gasket applications, and in the laboratory as a covering for magnetic stirrers, on stopcocks in liquid delivery devices, and on laboratory ware such as beakers, flasks, and condensers.

GLOSSARY

Ablative: A process in which the surface of the plastic is degraded and removed.

ABS: A tough copolymer containing repeating units of acrylonitrile, butadiene, and styrene.

Arylate: Trade name for the reaction product of bisphenol A and terephthalic acid.

Astrel: Trade name for a poly(aryl sulfone).

Bayblend: Trade name for a blend of PC and flexibilizing polymers.

Bisphenol A:

Cadon: Trade name for a heat-resistant terpolymer of styrene, acrylonitrile, and maleic anhydride.

Capping: Reaction with end groups in a polymer.

Carbonyl group: $C{=}O$.

Celcon: Trade name for an acetal copolymer.

Delrin: Trade name for POM.

Engineering plastic: A plastic with a high modulus and high melting point that can be used in place of metals in some applications.

Heat deflection temperature (HDUL): The temperature at which a heated beam deflects a specific distance.

Heterocyclic: A cyclic compound consisting of carbon and other atoms such as nitrogen.

Impact resistance: Toughness, the ability to withstand mechanical shock.

Inorganic polymer: A polymer, such as siloxanes and phosphazenes, that does not have carbon atoms in its backbone.

Kaptan: Trade name for PI.

Kevlar: Trade name for an aromatic nylon.

Kinel: Trade name for PI.

Ladder polymer: A polymer in which the backbone is a double chain.

Noryl: Trade name for a blend of polystyrene and PPO.

Nylon: A polyamide produced by the condensation of a diamine and a dicarboxylic acid or by the polymerization of a lactam.

Nylon 6: The product of the polymerization of caprolactam.

Nylon 6,6: The reaction product of hexamethylenediamine and adipic acid.

Phosgene: $COCl_2$.

Polyacetal (POM): A polymer of formaldehyde.

Polybenzimidazole (PBI): A heat-resistant heterocyclic polymer.

Polybutylene terephthalate (PBT): The reaction product of butylene glycol and terephthalic acid.

Polycarbonate (PC): The reaction product of phosgene and bisphenol A.

Poly(ether ether ketone) (PEEK): An aromatic thermoplastic having ether and carbonyl groups.

Poly(ethylene terephthalate) (PET): The reaction product of ethylene glycol and terephthalic acid.

Polyimide (PI): A polymer produced by the condensation of pyromellitic anhydride and diamines.

Polyoxymethylene (POM): Polyacetal.

Poly(phenylene oxide) (PPO): A polymer produced by the copper chloride-catalyzed oxidative coupling of 2,6-xylenol.

Poly(phenylene sulfide) (PPS): A polymer produced by the condensation of sodium sulfide (NaS) and *p*-dichlorobenzene.

Polyurethane (PUR): The reaction product of a diisocyanate and a dihydric alcohol.

Quartz: Silicon dioxide.

Reaction injection molding (RIM): A process in which the reactants are introduced and polymerized in the mold.

Ryton: Trade name for PPS.

Silicone: Siloxane.

Siloxane:

$$-O-[Si-O]_n$$

(with R groups above and below Si)

Tolylene diisocyanate (TDI):

(structure with H_3C, NCO, and NCO groups on benzene ring)

Torlon: Trade name for a poly(amide imide).
Udel: Trade name for a poly(aryl sulfone).
Ultem: Trade name for a poly(ether imide).
Victrex: Trade name for a poly(aryl sulfone).
Xenoy: Trade name for a blend of PC and PBT.

REVIEW QUESTIONS

1. Which of the following are engineering resins: polystyrene, polyimide, nylon, polycarbonate?

2. Which has the higher melting point: nylon 6,6 or Kevlar?

3. What is the big advantage of RIM?

4. Which has the higher melting point: PET or PBT?

5. Why must polyacetal (POM) be capped?

6. What is the advantage of a clear polycarbonate sheet over a sheet of poly(methyl methacrylate)?

7. What is the advantage of polymer blends?

8. Which of the following is a stiffening group in a polymer chain: SO_2, CO, CH_2, O?

9. What polymer is used in prototype automobile combustion engines?

10. What is a ladder polymer?

11. What is the principal difference in the structure of a quartz or silica sand and a siloxane polymer (silicone)?

12. Why is polyphosphazene preferred over natural rubber in the Alaskan oil fields?

BIBLIOGRAPHY

Boeke, P. V. (1983). Polyphenylene sulfide. *Modern Plastics*, **61**(10A), 78.

Caughey, E. C., *et al.* (1984). Nylon. *Modern Plastics*, **61**(10A), 32.

Cekis, G. V. (1984). Polyamide-imide. *Modern Plastics*, **61**(10A), 36.

Dickinson, B. L. (1984). Polyarylate. *Modern Plastics*, **61**(10A), 45.

Epel, J. N., Margolis, J. M., Newman, S., and Seymour, R. B. (Eds.). (1988). *Engineering Plastics*. Metals Park, OH: ASM International.

Feth, G. (1984). Polyphenylene oxide. *Modern Plastics*, **61**(10A), 27.

Floyd, D. E. (1958). *Polyamide Resins*. New York: Reinhold.

Fox, D. W. (1975). In J. K. Craver and R. W. Tess (Eds.), *Applied Polymer Science* (Chap. 30). Washington, DC: American Chemical Society.

Jackson, K. A., *et al.* (1984). Acetals. *Modern Plastics*, **61**(10A), 8.

Kohan, M. L. (1973). *Nylon Plastics*. New York: Wiley.

Nitschke, C. (1984). Thermoplastic polyesters. *Modern Plastics*, **61**(10A), 50.

Page, S. L. (1984). Polycarbonate. *Modern Plastics*, **61**(10A), 40.

Rigby, R. B. (1984). Polyether sulfone. *Modern Plastics*, **61**(10A), 99.

Seymour, R. B. (1975). *Modern Plastics Technology* (Chap. 15). Reston, VA: Reston.

Seymour, R. B. (1982). *Plastics vs. Corrosion*. New York: Wiley.

Seymour, R. B. (1987). *Polymers for Engineering Applications*. Metals Park, OH: ASM International.

Seymour, R. B. (1989). *Engineering Polymer Source Book*. New York: McGraw–Hill.

Seymour, R. B. (1989). *Polymeric Composites*. Utrecht, Holland: VNU Science Press.

Seymour, R. B., and Deanin, R. D. (1987). *History of Polymer Composites*. Utrecht, Holland: VNU Science Press.

Seymour, R. B., and Kirshenbaum, G. S. (1986). *High Performance Polymers: Their Origin and Development*. New York: Elsevier.

Sheh, T. M. (1984). Polyimide. *Modern Plastics*, **61**(10A), 75.

Sinker, S. M. (1984). Acetal copolymers. *Modern Plastics*, **61**(10A), 12.

ANSWERS TO REVIEW QUESTIONS

1. Polyimide, nylon, and polycarbonate.

2. Kevlar, it is an aromatic nylon.

3. Large molded parts can be made in relatively inexpensive molds in one step.

4. PET. (PBT has more methylene flexibilizing groups in its repeating unit.)

5. The uncapped polymer decomposes to formaldehyde when heated.

6. Polycarbonate is tougher.

7. They are more readily molded and produced in available processing equipment.

8. SO_2 and CO; CH_2 and O are flexibilizing groups.

9. Poly(amide imide).

10. A polymer with a double polymer chain.

11. They have similar backbones but the siloxane polymer has organic pendant groups on the silicon atoms.

12. The temperature in the Alaskan oil fields is often below the T_g of natural rubber but above the T_g of polyphosphazenes. The latter are flexible at very low temperatures at which natural rubber is brittle.

Inorganic Polymers

15.1 INTRODUCTION

Just as polymers are abundant in the world of organic chemistry, they also abound in the world of inorganic chemistry. Inorganic polymers are the major components of soil, mountains, cement (concrete), and glass.

The first man-made, synthetic polymer was probably alkaline silicate glass, used in the Badarian period in Egypt (about 12,000 B.C.) as a glaze that was applied to steatite after it had been carved into various animal or ornamental shapes. Faience, a composite containing a powdered quartz or steatite core covered with a layer of opaque glass, was employed from about 9000 B.C. to make decorative objects. The earliest known piece of regular (modern type) glass has been dated at 3000 B.C. and is a lion's amulet found at Thebes and now housed in the British Museum. It is a blue opaque glass partially covered with a dark-green glass. Transparent glass appeared about 1500 B.C. Several fine pieces of glass jewelry were found in Tutankhamen's tomb (about 1300 B.C.), including two birds' heads of light-blue glass incorporated into the gold pectoral worn by the Pharaoh.

15.2 PORTLAND CEMENT

Portland cement is the least expensive and most widely used synthetic inorganic polymer. It is employed as the basic nonmetallic, nonwoody material in construction. Concrete highways and streets span our countryside and concrete skyscrapers crowd the urban skyline. Less spectacular uses include sidewalks, fence posts, and parking bumpers.

The name portland is derived from the cement having the same color as the natural stone quarried on the Isle of Portland, south of Great Britain. The word cement comes from the Latin word *caementum*, which means pieces of rough, uncut stone. Concrete comes from the Latin word *concretus*, meaning to grow together. Common cement consists of anhydrous, crystalline calcium silicates, the major ones being tricalcium silicate (Ca_3SiO_5) and β-dicalcium silicate (Ca_2SiO_4); lime (CaO, 60%) and alumina (a complex aluminum-containing silicate, 5%) are also present.

When anhydrous (without water) cement mix is mixed with water, the silicates react by forming hydrates (compounds containing water and calcium hydroxide, $Ca(OH)_2$). Hardened portland cement contains about 70% cross-linked calcium silicate hydrate and 20% crystalline calcium hydroxide.

$$2Ca_3SiO_5 + 6H_2O \rightarrow Ca_3Si_2O_73H_2O + 3Ca(OH)_2$$

$$2Ca_3SiO_4 + 4H_2O \rightarrow Ca_3Si_2O_73H_2O + Ca(OH)_2$$

Calcium silicate   Calcium hydroxide
trihydrate

The manufacture of portland concrete consists of three basic steps—crushing, burning, and finish grinding. Most cement plants are located near limestone ($CaCO_3$) quarries, since this is the major source of lime. Lime may also come from oyster shells, chalk, and a type of clay called marl. The silicates and alumina are derived from clay, silica sand, shale, and blast-furnace slag.

The powdery mixture is fed directly into rotary kilns for burning at 2700°F. These cement kilns are the largest pieces of moving machinery used in industry and can be 25 ft in diameter and 750 ft long. The heat changes the mixture into particles called clinkers, which are about the size of a marble. The clinkers are cooled and reground, with the final grinding producing portland cement that is finer than flour. The United States produces over 60 million tons of portland cement a year.

15.3 OTHER CEMENTS

There are a number of cements that are specially formulated for specific uses. Air-entrained concrete contains small air bubbles that were formed by addition of soaplike resinous materials to the cement or to the concrete when it is mixed.

Lightweight concrete may be made through use of lightweight fillers such as clays and pumice in place of sand and rocks or through the addition of chemical foaming agents, which produce air pockets as the concrete hardens.

Reinforced concrete is made by casting concrete about steel bars or rods. Most large cement structures such as bridges and skyscrapers employ reinforced concrete.

Prestressed concrete is typically made by casting concrete about steel cables stretched by jacks. After the concrete hardens, the tension is released, resulting in the entrapped cables compressing the concrete. Steel is stronger when tensed and concrete is stronger when compressed. Thus prestressed concrete takes advantage of both of these factors. Archways and bridge connections are often made from prestressed concrete.

There are non-portland cements as well. Calcium-aluminate cement has a much higher percentage of alumina than portland cement. Its active ingredients are lime and alumina. In the United States, it is manufactured under the trademark of Lumnite. Its major advantages are the rapidity of hardening and high strength within a day or two.

Magnesia cement is largely composed of magnesium oxide (MgO). In practice, the magnesium oxide is mixed with fillers and rocks and an aqueous solution of magnesium chloride. This cement sets up (hardens) within 2 to 8 h and is employed for flooring.

Gypsum, or hydrated calcium sulfate ($CaSO_4 \cdot 2H_2O$), serves as the base of a number of products, including plaster of Paris (also known as molding plaster, wall plaster, and finishing plaster), Keen's cement, Parisian cement, and Martin's cement.

15.4   SILICON DIOXIDE (AMORPHOUS)—GLASS

Silicon dioxide (SiO_2) is the repeating general structural formula for most rock and sand, and for the material we refer to as glass. The term glass can refer to many materials but here we will use the ASTM definition: glass is an inorganic product of fusion that has been cooled to a rigid condition without crystallization. We will consider silicate glasses and the common glasses used for electric light bulbs, window glass, drinking glasses, glass bottles, glass test tubes and beakers, and glass cooking ware.

Glass has many desirable properties. It ages (changes chemical composition and/or physical property) slowly, typically retaining its fine optical and hardness-related properties for centuries. Glass is referred to as being a supercooled liquid, or a very viscous liquid. Indeed it is a slow-moving liquid as confirmed by sensitive measurements carried out in many laboratories. This information is corroborated by the observation that the old stained glass windows adorning European cathedrals are a little thicker at the bottom of each small, individual piece than at the top of the piece. For our purposes, though, we should treat glass as a brittle solid that shatters on sharp impact.

Glass is mainly silica sand and is made by heating silica sand and powdered additives together in a specified manner and proportion, much as one bakes a cake. This recipe describes the items to be included, amounts, mixing procedure (including sequence), oven temperature, and heating time. The amounts and nature of additives all affect the physical properties of the final glass.

Typically, cullet—recycled or waste glass—is added (from 5 to 40%) along with the principal raw materials. The mixture is thoroughly mixed and then added to a furnace, where the mixture is heated to near 2725°F to form a viscous, syruplike liquid. The size and nature of the furnace correspond to the intended uses. For small, individual items the mixture may be heated in small clay (refractory) pots.

Most glass is melted in large (continuous) tanks that can melt 400 to 600 tons a day for the production of glass products. The process is continuous, with raw materials fed into one end as molten glass is removed from the other. Once the process (called a campaign) begins, it is continued indefinitely, night and day, often for several years until the demand is met or the furnace breaks down.

A typical window glass will contain 95–99% silica sand with the remainder being soda ash (Na_2CO_3), limestone ($CaCO_3$), feldspar, borax, or boric acid, along with the appropriate coloring, decolorizing, and oxidizing agents.

Processing of glass includes shaping and retreatments of glass. Since shaping may create undesirable sites of amorphous structure, most glass objects are again heated to *near* their melting point. This process is called annealing. Since many materials tend to form more ordered structures when heated and recooled slowly, the effect of annealing is to "heal" these sites of major dissymmetry.

Four main methods are employed for shaping glass: drawing, pressing, casting, and blowing. Most flat glass is shaped by drawing a sheet of molten (heated so it can be shaped but not so it freely flows) glass onto a surface of molten tin. Since the glass literally floats on the tin, it is called "float glass." Its temperature is carefully controlled. The glass from a "float bath" typically has both of its sides quite smooth with a brilliant finish that requires no polishing.

Glass tubing is made by drawing molten glass around a rotating cylinder of appropriate shape and size. Air can be blown through the cylinder or cone to make glass tubing like that used in laboratories. Fiberglass is made by drawing molten glass through tiny holes, with the drawing process helping to align the tetrahedral clusters.

Pressing is accomplished by simply dropping a portion of molten glass into a form and then applying pressure to ensure that the glass takes the form of the mold. Lenses, glass blocks, baking dishes, and ashtrays are examples of objects commonly made by pressing.

The casting process involves filling molds with molten glass much the same way cement and plaster of Paris molded objects are produced. Art glass pieces are typical examples of articles produced by casting.

Glassblowing is one of the oldest arts known to human culture. For art or tailor-made objects, the working and blowing of the glass are done by a skilled worker who blows into a pipe intruded into molten glass. The glass must be maintained at a temperature that permits working but not free flow and the blowing must be at a rate and force to give the desired result. Mass-produced items are made using machine blowers, which often blow the glass to fit a mold, much like the blow molding of plastics.

In the following we give brief summaries describing different kinds of glass. The type and properties of glass can be readily varied by changing the relative amounts and nature of ingredients. Soda-lime glass is the most common of all glasses, accounting for about 90% of the glass made. Window glass, glass for bottles and other containers, glass for light bulbs, and many art glass objects are all soda-lime glass. Soda-lime glass typically contains 72% silica, 15% soda (sodium oxide, Na_2O), 9% lime (calcium oxide, CaO), and the remaining 4% is minor ingredients. Its relatively low softening temperature and thermal shock resistance limit its high-temperature applications.

Vycor, or 96% silicon, glass is made using silicon and boron oxide. Initially, the alkali–borosilicate mixture is melted and shaped using conventional procedures. The article is then heat-treated, resulting in the formation of two separate phases—one high in alkali and boron oxide and the other phase containing 96% silica and 3% boron oxide. The alkali–boron oxide phase is soluble in strong acids and is leached away by immersion in hot acid. The remaining silica–boron oxide phase is quite porous. This porous glass is again heated to about 2200°F, resulting in a 14% shrinkage as the remaining portion fills the porous voids. The best variety is "crystal" clear and is called fused quartz. The 96% silica glasses are more stable and exhibit higher melting points (2725°F) than soda-lime glass. Crucibles, ultraviolet filters, range-burner plates, induction-furnace linings, optically clear filters and cells, and super-heat-resistant laboratory ware are often 96% silicon glass.

Borosilicate glass contains about 80% silica, 13% boric oxide, 4% alkali, and 2% alumina. It is more resistant to heat shock than most glasses because of its unusually small coefficient of thermal expansion (typically between 2 and 5×10^{-6} cm/cm °C; the value for soda-lime glass is about $8–9 \times 10^{-6}$ cm/cm °C). Borosilicate glass is better known by trade names such as Kimax and Pyrex. Bakeware and glass pipelines are often made of borosilicate glass.

Lead glasses (also called heavy glass) are made by replacing some or all the calcium oxide by lead oxide (PbO). Very high amounts of lead oxide can be incorporated—up to 80%. Lead glasses are more expensive than soda-lime glasses, but they are easier to melt and work with. They are more easily cut and engraved and give a product with high sparkle and luster (due to higher refractive indexes). Fine art glass and tableware are often made of lead glass.

Glazes are thin, transparent coatings (colored or colorless) fused on ceramic materials. Vitreous enamels are thin, normally opaque or semi-opaque, colored coatings fused on metals, glasses, or ceramic materials. Both are special glasses but may contain little silica. They are typically low melting and often are not easily mixed in with more traditional glasses.

There are also special glasses for specific applications. Laminated automotive safety glass is a sandwich form made by combining alternate layers of poly(vinyl butyral) (containing about 30% plasticizer) and soda-lime glass. This sticky organic polymer layer acts both to absorb sudden shocks (like hitting another car) and to hold broken pieces of the glass together. Bulletproof or, more correctly stated, bullet-resistant glass is a thicker, multilayer form of safety glass.

Tempered safety glass is a single piece of specially heat-treated glass often used for industrial glass doors, laboratory glass, lenses, and side and rear automotive windows. Because of the tempering process, the material is much stronger than normal soda-lime glass. Optical fibers are glass fibers that are coated with a highly reflective polymer coating such that light entering one end of the fiber is transmitted through the fiber (even around curves and corners, as when inserted into a person's stomach) to emerge from the other end with little loss of light energy. Optical fibers can also be made to transmit sound and serve as the basis for the transmission of television and telephone signals over great distances.

15.5 SILICON DIOXIDE (CRYSTALLINE)—QUARTZ

Silicon crystallizes in mainly three forms—quartz, tridymite, and cristobalite. After the feldspars, quartz is the most abundant mineral in the earth's crust, being a major component of igneous rocks and one of the commonest sedimentary materials in the form of sandstone and sand. Quartz can occur as large (several pounds) single crystals but is normally present as much smaller components of many of the common materials around us. The structure of quartz is a three-dimensional network of 6-membered Si—O rings (three SiO_4 tetraheda) connected such that every six rings enclose a 12-membered Si—O ring (six SiO_4 tetrahedra).

15.6   ASBESTOS

Asbestos has been known and used for over 2000 years. Egyptians used asbestos cloth to prepare bodies for burial. The Romans called it *amiantus* and used it as a cremation cloth and for lamp wicks. Marco Polo described its use in the preparation of fire-resistant textiles in the thirteenth century. Asbestos is not a single mineral but rather a grouping of materials that yield soft, threadlike fibers. These materials are examples of two-dimensional sheet polymers containing two-dimensional silicate ($Si_4O_{10}^{4-}$) anions bound on either one or both sides by a layer of aluminum hydroxide ($Al(OH)_3$, gibbsite) or magnesium hydroxide ($Mg(OH))_2$, brucite). The aluminum and magnesium are present as positively charged ions, that is, cations. These cations can also have a varying number of water molecules (waters of hydration) associated with them. The spacing between silicate layers varies with the nature of the cation and the amount of its hydration.

These fibrous silicates are generally divided into the serpentine and amphibole groups of minerals. Chrysotile is the most abundant and widely used type of asbestos and is a member of the serpentine mineral group. It consists of alternate sheets of magnesia ($Mg(OH)_2$) and silica with an overall empirical formula of $Mg_3Si_2O_5(OH)_4$. The chrysotile fibers are coiled and exist as bundles of hollow tubes called fibrils. Chrysotile is mined in Canada, the United States, South Africa, Zimbabwe, and the U.S.S.R., and accounts for 90% of the world asbestos market.

Asbestosis is a disease that blocks the lungs with thick, fibrous tissue, causing shortness of breath and swollen fingers and toes. Bronchogenic cancer, or cancer of the bronchial tubes, is prevalent among asbestos workers who also smoke cigarettes. Asbestos also causes mesothelioma, a fatal cancer of the lining of the abdominal chest. These diseases may lie dormant for many years after exposure.

The exact causes of these diseases are unknown but appear to be characteristic of particles (whether asbestos or other particulants) about 5 to 20 μm in length (about 2×10^{-4} in.), corresponding to the approximate sizes of the mucous membrane openings in the lungs. Further, the sharpness of asbestos fibers intensifies their toxicity since these fibers actually cut the lung walls; even though the walls heal, the deposited asbestos, if not flushed from the lungs, will again cut the lung walls when the individual coughs, thus causing more scar tissue to form. The cycle continues until the lungs are no longer able to function properly.

15.7   POLYMERIC CARBON—DIAMOND

Just as carbon serves as the basic building element for organic materials, it is also a building block in the world of inorganic materials. Elemental carbon exists in two distinct crystalline forms—graphite and diamond.

Although diamonds may not truly be "a girl's best friend," they are an important industrial mineral because they are the hardest naturally occurring substance.

Table 15.1   Mohs' Scale of Hardness[a]

1. Talc	6. Feldspar
2. Gypsum	7. Quartz
3. Calcite	8. Topaz
4. Fluorite	9. Corundum
5. Apatite	10. Diamond

[a]Hardness of other materials: fingernail, 2; copper penny, 3; knife blade, 5.5; window glass, 5.5.

Hardness is a relative term. In the world of rocks and minerals, hardness is measured on the basis of a 10-point scale (Table 15.1) that is dependent on the ability of a material to scratch a mark on a member of this scale. Each mineral on the scale scratches the ones with lower values. The hardness is more accurately measured today using an instrument called a sclerometer, which records the force required to scratch the material with a diamond.

Diamonds are almost pure carbon that occurs in a tetrahedral structure in which each carbon atom is at the center of a tetrahedra composed of four other carbon atoms (Figure 15.1). Natural diamonds were formed millions of years ago when concentrations of pure carbon were subjected by the earth's mantle to great pressure and heat. The majority of diamonds (nongems) are now man-made. The first synthetic diamonds were made by a team of scientists in 1955 at the General Electric Research Laboratory by compressing pure carbon under extreme pressure and heat. The majority of the synthetic diamonds are no larger than a grain of common sand. By 1970, G.E. was manufacturing diamonds of gem quality and size. These diamonds, available at a cost much higher than that of natural diamonds, are used for research. For instance, it was discovered that the addition of small amounts of boron to the diamonds causes the diamonds to become semiconductors. Today such doped diamonds are used to make transistors.

The major uses of diamonds are in industry as shaping agents to cut, grind, and bore (drill) holes in metals and ceramics. Most turntable cartrides employ a diamond needle to transmit differences in the record grooves into sound.

15.8   POLYMERIC CARBON—GRAPHITE

Although diamonds are the hardest naturally occurring material, the most common form of crystalline carbon is the softer graphite. Graphite occurs as sheets of hexagonally fused benzene rings (Figure 15.1) or "hexa-chicken wire." The bonds holding the fused hexagons together are traditional primary, covalent bonds. In contrast, the bonding between the sheets of fused hexagons consists of a weak overlapping of pi electrons and is considerably weaker than the bonding within the sheet. Thus, graphite exhibits many properties that are dependent on the angle with which they are measured. Graphite has some strength when measured along the sheet but very little strength if the layers are allowed to "slide over one another." Furthermore, the fused hexagons are situated such that the atoms in each layer lie opposite to the centers of the six-membered rings in the next layer. This arrangement

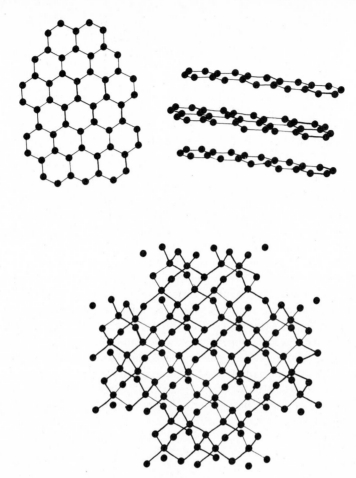

Figure 15.1 Structures of graphite (top) emphasizing the sheet nature (left) and emphasizing the layered nature (right), and diamond (bottom).

further weakens the overlapping of pi electrons between layers such that the magnitude of layer-to-layer attraction is on the order of ordinary secondary forces. The "slipperiness" of the layers relative to one another accounts for graphite's ability to be a good lubricant.

As with diamonds and most other natural materials, graphite's discovery and initial use by humankind is lost in antiquity. Graphite was long confused with molybdenite (MoS_2), and at different times it was known as plumbago ("like lead"), crayon noir, silver lead, black lead, or carbo mineralis. In 1789 Werner first named it graphite, from the Greek *graphein*, meaning to write.

Although graphite has been extensively mined in China, Mexico, Austria, North and South Korea, Russian, and Madagascar, the majority of graphite used in the United States is manufactured from coke.

Graphite's properties led directly to its many uses in today's society. Because of its tendency to mark, hardened mixtures of clay and graphite are the "lead" in today's lead pencils. Graphite conducts electricity and is not easily burned, thus many

Table 15.2   Comparison of Properties of Diamond and Graphite

Property	Diamond	Graphite
Density (g/cc)	3.5	2.3
Electrical resistance	Increases with temp.	Decreases with temp.
Mohs' hardness	10	0.5–1.5
C—C bond length (Å)	1.54	1.42
Stability temperature (°F)	to ca. 4000	6300

industrial electrical contact points (electrodes) are made of graphite. Graphite is a good conductor of heat and is chemically quite inert even at high temperatures, and so many crucibles for melting metals are graphite lined. Graphite has good stability to even strong acids, thus it is employed to coat acid tanks. It also is effective at slowing down neutrons, and thus composite bricks and rods (often called carbon rods) are employed in nuclear generators to regulate the progress of the nuclear reaction. Its "slipperiness" accounts for its use as a lubricant for clocks, door locks, and hand-held tools. Graphite is also the major starting material for the synthesis of synthetic diamonds. Dry cells and some types of alkali storage batteries also employ graphite.

At ordinary pressures and temperatures, both graphite and diamonds are stable. At high temperatures (about 3250°F), a diamond is readily transformed to graphite. The reverse transformation of graphite to a diamond occurs only with application of great pressure and high temperatures. Thus, the naturally more stable form of crystalline carbon is not diamond but rather graphite. A comparison of some physical properties of diamond and graphite appear in Table 15.2.

GLOSSARY

Alumina: Al_2O_3.

Anisotropic: Dependent on direction; directionally dependent.

Annealing: Subjecting materials to heat near their melting point.

Asbestos: Grouping of silica-intensive materials containing aluminum and magnesium that give soft, threadlike fibers.

Asbestosis: Disease that blocks lungs with thick, fibrous tissue.

Borosilicate glass: Relatively heat-shock-resistant glass with a small coefficient of thermal expansion (Kimax and Pyrex).

Calcium-aluminate cement: Cement with more alumina than portland cement.

Chrysotile: The most abundant and widely used type of asbestos.

Colored glass (stained glass): Glass containing coloring agents such as metal salts and oxides.

Concrete: Combination of cement, water, and filler material such as rocks and sand.

Diamond: Polymeric carbon in which the carbon atoms are at the centers of tetrahedra composed of four other carbon atoms; the hardest known natural material.

Feldspar: A derivative of silica in which one-half to one-quarter of the silicon atoms are replaced by aluminum atoms.

Fiberglass: Fibers of drawn glass.

Float glass: Glass made by cooling sheets of molten glass in a tank of molten tin; most common window glass is of this type.

Glass: An inorganic product of fusion that has been cooled to a rigid condition without crystallization; most glasses are based on amorphous SiO_2.

Glaze: Thin, transparent coatings fused on ceramic materials.

Graphite: Polymeric carbon consisting of sheets of hexagonally fused rings in which the sheets are held together by weak overlapping pi electron orbitals; anisotropic in behavior.

Gypsum: $CaSO_2 \cdot 2H_2O$; serves as the basis of plaster of Paris, Martin's cement, Keen's cement, and Parisian cement; shrinks very little on hardening; rapid drying.

Hole: Unoccupied site in a material.

Inorganic polymer: A polymer containing no organic portions.

Kaolinite: An important type of asbestos clay.

Lead glass (heavy glass): Glass in which some or all the calcium oxide is replaced by lead oxide.

Lime: $CaCO_3$; derived from oyster shells, chalk, and marl.

Magnesia cement: Cement composed mainly of magnesium oxide; rapid hardening.

Optical fiber: Glass fibers coated with highly reflective polymer coatings; allows light entering one end of the fiber to pass through to the other end with little loss of energy.

Piezoelectric material: Materials that develop net electronic charges when pressure is applied; sliced quartz is piezoelectric.

Portland cement: A major three-dimensional inorganic polymer construction material consisting of calcium silicates, lime, and alumina.

Precast concrete: Portland concrete cast and hardened prior to being taken to the site of use.

Prestressed concrete: Portland concrete cast around steel cables stretched by jacks.

Quartz: Crystalline forms of silicon dioxide; basic material of many sands, soils, and rocks.

Reinforced concrete: Portland concrete cast around steel rods or bars.

Safety glass: Laminated glass; sandwich form containing alternate layers of poly(vinyl butyral) and soda-lime glass.

Sandstone: Granular quartz.

Silicon glass: Glass made by fusing pure quartz crystals or glass sand; high melting.

Soda: Na_2O.

Soda ash: Na_2CO_3.

Soda-lime glass: Most common glass; based on silica, soda, and lime.

Tempered safety glass: A single piece of specially heat-treated glass.

Tempering: A process of rapidly cooling glass, resulting in an amorphous glass that is weaker but less brittle.

Vitreous enamel: Thin, normally somewhat opaque-colored inorganic coatings fused on metals.

Vycor: 96% silicon glass; made from silicon and boron oxide; best variety is called fused quartz.

REVIEW QUESTIONS

1. Compare window glass with organic thermoplastics.

2. Why is portland cement an attractive basic building material?

3. Name five important natural inorganic giant molecules.

4. Is window glass a thermoset polymer?

5. Why are specialty cements and concretes necessary?

6. What is meant by the comment that "glass is a supercooled liquid"?

7. Is quartz a thermoset polymer?

8. Why are specialty glasses important in today's society?

9. Where does sand come from?

10. Which is the most brittle: window glass, quartz, fibrous glass, asbestos fiber, polypropylene?

11. Compare the structures of graphite and diamond.

BIBLIOGRAPHY

Adams, D. W. (1974). *Inorganic Solids*. New York: Wiley.

Allcock, H. R. (1972). *Phosphorus–Nitrogen Compounds*. New York: Academic Press.

Carraher, C. E. (1977). Organometallic polymers. *Coatings and Plastics Preprints*, **37**(1), 59.

Carraher, C. E. (1981). *Metallo-organic Polymers*. Moscow: Mir.

Carraher, C. E. (1982). *Advances in Organometallic and Inorganic Polymer Science*. New York: Dekker.

Carraher, C. E., and Preston, J. (1982). *Interfacial Synthesis*, Vol. III, *Recent Advances*. New York: Dekker.

Carraher, C. E., and Reese, D. R. (1977). Lead polyesters. *Coatings and Plastics Preprints*, **37**(1), 162.

Carraher, C. E., Sheats, J., and Pittman, C. U. (Eds.). (1978). *Organometallic Polymers*. New York: Academic Press.

Deer, W., Howie, R., and Zussman, J. (1982). *Rock-Forming Minerals*. New York: Longmans.

Eitel, W. (1964). *Silicate Science*. New York: Academic Press.

Holliday, L. (1975). *Ionic Polymers*. New York: Halsted.

Klein, C., and Hurlbut, C. (1985). *Manual of Mineralogy*. New York: Wiley.

MacGregor, E., and Greenwood, C. (1980). *Polymers in Nature*. New York: Wiley.

Mineralogical Society of America *Reviews in Mineralogy* (ongoing series). Washington, DC: Mineralogical Society of America.

Neuse, E. W., and Rosenberg, H. (1970). *Metallocene Polymers*. New York: Dekker.

Ray, N. H. (1978). *Inorganic Polymers*. New York: Academic Press.

Sheats, J. E., Carraher, C. E., and Pittman, C. U. (1985). *Metal-Containing Polymeric Systems*. New York: Plenum.

Zoltai, T., and Stout, J. (1984). *Mineralogy*. Minneapolis: Burgess.

ANSWERS TO REVIEW QUESTIONS

1. Window glass is held together by directional covalent bonds; acts physically like many organic polymers that are above their glass transition temperature in being very viscous; acts like a solid on rapid impact but like a liquid on a much elongated time scale.

2. It is readily available on a large scale; inexpensive; relatively nontoxic; stands up well to most natural elements such as rain, cold, heat, and mild acids and bases; light and strong.

3. Quartz, asbestos, alumina, graphite, and diamond.

4. No.

5. To perform for a wide variety of conditions and applications.

6. It flows like a liquid, but the flow rate is very low.

7. Yes.

8. A wide variety of applications require materials that may possess glasslike properties; glasses have generally good resistance to natural elements, are easily shaped, polished, and cut, and many transmit light and can be colored.

9. From breakup of larger silicon dioxide-intense rocks.

10. Quartz.

11. Both are made from carbon but graphite is a sheet and diamond is a three-dimensional tetrahedra.

CHAPTER XVI

Specialty Polymers

16.1 WATER-SOLUBLE POLYMERS

Because of the nonpolar structure of some polymers, such as polystyrene and polyethylene, they repel water. Some polymers are polar and a few are polar enough to be soluble in water. These polymers are said to be hydrophilic, that is, water-loving.

Most of us are familiar with water-soluble starch, which is used both as a food and as a textile and paper coating. However, few of us are aware of water-soluble derivatives of cellulose, such as hydroxyethylcellulose, methylcellulose, and carboxymethylcellulose, which are used to increase the viscosity (resistance to flow) of water and to provide water-soluble coatings. These and other water-soluble polymers, such as guar gum, have been used to enhance the recovery of oil from wells and thus increase oil production.

Aqueous solutions of other water-soluble polymers, such as polyacrylamide,

$$\begin{array}{c} COONH_2 \\ | \\ \text{---}CH_2\text{---}CH\text{---}_n \end{array}$$

the sodium salt of polyacrylic acid,

$$\begin{array}{c} COONa \\ | \\ -CH_2-CH-_n \end{array}$$

and copolymers of vinyl acetate and maleic anhydride

$$\begin{array}{c} O \\ \diagup \diagdown \\ C=O \quad C=O \\ | \quad\quad | \\ -CH_2-CH-CH-CH-_n \\ | \\ COCCH_3 \end{array}$$

have also been used for enhanced oil recovery. Maleic anhydride copolymers must be hydrolyzed to produce a water-soluble salt.

$$\begin{array}{c} COONa \quad COONa \\ | \quad\quad\quad | \\ -CH_2-CH-CH-CH-_n \\ | \\ COOCH_3 \end{array}$$

Sodium salt of copolymor of vinyl
acetate and maleic acid

Water-soluble polymers such as polyacrylamide are used as flocculants in papermaking. Salts of polyacrylic acid, polyacrylamide, polyvinyl alcohol, and polyethylene oxide are also used as flocculants to improve the efficiency of municipal waste treatment. Polyethylene oxide, $-(CH_2CH_2-O)_n$, has been added to water to increase the flow when it is used to extinguish fires. Water-soluble polymers such as hydrolyzed maleic anhydride copolymers are also used to increase the utility of floor polishes.

16.2 OIL-SOLUBLE POLYMERS

Nonpolar polymers are used to assist the flow of crude oil in pipelines, and they are also used to control the viscosity of lubricating oils. The 10-40 oil that is used in automobiles contains a small amount of an oil-soluble polymer, such as polyisobutylene, polybutyl methacrylate, or polycyclohexylstyrene. Since these polymers are not particularly soluble in cold oil, their chains tend to form tight coils, which have little effect on the viscosity of the oil. However, these coils extend when the oil is heated and these extended chains increase the viscosity of the oil so that it is more viscous at engine operating temperatures than at room temperature. This increased viscosity helps the oil reduce the friction between the moving parts in the engine.

$$\begin{array}{c} CH_3 \\ | \\ CH_2=C \\ | \\ CO_2C_4H_9 \end{array} \longrightarrow \begin{bmatrix} \quad CH_3 \quad \\ | \\ -CH_2-C- \\ | \\ CO_2C_4H_9 \end{bmatrix}_n$$

n-Butyl methacrylate   PMBA

16.3 POLYMERIC FOAMS

The insulation properties of nonconductive polymers may be enhanced by the addition of hollow glass or polymer spheres to produce syntactic foams. Foams may also be produced by the addition of volatile liquids or gases to molten polymers. Thus, foamed polystyrene (PS, Styrofoam) is produced by the extrusion of a mixture of polystyrene and pentane.

The propellant pentane may also be incorporated into PS beads before molding. Coffee cups produced by molding these expandable beams have been widely used by fast-food establishments, but their use is being curtailed because of the alleged adverse environmental impact of these chemicals.

Most rigid foam is also made from polyurethane derived from methylenediphenyl isocyanate (MDI) and difunctional polyether polyols. The rigid foams are typically blown using fluorocarbons as the gas to produce a closed cell foam with outstanding insulation properties. Major uses include in building and construction (60%), cryogenic (low-temperature) transport of materials, furniture, packaging, and molded structural parts such as solar panels.

Bedding and furniture (50%) account for the major end use of flexible PUR foams. Most furniture and pillow filling (30%) is PUR foam, as is much of the automotive seating material. PUR foam is also employed as carpet underlay and in the packaging of breakable items.

16.4 POLYMER CONCRETE

The use of polymers for structural applications has usually been restricted because of the relatively high costs of these polymers. However, their lower specific gravity, good tensile strength, and the increasing cost of competitive products, such as hydraulic cement, have sparked considerable interest in polymer concrete.

The original polymer concrete was a filled resole phenolic resin that was converted to a hard polymer *in situ* by the presence of a compatible acid, such as *p*-toluene sulfonic acid,

$$H_3C-\!\!\left\langle\!\bigcirc\!\right\rangle\!\!-SO_3H$$

The phenolic polymer concrete has been displaced to some extent by a more alkaline-resistant cement based on furfuryl alcohol,

$$\text{furfuryl alcohol structure: }\overset{H}{\underset{H}{|}}\text{—C—OH}$$

epoxy, and polyester resins.

The most widely used polymer cement is based on a filled polyester prepolymer and is called "cultured marble." It is used for casting bathroom sinks and tubs.

Because of its high strength, the epoxy concrete is competitive with portland cement and is used for patching roads and surfacing bridges. Unlike patching with portland cement, where appreciable bonding does not occur between the patched portion and the newly deposited patch material, considerable bonding does occur between the polymer cement and the old section, thus making such patches more permanent.

16.5   XEROGRAPHY

Photoconductive materials form the basis for xerography. The two major photoconductive materials are selenium and polyvinylcarbazole, which are applied to a backing. The surface is made light sensitive by electrostatic charging in darkness. The surface is then exposed to the desired image, which in turn is developed by application of a toner. The image is then transferred to a paper and set or fixed by heat, and the xerography paper emerges from inside the xerography machine. The drum is then automatically cleared and ready to accept another image. Improvements are continuing, with the giant molecule polyvinylcarbazole playing a major role.

$$CH_2{=}CH \qquad\qquad {+}CH_2{-}CH{)}_{\overline{n}}$$

N-Vinylcarbazole   Poly(N-vinylcarbazole)

16.6   PIEZOELECTRIC MATERIALS

A slice of quartz develops a net positive charge on one side and a negative charge on the other side when pressure is applied. The same effect is found when pressure is applied by means of an alternating electric field. Such crystals are employed for quartz watches and clocks, and for TVs, radios, hearing aids, ignition of flash bulbs, and in telephone receivers. Rochelle salt and tourmaline also exhibit piezoelectric properties.

Several organic polymers are also effective piezoelectric materials. The most widely used is polyvinylidene fluoride, which is employed in loudspeakers, fire and burglar alarm systems, microphones, and earphones. Nylon 11 is also piezoelectric and can be aligned when placed in a strong electrostatic field, giving films used in infrared-sensitive TV cameras, in underwater detection devices, and as part of electronic devices because nylon 11 films can be overlaid with printed circuits.

$$+CH_2CF_2{]}_{\overline{n}}$$
Polyvinylidene fluoride

$$H_2N(CH_2)_{10}CO_2H \xrightarrow{\text{heat}} \left[NH(CH_2)_{10}\overset{\overset{\displaystyle O}{\|}}{C} \right]_{\overline{n}} + H_2O$$

11-Aminoundecanoic acid   Nylon 11

16.7   CONDUCTIVE AND SEMICONDUCTIVE MATERIALS

Numerous conductive (copper, aluminum) and semiconductive materials have been utilized. Whereas most giant molecules have been employed because of their lack of electrical conductivity, today some are being considered for their conductive properties. Advantages of polymeric materials are the possibilities of flexible, finer, and more easily processed avenues of conductivity. The major polymeric candidates are polyacetylene, polythiazyl, polyphenylene, and polyphenylene sulfide (Table 16.1). It is customary to add a dopant, such as SbF_3, to enhance the conductivity of these polymers.

16.8   SILICON CHIPS

The silicon chip is essential for transmission, but the photoresist polymers supply the message. Thus, there are many sophisticated polymers that are coated on the chip and then selectively degraded to produce an effective chip.

Since polymethyl methacrylate (PMMA) is easy to thermally degrade, it was one of the first coatings used on the chip. However, PMMA has been displaced by other polymers designed specifically for this end use.

16.9   ION-EXCHANGE RESINS AND ANCHORED CATALYSTS

Just as the calcium ion forms an insoluble compound through reaction with the carbonate ion (namely, calcium carbonate), so also will it form a complex with a carboxyl group attached to a polymer. This concept forms the basis for many analysis, separation, and concentration techniques. Many of these are based on benzene and divinylbenzene, silicon dioxide, and dextran-based resins. These resins are almost always cross-linked and then functionalized, that is functional groups are added onto the surface of the resin beads. Functional groups such as carboxylic acid and sulfonic acid attract and retain positively charged ions and are appropriately called cation-exchange resins. Functional groups such as amines attract anions and are called anion-exchange resins (Table 16.2).

Table 16.1   Structures of Some Conductive Polymers

Name	Structure	σ (reciprocal ohm, cm)
Poly (p-phenylene)		5000
Polythiazyl		3700
Polyacetylene		3000
Polyphenylene sulfide (PPS)		1.0

Table 16.2   Active Functional Groups on Ion-Exchange Resins

Active Group	Structure
Cation-Exchange Resins Sulfonic acid	benzene ring $-SO_3H$
Carboxylic acid	$-CH_2CHCH_2-$ $\quad\quad\vert$ $\quad\quad COOH$
Anion-Exchange Resins	
Quaternary ammonium salt	benzene ring $-CH_2N(CH_3)_3]^+Cl^-$
Secondary amine	benzene ring $-CH_2NHR$
Tertiary amine	benzene ring $-CH_2NR_2$

Applications of these resins are extremely varied, including the purification of sugar, identification of drugs and biomacromolecules, concentration or uranium, and as therapeutic agents for the control of bile acid and gastric acidity. In the latter use a solid polyamide (Colestid) is diluted to be taken with orange juice to help in the body's removal of bile acids. Removal of bile acids causes the body to produce more bile acid from cholesterol, thus effectively reducing the cholesterol level. Recent catalysts and stereoregulating groups have been attached to the backbones of polymeric materials.

The Merrifield protein synthetic (using chloromethylated polystyrene) makes use of ion-exchange resins as do many of our industrial and home water purifiers. Water containing calcium, iron, or magnesium ions is called hard water. These ions are normally from natural sources, for example, calcium ions are generally derived from the passage of water over and through limestone ($CaCO_3$). Hardness in water is objectionable since the metal ions generally form insoluble salts when the water is heated or when soap is added. The precipitate (formation of insoluble materials) may be deposited in the pipes and water heater, forming boiler scale. The ions also lower the efficiency of added soaps and the precipitate forms "curds" that are often "captured" in the laundery. Furthermore, the precipitate forms bathtub rings and it adheres to us as we are taking a shower or bath.

Most home water softeners are based on ion-exchange resins. The first ion-exchange materials used in softening water were naturally occurring polymeric aluminum silicates called zeolites ($NaAlSi_2O_6$), which exchanged their ions for calcium, iron, and magnesium ions. Synthetic zeolites were later developed. Today most ion-exchange materials are based on styrene and divinylbenzene (vinylstyrene) resins, which are then sulfonated. When the resin system is ready for use, sodium ions generated from rock salt (sodium chloride) are passed through the resin "bed" replacing the hydrogen ions (protons). Then water to be used for drinking, cooking,

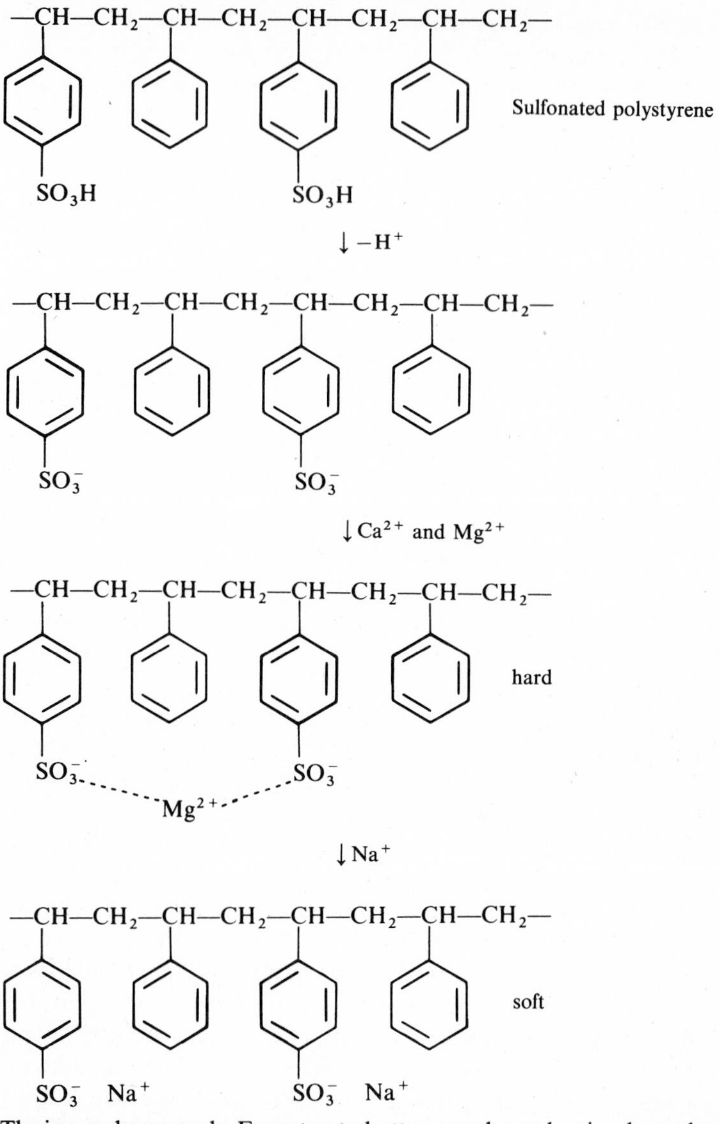

Figure 16.1 The ion-exchange cycle. From top to bottom: uncharged resin, charged resin, resin with complexed "hard ions," and recharged resin. After initial use, the usual cycle will involve the two bottom steps.

washing, and bathing is passed through the resin bed. The sulfonate functional groups have a greater affinity for calcium, iron, and magnesium ions so these displace the sodium ions, resulting in water that has few if any "hard ions," but with a few more sodium ions. Eventually, the sulfonate sites on the resin become filled and the resin bed must be recharged by adding large amounts of dissolved sodium ions from sodium chloride, which displace the more tightly bound, but overwhelmingly outnumbered, "hard ions" after the system is flushed free of these "hard ions" the resin bed is again ready to deliver "soft water" for our use (Figure 16.1).

16.10   PHOTOACTIVE MATERIALS

Photo-cross-linking and photopolymerization have been employed to produce permanent images. For duPont's Dyoryl and Lydel systems, soluble linear giant molecules are made insoluble through cross-linking by application of ultraviolet radiation.

Three major approaches have been studied in developing photoactive materials for imaging. An Eastman–Kodak photo-cross-linkable system employs polyvinyl cinnamates. The photoactive group in this case is the cinnamate unit, which is attached to polyvinyl alcohol and is thus located as a pendant group. Another approach involves the photoactive group as part of the main chain. A third approach employs a difunctional reactant such as diazide, which when activated acts as a cross-linking agent. These photosensitive systems are called negative imaging systems, similar to the more traditional photographic systems. Since the exposed areas are cross-linked, these insoluble exposed areas are left after the unexposed areas are washed away, creating negative images. Positive images can be obtained through use of special solvent systems that render the cross-linked portions soluble through disruption of the cross-linking.

These systems are much superior to the classical silver halide systems in which the image sharpness was limited by the size of the silver halide grains, whereas the photo-cross-linking systems are limited only by the size of the individual giant molecules. Today, microcircuits and most of our newspapers are printed using photopolymers.

$$CH_2{=}CH$$

$$OCH_2CH_2OCCH{=}CH- \text{(ring)} \quad \xrightarrow[\text{toluene}]{BF_3 \text{ etherate}}$$

β-Vinyloxyethyl cinnamate

$$\{CH_2CH\}_n$$

$$OCH_2CH_2OCCH{=}CH- \text{(ring)}$$

Poly β-vinyloxyethyl cinnamate

16.11   CONTROLLED-RELEASE POLYMERS

Numerous examples of controlled-release systems are known. For many cases, the to-be-delivered group is incorporated within the polymer chain or as a side group. The controlled-release polymer is then allowed to work—in a book as a scratch and smell, in the body to treat drug addiction, on a flea collar to ward off fleas, in waterways to control undesirable plants, in chewing gum to prolong the flavor, on an adhesive bandage to extend the antiseptic period, in cattle to promote weight gain, or in a diabetic patient to maintain a balanced level of insulin.

GLOSSARY

Antioxidant: A stabilizer for polymers.

Benzophenone: Diphenyl ketone.

Cultured marble: A polymer cement usually based on a filled polyester.

Dopant: An additive that enhances the electrical conductivity of a polymer.

Guar gum: The endosperm of *Cyanopsis tetraganoloba*, cultivated in India. The water-soluble portion (85%) is called guaran.

Polyacetylene:

$$\begin{array}{cccc} H & H & H & H \\ | & | & | & | \\ \left[\!\!\!\!\!\!\begin{array}{c} \\ C \end{array}\right.\!\!\!\!=C-C=C\!\!\!\!\left.\!\!\!\!\right]_n \end{array}$$

Polymer concrete: A composite produced by the *in situ* polymerization of a mixture of polymer and filler.

Polymethyl isopropenyl ketone (PMIK):

$$\begin{array}{cccc} H & & H & H \\ | & & | & | \\ \left[\!\!\!\!\! C\right.\!\!-\!\!C\!\!-\!\!C\!\!-\!\!C\!\!\left.\right]_n \\ | & \| & | & | \\ H & O & H & CH_3 \end{array}$$

Resole phenolic resin: The reaction product of phenol and formaldehyde under alkaline conditions, usually used as a prepolymer.

Solubility parameter: A scale of solubility equal to the square root of the cohesive energy density, $(CED)^{1/2}$. The values range from about 4H (Hildebrands) to 22.5 H (for water).

Specialty polymer: A polymer that is used in applications other than molding, extrusion, or as a fiber, coating, adhesive, or elastomer.

Structural foam: A molded cellular plastic with a solid surface.

Styrofoam: Trade name for cellular polystyrene.

Syntactic foam: A cellular product consisting of a polymer and a hollow glass bead filler.

REVIEW QUESTIONS

1. Why do some toys break when dropped on concrete surfaces?

2. Do the principles outlined in previous chapters apply to biopolymers as well as to elastomers, coatings, fibers, and plastics?

3. What is the requirement for a polymer to be water soluble?

4. Will a hydrophilic polymer like polyvinyl alcohol be soluble in gasoline?

5. What happens if sodium chloride (rock salt) is not added to home ion-exchange water purifiers?

6. Is starch hydrophilic or hydrophobic?

7. Is polyethylene hydrophilic or hydrophobic?

8. Cellulose is not soluble in water but hydroxyethylcellulose is water soluble. Why?

9. What is the difference in the shape of polyisobutylene in hot and cold lubricating oil?

10. Name one application in which biodegradable polymers are useful?

11. What is the advantage of a glyceryl nitrate controlled-release patch (nitro patch) over the glyceryl nitrate pill for control of angina pectoris?

12. Why are polymers essential for computers?

13. What is the advantage of a conductive polymer over a copper wire?

14. What is the principal use of flexible foams?

15. What is the principal use of rigid foams?

16. Which is more expensive: portland cement or polymer concrete?

17. What is the advantage of epoxy concrete over portland cement for patching cracks in highways?

18. Name some applications of synthetic polymers in the human body.

BIBLIOGRAPHY

Banks, W., and Greenwood, A. T. (1925). *Starch and Its Components*. Edinburgh, Scotland: Edinburgh University Press.

Deanin, R. D. (1974). *New Industrial Polymers*, ACS Symposium Series. Washington, DC: American Chemical Society.

Fettes, E. M. (1964). *Chemical Reactions of Polymers*. New York: Interscience.

Radley, J. A. (1923). *Industrial Uses of Starch and Its Derivatives*. London: Applied Science Publishers.

Seymour, R. B. (1979). *Plastic Mortars, Sealants and Caulking Compounds*, ACS Symposium Series. Washington, DC: American Chemical Society.

Seymour, R. B. (1981). *Conductive Polymers*. New York: Plenum.

Seymour, R. B., and Harris, F. (1927). *Polymer Structure–Solubility Relationships*. New York: Academic Press.

Seymour, R. B., and Mark, H. F. (1988). *Applications of Polymers*. New York: Plenum.

Seymour, R. B., and Stahl, G. A. (1981). *Macromolecular Solutions*. Elmsford, NY: Pergamon.

Ulrich, H. (1982). *Introduction to Industrial Polymers*. Munich: Hansen Publishers.

ANSWERS TO REVIEW QUESTIONS

1. Toys molded from inexpensive brittle polymers will break but there are many tough plastics that can be used. Unfortunately, these tough plastics are more expensive.

2. Yes.

3. It must have polar pendant groups such as hydroxyl, amino, or carboxyl or have polar groupings in the polymer backbone.

4. No.

5. The resins will not be recharged and the resins will not be able to remove the "hardness" from the water.

6. Hydrophilic.

7. Hydrophobic.

8. The pendant group is a bulky group that reduces intermolecular hydrogen bonding.

9. Polyisobutylene is a tight coil in cold lubricating oil but is an extended chain in hot oil.

10. As a polyolefin agricultural mulch.

11. The glyceryl nitrate is released in small amounts from the patch over a long period of time.

12. The message is supplied by photoresist polymers.

13. Lighter weight and usually more flexible.

14. Upholstery, mattresses, and filling.

15. Insulating and packaging.

16. Polymer concrete on a weight basis but not always on a performance basis.

17. The epoxy concrete bonds adhere to clean concrete.

18. Lens implants, tooth filling, false teeth, wigs, body parts, such as ears and noses, and joint implants.

CHAPTER XVII

Additives for Polymers

17.1   INTRODUCTION

Some natural polymers, such as wool, silk, or cotton fibers, and some natural coatings, such as shellac and gutta-percha, may be used without additives. However, plant leaves consist of cellulose and pigments, natural rubber contains stabilizers (antioxidants), and wood is a reinforced polymer consisting of a continuous phase (lignin) and a discontinuous phase (cellulose). Likewise, most synthetic plastics, elastomers, and coatings consist of polymers and functional additives.

The types and purposes of additives (the word "additives" is derived from "addition" and simply means material or materials added) are varied and the exact proportions and nature of the additives are as much an art as a science. It is important to remember that (a) the addition of additives often requires extra processing steps, thus increasing the cost of the item; (b) the additives themselves may vary in cost from clay fillers and sulfur, which cost pennies per pound, to bioactive additives to prevent rot and mildew, which may cost dollars per pound; and (c) the "additive industry" is also a major contributor to the polymer industry and to our industrial complex.

17.2   FILLERS

Charles Goodyear patented the use of small quantities of carbon black as a pigment for natural rubber in the 1840s. However, the advantageous use of larger quantities (50%) of carbon black as a reinforcement for rubber tires was not recognized until 1920. Wood-flour-filled shellac (Florence Compound) was formulated by A. P. Critchlow in 1845. Wood flour is made up of fibrouslike wood particles obtained by the attrition grinding of wood.

Heming and Baekeland used asbestos as reinforcements for cold-molded bituminous composites and for phenolic resins, respectively, in the early 1900s. α-Cellulose was used as a filler in urea and melamine plastics in the early 1930s and these formulations are still in use today. Over 2 million tons of fillers are used annually by the American plastics industry.

The most widely used inorganic filler is calcium carbonate, which is used at an annual rate of over 1 million tons. Asbestos continues to be used in moderate amounts (250,000 tons) but, because of its toxicity, it is being displaced by other fillers. Among the naturally occurring filler materials are cellulosics, such as wood flour, α-cellulose, shell flour, and starch, and proteinaceous fillers, such as soybean residues. Approximately 40,000 tons of cellulosic fillers are used annually by the American polymer industry.

Wood flour, which is produced by the attrition grinding of wood wastes, is used as a filler for phenolic resins, dark-colored urea resins, polyolefins, and PVC. Shell flour, which lacks the fibrous structure of wood flour, has been made by grinding walnut and peanut shells. It is used as a replacement for wood flour.

Cellulose, which is more fibrous than wood flour, is used as a filler for urea and melamine plastics. Melamine dishware is a laminated structure consisting of molded resin-impregnated paper. Starch and soybean derivatives are biodegradable, and the rate of disintegration of resin composites may be controlled by the amount of these fillers present.

Many incompatible polymers are added to increase the impact resistance of other polymers, such as polystyrene. Other comminuted resins, such as silicones or polyfluorocarbons, are added to increase the lubricity of some plastics. For example, a hot melt dispersion of polytetrafluoroethylene in polyphenylene sulfide is used as a coating for antistick cookware.

Carbon black, which was produced by the smoke impingement process by the Chinese over a thousand years ago, is now the most widely used filler for polymers. Much of the 1.5 million tons produced annually in the United States is used for the reinforcement of elastomers. The most widely used carbon black is furnace carbon black.

Carbon-filled polymers, especially those made from acetylene black, are fair conductors of heat and electricity. Polymers with fair conductivity have also been obtained by embedding carbon black in the surfaces of nylon or polyester filament reinforcements. The resistance of polyolefins to ultraviolet radiation is also improved by the incorporation of carbon black.

Although glass spheres are classified as nonreinforcing fillers, the addition of 40 g of these spheres to 60 g of nylon 6,6 increases the flexural modulus, compressive strength, and melt index. However, the tensile strength, impact strength, creep resistance, and elongation of these composites are less than those of the unfilled nylon 6,6.

Zinc oxide is used to a large extent as an active filler in rubber and as a weatherability improver in polyolefins and polyesters. Titanium dioxide is used as a white pigment and as a weatherability improver in many polymers.

The addition of finely divided calcined alumina, corundum, or silicon carbide produces abrasive composites. Alumina trihydrate (ATH) serves as a flame-retardant filler in plastics. Ground barytes ($BaSO_4$) yield x-ray-opaque plastics with controlled density. Zirconia, zirconium silicate, and iron oxide, which have specific gravities greater than 4.5, are also used to produce plastics with controlled densities.

Clay is used as a filler in making synthetic paper and rubber. Talc, a naturally occurring, fibrouslike, hydrated magnesium silicate, is used with polypropylene (Figure 17.1). Since talc-filled polypropylene is much more resistant to heat than PP, it is used in automotive accessories subject to high temperatures. Over 40 million tons of talc are used annually as a filler.

Silica, which has a specific gravity of 2.6, is used as naturally occurring and synthetic amorphous silica, as well as in the form of large crystalline particulates, such as sand and quartz. Diatomaceous earth, also called infusorial earth, fossil flour, or Fuller's earth, is a finely divided amorphous silica consisting of the skeletons of diatoms. Diatomaceous earth is used to prevent rolls of film from sticking to themselves (antiblocking) and to increase the compressive strength of polyurethane foams.

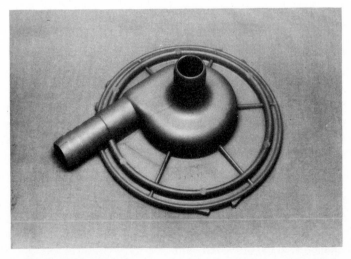

Figure 17.1. A washing machine pump housing made from 40% talc and 60% polypropylene, the latter containing 0.5% titanate.

Pyrogenic or fumed silica is a finely divided filler obtained by heating silicon tetrachloride in an atmosphere of hydrogen and oxygen. This filler is used as a thixotrope to increase the viscosity of liquid resins. Finely divided silicas are also produced by the acidification of sodium silicate solutions and by the evaporation of alcoholic solutions of silicic acid.

Sharp silica sand is used as a filler in resinous cement mortars. Reactive silica ash, produced by burning rice hulls, and the lamellar filler novaculite, from the novaculite uplift in Arkansas, are also used as silica fillers in polymers.

Conductive composites are obtained when powdered metal fillers or metal-plated fillers are added to resins. These composites have been used to produce forming tools for the aircraft industry. Powdered lead-filled polyolefin composites have been used as shields for neutron and gamma radiation.

It is of interest to note that much of the theory on property enhancement by fillers is based on an equation developed by Einstein, which states that the viscosity (η) of a liquid is increased as the concentration (C) of spherical particles is increased in accordance with the equation

$$\eta = \eta_0(1 + K_E C)$$

The constant K_E is a universal constant that is equal to 2.5 for spherical particles, that is, those in which the ratio of length (l) to the diameter (d) is equal to 1. This constant increases as the aspect ratio (l/d) increases.

The stiffness (modulus) and other strength properties of composites (mixtures of polymers and fillers) are related to the viscosity (i.e., the resistance to flow). Many properties of the composites may be estimated from the rule of mixtures in which the volume percentage of the additive and that of the polymer are important factors in determining the properties of the composite. For example, the specific heat of a composite is equal to that of the polymer multiplied by its fractional volume plus the specific heat of the filler multiplied by its specific volume.

Fillers such as glass beads are spherical, those like wood flour and α-cellulose are fibrous, and those like mica are platelike. The properties of the composites are usually enhanced when the surface of the filler is treated with a coupling agent that makes the filler more compatible with the resin. See Table 17.1 for a listing of common fillers.

17.3 REINFORCEMENTS

By definition, reinforcements are fillers with aspect ratios (l/d) greater than 100. Fiberglass, which is the most widely used reinforcement, was produced commercially from molten glass by Slayter in 1938. In spite of the potential of this fiber as a reinforcement for unsaturated polyester resins, little enhancement in the properties of glass-reinforced plastics was noted until the glass surface was treated with coupling agents.

Alkylsilane ($R_2Si(OCH_3)_2$) coupling agents are widely used today for the surface treatment of fiberglass. Organotitanates and organozirconates have also been used successfully as coupling agents. The fiberglass may be used as chopped strands, continuous filaments, mats, or woven cloth.

Table 17.1   Fillers for Polymers

I. ORGANIC
  A. Cellulosic products
    1. Wood
    2. Comminuted cellulose
    3. Fibers (cellulose, cotton, jute, rayon)

  B. Lignin based
  C. Synthetic fibers
    1. Polyesters
    2. Nylons
    3. Polyacrylonitrile
  D. Carbon
    1. Carbon black
    2. Graphite whiskers and filaments
    3. Ground petroleum coke

II. INORGANIC
  A. Silicates
    1. Minerals (asbestos, mica, China clay (kaolinite), talc, wollastonite)
    2. Synthetics (calcium silicate, aluminum silicate)
  B. Silica based
    1. Minerals (sand, quartz, diatomaceous earth, tripoli)
    2. Synthetics
  C. Metals
  D. Boron filaments
  E. Glass
    1. Solid and hollow glass spheres
    2. Milled fiber
    3. Flakes
    4. Fibrous glass (woven, roving, filament, yarn, mat, fabric)

  F. Metallic oxides
    1. Ground (zinc oxide, titania, magnesia, alumina)
    2. Whiskers (aluminum oxide, magnesium oxide, thorium oxide, zirconium oxide, beryllium oxide)

    Polyfluorocarbons

  G. Calcium carbonate
    1. Limestone
    2. Chalk
    3. Precipitated calcium carbonate

  H. Other fillers
    1. Whiskers (nonoxides; aluminum nitride, boron carbide, silicon nitride, tungsten carbide, beryllium carbide)
    2. Barium sulfate
    3. Barium ferrite

Graphite is an excellent but expensive reinforcement for plastics. Aramid (aromatic polyamide), polyester, and boron filaments are also used as reinforcements.

Polyester resin-impregnated fibrous glass is used as a sheet molding compound and bulk molding compound. The former is used like a molding powder and the latter is hot pressed in the shape of the desired object, such as one-half of a suitcase. Chopped fibrous glass roving may be impregnated with resin and sprayed, and glass mats may be impregnated with resin just prior to curing.

Table 17.2   Properties of Reinforcements

Reinforcement	Specific Gravity	Tensile Strength (psi)[a]	Tensile Modulus (psi)	Specific Modulus (psi)[b]
E-glass	2.55	450,000	10,000,000	4,000,000
S-glass	2.48	650,000	12,500,000	5,000,000
Graphite—whiskers	1.74	400,000	40,000,000	23,000,000
Alumina	4.0	4,100,000	103,000,000	26,000,000
Silicon carbide	3.2	2,000,000	70,000,000	22,000,000
Silicon nitride	3.2	20,000,000	57,000,000	19,000,000
Potassium titanate	3.2	10,000,000	40,000,000	12,000,000

[a]68,948 psi = kPa.
[b]Specific modulus = tensile modulus/specific gravity.

The strongest composites are made from continuous filaments impregnated with resin before curing. These continuous filaments are wound around a mandrel in the filament winding process, gathered together, and forced through an orifice in the pultrusion molding process.

The first continuous filaments were rayon and these, as well as polyacrylonitrile fibers, have been pyrolyzed to produce graphite fiber. High-modulus reinforcing filaments have also been produced by the deposition of boron atoms from boron trichloride vapors on tungsten or graphite filaments.

Small single crystals, such as potassium titanate, are being used at an annual rate of over 10,000 tons for the reinforcement of nylon and other thermoplastics. These composites are replacing die-cast metals in many applications. Another microfiber, sodium hydroxycarbonate (called Dawsonite), also improves the physical properties and flame resistance of many polymers. Many other single crystals, called whiskers, such as alumina, chromia, and boron carbide, have been used for making high-performance composites.

Microfibers that have an aspect ratio of at least 60 to 1 also have good reinforcing properties in composites. The principal microfibers are processed mineral fiber (PMF, slag), Franklin fiber (gypsum, $CaSO_4$), Dawsonite and Fybex (potassium titanate, K_2TiO_3). Over 1 million tons of reinforcements are used annually by the American plastics industry. The properties of reinforcing fibers are shown in Table 17.2.

17.4   COUPLING AGENTS

Although natural rubber bonds well to carbon black, in 1956, H. M. Leeper discovered that the adhesion could be enhanced by the addition of small amounts of Elastopar (*N*-4-dinitroso-*N*-methylaniline).

In addition to improving the properties of carbon black-filled butyl rubber, this additive also dramatically reduces the time of milling of the filler and butyl rubber.

There are many different proprietary alkylsilanes, organozirconates, and organotitanates that have been developed for use with specific composites. These coupling agents have two different functional groups, one that is attracted to the resin and the other to the surface of the filler. For example, dialkyldimethoxysilanes are hydrolyzed to produce dialkyldihydroxysilanes *in situ.* As shown in the following equation, the hydroxyl groups bond with the filler surface and the alkyl groups are attracted to the resin.

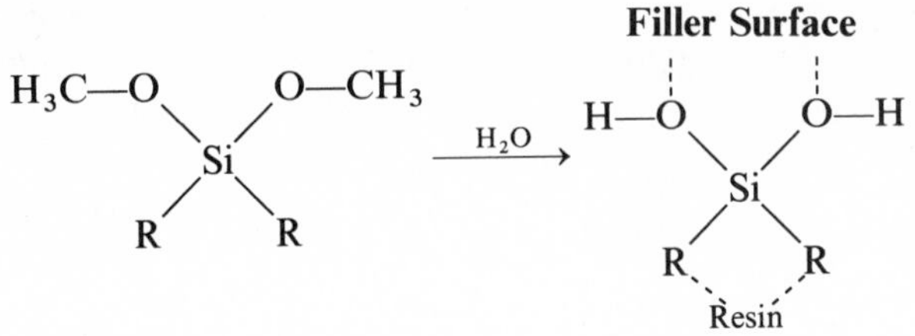

17.5 ANTIOXIDANTS

Polypropylene cannot be used out-of-doors unless a small amount of stabilizer (antioxidant) is present. It is known that the tertiary hydrogen atoms on every other carbon atom in the repeating units may be readily cleaved to form free radicals (R·):

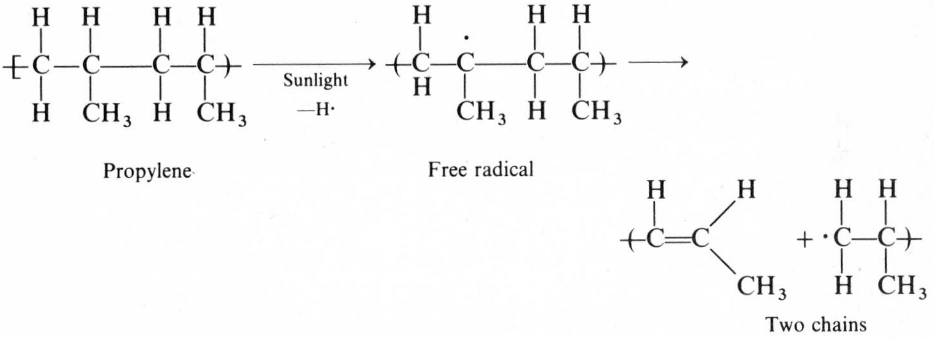

The deterioration resulting from the formation of these free radicals is lessened when antioxidants, such as hindered phenols, alkyl phosphites $((ArO)_3P{=}O)$, thioesters

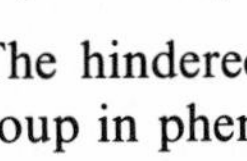

$$\underset{(RC-SR)}{\overset{\displaystyle S}{\overset{\displaystyle \|}{}}}$$

or hindered amines, are present. The hindered phenols have relatively large alkyl groups adjacent to the hydroxyl group in phenols.

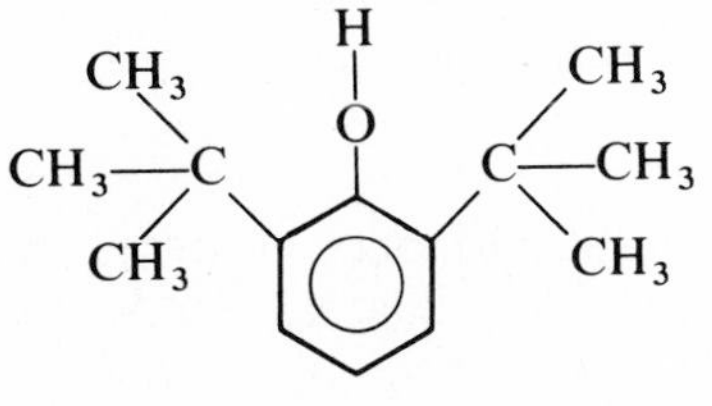

2,6-di-*tert*-butylphenol

Aromatic amines, such as phenyl β-naphthylamine (Ar_2NH), are used as antioxidants in the rubber industry, but some tests have shown that these antioxidants cause cancer in rats. In contrast, the hindered phenols extend the life of rats.

17.6   HEAT STABILIZERS

As mentioned previously, polyvinyl chloride (PVC) is difficult to process since it decomposes at temperatures below the processing temperature. However, stabilizers such as barium, cadmium, calcium, and zinc salts of moderately high molecular weight carboxylic acids, organotin, and organoantimony compounds are effective stabilizers. Presumably these heavy metal cations react with hydrogen chloride to produce insoluble salts. The general reaction may be represented by

$$2HCl + M(OOCR)_2 \longrightarrow 2HOOCR + MCl_2$$

17.7   ULTRAVIOLET STABILIZERS

Sunlight, which has an energy corresponding to 72 to 100 kcal, may break bonds having similar energy. This degradation of polymers may be minimized when compounds that absorb this high-UV energy are present. Since the reaction is equivalent to that which occurs in the sunburning of skin, the additives present in sunburn lotions, such as phenyl salicylate, are effective UV stabilizers for polymers. Phenyl salicylate, which has been used as a medicinal for years, rearranges to 2,2'-hydroxybenzophenone in the presence of ultraviolet light.

Phenyl salicylate   2,2'-Dihydroxybenzophenone

$\longrightarrow$ Quinone + $h\nu$

Chelate

Stabilizers such as 2,2'-hydroxylbenzophenone produce cyclic compounds (chelates) that absorb the UV energy and release it at a lower, less destructive energy level. A chelate is a five- or six-membered ring that may be formed by intramolecular attraction of a hydrogen atom to an oxygen atom in the same compound. Thus, when the hydroxyl group is in the proper (2) position, it will form a chelate with the carbonyl oxygen. Hindered amine light stabilizers (HALS), which, in the presence of sunlight, produce nitroxyl radicals ($\mathrm{=NO\cdot}$), are excellent UV stabilizers.

17.8 FLAME RETARDANTS

General-purpose plastics, such as polyethylene and polystyrene, are readily combustible. However, when polymers are used for furniture or in construction, it is essential that they not be combustible under conditions that may exist in a burning building. Polymers such as polytetrafluoroethylene (PTFE) are usually considered to be flame resistant. However, they will burn in the high-oxygen/low-nitrogen atmosphere present in aerospace capsules.

Other halogen (Cl, Br) and phosphorus-containing polymers are also flame resistant in air. However, it is usually essential that flame retardants be added to most flame-resistant polymers. Alumina hydrate (ATH, $Al_2O_3\cdot3H_2O$) is a colorless filler that is readily used as a flame retardant. This filler releases water, which helps to quench the flames at the burning temperature.

Most flame retardants consist of mixtures of aliphatic chlorides (RCl) and antimony oxide (Sb_2O_3). The antimony chloride ($SbCl_3$) produced at the temperature of the burning plastic is the flame retardant.

Char, formed in some combustion processes, also shields the reactants from oxygen and retards the outward diffusion of volatile combustible products. Aromatic polymers tend to char, and some phosphorus and boron compounds tend to catalyze char formation.

Synergistic flame retardants, such as a mixture of antimony trioxide and an organic bromo compound, are much more effective than single flame retardants. Thus, whereas a polyester containing 11.5% tetrabromophthalic anhydride burned without charring at high temperatures, charring but no burning was noted when 5% antimony oxide was added.

Since combustion is subject to many variables, tests for flame retardancy may not predict flame resistance under unusual conditions. Thus, a disclaimer stating that flame-retardant tests do not predict performance in an actual fire must accompany all flame-retardant polymers. Flame retardants, like many other organic compounds, may be toxic or they may produce toxic gases when burned. Hence, extreme care must be exercised when using fabrics or other polymers treated with flame retardants.

17.9 PLASTICIZERS

W. Semon lowered the processing temperature of PVC by the addition of tricresyl phosphate as a plasticizer in the early 1930s. This somewhat toxic plasticizer has been replaced by esters of phthalic acid, such as diethylhexylphthalate (DOP, DEHP). It is believed that plasticizers weaken the intermolecular attractions between molecules in PVC and allow the semicrystalline polymer to flow at lower than normal temperatures. It should be noted that tests with massive doses of DEHP have shown it to be toxic to laboratory animals.

Dioctyl phthalate (DOP)

Synthetic plasticizers are fairly large molecules. Natural plasticizers, such as some proteins, are also large but molecular water is small—yet it probably acts as a larger unit through its hydrogen bonding. Most plasticizers work on the basis of solubilizing polymer units to permit segmental movement but not wholesale chain movement.

The annual worldwide production of plasticizers is 3.2 million tons, and the U.S. production is in excess of 1 million tons. In fact, plasticizers are major components of a number of polymer-containing products. For instance, automobile safety glass is composed mainly of polyvinyl butyral and about 30% plasticizer.

The development of plasticizers has been plagued with toxicity problems. For example, the use of highly toxic polychlorinated biphenyls (PCBs) has been discontinued. Phthalic acid esters, such as DOP, may be extracted by blood stored in plasticized PVC blood bags and tubing. These aromatic esters are also extracted from PVC tubing and are distilled from PVC upholstery in closed automobiles in hot weather. These problems have been solved by using oligomeric polyesters instead of DOP as nonmigrating plasticizers.

Many copolymers, such as polyvinyl chloride-*co*-vinyl acetate, are internally plasticized because of the flexibilization brought about by the change in structure of the polymer chain. In contrast, DOP and others are said to be external plasticizers. The presence of bulky groups on the polymer chain increases segmental motion. Thus, the flexibility increases as the size of the pendant group increases. However, linear bulky groups with more than 10 carbon atoms will reduce flexibility because of side chain crystallization when the groups are regularly spaced.

Plasticizer containment still remains a major problem, particularly for periods of extended use. For instance, most plastic floor tiles become brittle with age, mainly due to the leaching out of plasticizer. This may be overcome through many routes, including treatment of polymer product surfaces to reduce porosity and use of branched polymers that can act as plasticizers to themselves. These highly branched polymers are slow to leach because of physical entanglements within the total polymer matrix.

17.10   IMPACT MODIFIERS

Polystyrene (PS) is brittle and not suitable for applications that require good resistance to impact. This objection has been overcome, to some extent, by blending PS with SBR to produce high-impact polystyrene (HIPS). HIPS was patented by Seymour in the early 1940s. The impact resistance of PS and other brittle polymers may be overcome by the addition of ABS or polyalkyl vinyl ethers. Of course, specific copolymers, such as those of methyl methacrylate and ethyl acrylate, have better resistance to impact than polymethyl methacrylate (PMMA).

17.11   COLORANTS

Since cosmetic effects are important to the consumer, colorants are added to plastics, fibers, and coatings to meet aesthetic requirements. Some polymeric objects, such as rubber tires, are black because of the presence of high proportions of carbon black filler. Many other products, including some paints, are white because of the presence of titanium dioxide, the most widely used inorganic pigment. Over 50,000 tons of colorants are used annually by the American polymer industry.

Pigments are classified as organic or inorganic. The former are brighter, less dense, and smaller in size than the more widely used, more opaque inorganic colorants. Iron oxides or ochers, available as yellow, red, black, brown, and tan, are the most common pigments.

Carbon black is the most widely used organic pigment, but phthalocyanine blues and greens are available in many different shades and are also common. Other organic pigments are the azo dyestuffs, such as the pyrazolone reds, diarylide yellows, dianisidine orange, and tolyl orange; quinacridone dyestuffs, such as quinacridone violet, magenta, and red; the red perylenes; acid and basic dyes, such as rhodamine red and victoria blue; anthraquinones, such as flavanthrone yellow; dioxazines, such as carbazole violet; and isoindolines, available in the yellow and red range.

17.12   CATALYSTS AND CURING AGENTS

By definition, a catalyst is an additive that accelerates the velocity of reaction but remains unchanged, thus most so-called catalysts used in polymer reactions are actually initiators and not catalysts. One of the most important of this type of additive is 2-mercaptobenzothiazole (Captax), which accelerates the rate of cross-linking of rubber with sulfur. The first rubber accelerator was developed by G. Oenslager in 1906.

Hexamethylenetetramine, which is produced by the condensation of ammonia and formaldehyde, is used as a source of formaldehyde in the curing of phenolic (novolak) resins. Large quantities of peroxides, such as benzoyl peroxide (BPO), and azo compounds, such as azobisisobutyronitrile (AIBN), are used as initiators for the polymerization of vinyl monomers.

Cyclic anhydrides, such as phthalic anhydride, are used for the elevated temperature curing of epoxy resins. These prepolymers are cured at ordinary temperatures in the presence of secondary amines. Tertiary amines and organic tin compounds are used as curing agents for polyurethanes.

17.13   FOAMING AGENTS

Nitrogen gas, produced by the decomposition of azo compounds such as azodicarbonamide (ABFA), is used to form cellular plastics (foam). In addition to these chemical blowing agents (CBA), physical blowing agents (PBA) such as nitrogen, pentane, and volatile fluorocarbons have been used for foam formation.

17.14 BIOCIDES

Biocides are added to certain polymers such as paints to retard degradation by microorganisms. For example, wood is readily attacked under moist, humid conditions by bacteria and fungi, which cause mildew and rot. Antimicrobials such as tin, arsenic, antimony, and copper-containing organometallic salts and oxides are employed in the polymer industry to prevent or discourage the attack of bacteria or fungi on both the natural polymer and, in the case of coatings, the coated material. It must be remembered that most synthetic polymers are not attacked by microorganisms.

Tin-containing compounds have been widely employed as additives in coatings and sealants to inhibit mildew- and rot-causing microorganisms. The monomeric tin compounds have recently been outlawed for use in marine paints since such compounds dissolve in the water and kill marine plants and fish. Organotin-containing polymers have not been banned, are only slightly soluble, and are thus more environmentally acceptable.

17.15 LUBRICANTS AND PROCESSING AIDS

Lubricants, such as calcium or lead stearates, paraffin wax, or fatty acid esters, serve as lubricants and processing aids for plastics. Polybutylene and polystyrene promote flow of plastics in extrusion and injection molding, and thixotropes, such as colloidal silica, reduce the flow of prepolymers.

17.16 ANTISTATS

Since most polymers are nonconductors of electricity, they tend to store electrostatic charges and attract dust. These electrostatic charges can be reduced by the addition of organic compounds such as hydroxylated amines, which attract water, which in turn dissipates the charge. Since electromagnetic interference (EMI) is a problem with business machines with plastic housings, metallic flakes such as aluminum are added to solve this electrostatic problem. It is important to note that most of these additives are used in small amounts (1%). However, their presence is essential for the attainment of optimum properties in plastics.

GLOSSARY

α-Cellulose: High-molecular-weight cellulose, insoluble in 17.5% aqueous sodium hydroxide.

Accelerator: An additive that accelerates the cross-linking (vulcanization) of rubber by sulfur.

Additive: A substance added to a polymer.

AIBN: Azobisisobutyronitrile.

Alkylsilane: A coupling agent for polymer and filler.

Alumina hydrate (ATH): $Al_2O_3 \cdot 3H_2O$, a flame retardant.

Antioxidant: An additive that deters polymer degradation in air.

Antistat: An additive that aids in dissipating electrostatic charges in polymers.

Aramid: Aromatic polyamide.

Aspect ratio: l/d of a fiber.

Biocide: An additive that deters attack on polymers by bacteria, fungi, et cetera.

BPO: Benzoyl peroxide.

Calcium carbonate: $CaCO_3$.

Carbon black: Finely divided carbon produced by the incomplete combustion of natural gas or liquid hydrocarbons.

Catalyst: A substance that hastens the attainment of equilibrium. The term is often misused and applied to initiators.

CBA: Chemical blowing agent.

Chelate: A cyclic compound formed by intramolecular attractions, such as hydrogen bonding between atoms.

Colorant: Pigment or dye.

Continuous phase: The polymeric phase in a composite.

Coupling agent: A compound that contains resin-attracting and fiber-attracting functional groups, such as organosilane.

DEHP: Diethylhexylphthalate plasticizer.

Discontinuous phase: The noncompatible additive phase in a composite.

E-glass: Electrical-grade fiberglass.

Einstein, Albert: Nobel laureate who developed equations for the property enhancement by fillers.

Einstein equation: $\eta = \eta_0(1 + KC)$, where η = viscosity of composite, η_0 = viscosity of polymer, $K_E = 2.5$ for spherical particles, and C = concentration of filler.

Elastopar: N-4-Dinitroso-N-methylaniline, a coupling agent.

Electromagnetic interference (EMI): Interference resulting from storage of electrostatic charges by nonconductive polymers.

Filler: Originally considered a cost-reducing additive but has been found to be a functional additive that enhances the physical properties of polymers in many instances.

Flame retardant: An additive that retards combustion of polymers.

Franklin fiber: Calcium sulfate microfiber.

Fybex: Potassium titanate microfiber.

Graphite fiber: Carbon fiber obtained by the pyrolysis of polyacrylonitrile fibers or pitch.

HALS: Hindered amine light stabilizer.

Heat stabilizer: An additive that deters degradation of polymers, such as PVC, at elevated temperatures.

Impact modifier: An additive that reduces the brittleness of polymers.

Microfiber: A fiber with an aspect ratio (l/d) of at least 60 to 1.

Modulus: Stiffness.

Novolak: A resin obtained by the condensation of phenol with an inadequate amount of formaldehyde under acidic conditions.

Organotitanates: Coupling agents for polymers and fillers.

PBA: Physical blowing agent.

Plasticizer: A flexibilizing additive for polymers.

PMF: Slag microfiber.

Reinforcing fiber: One in which $l/d > 100$.

Rule of mixtures: Each component contributes in accordance to its concentration.

S-glass: A high-tensile-strength fibrous glass.

Tertiary amine: Trisubstituted amine, R_3N.

Ultraviolet light stabilizer: An additive that deters degradation of polymers in sunlight.

Viscosity: Resistance to flow.

Whiskers: Finely divided single crystals.

Wood flour: Fibrouslike wood particles produced by the attrition grinding of debarked wood.

REVIEW QUESTIONS

1. Name a commercial polymer that does not contain an additive.

2. Which is the continuous phase in a polymeric composite?

3. What function does carbon black have in addition to its reinforcing property?

4. What is the difference between wood flour and sawdust?

5. How is α-cellulose separated from lower-molecular-weight cellulose?

6. What is the aspect ratio of a sphere?

7. Name an antioxidant.

8. Name an ultraviolet light stabilizer.

9. Why are flame retardants important?

10. How does ATH function as a flame retardant?

11. Name a widely used rubber accelerator.

12. Why should an antistat be added to the polymer in plastic bottles?

BIBLIOGRAPHY

Berry, R. M. (1972). *Plastics Additives.* West Port, CT: Technomic Publishing Co.

Dearman, R. D., and Schott, N. R. (1974). *Fillers and Reinforcements for Plastics,* ACS Advances in Chemistry Series. Washington, DC: American Chemical Society.

Mascia, L. (1974). *The Role of Additives in Plastics.* London: Edward Arnold.

Milewski, J. V., and Katz, H. W. (Eds.). (1978). *Handbook of Fillers and Reinforcements*. New York: Van Nostrand–Reinhold.

Morton, M. (Ed.). (1959). *Introduction to Rubber Technology*. New York: Reinhold.

Plueddemann, E. P. (1982). *Silane Coupling Agents*. New York: Plenum.

Ritchie, P. D., Catchley, S. W., and Hill, A. (1972). *Plasticizers, Stabilizers and Fibers*. London: Wolfe Books.

Seymour, R. B. (1978). *Additives for Plastics* (Vols. I and II). New York: Academic Press.

Seymour, R. B. (1980). *History of Polymer Science and Technology*. New York: Dekker.

Seymour, R. B. (1985). Fillers and reinforcements (Chap. 5). In J. L. Craft and T. Whelan (Eds.), *Developments in Plastics Technology*. London: Applied Science.

Solomon, D. H., and Hawthorne, D. G. (1983). *Chemistry of Pigments and Fillers*. New York: Wiley.

Tess, R., and Pohlein, L. (1985). *Applied Polymer Science*, ACS Symposium Series. Washington, DC: American Chemical Society.

ANSWERS TO REVIEW QUESTIONS

1. If you had trouble, it's understandable since almost all commercial polymers have additives. Nondyed textile yarn is one.

2. The resinous phase.

3. It is a black pigment.

4. Wood fiber is more fibrous than sawdust.

5. Lower-molecular-weight cellulose is soluble in 17.5% aqueous sodium hydroxide.

6. 1.

7. 2,6-Di-*tert*-butylphenol.

8. Phenyl salicylate (salol).

9. Because polymers used in buildings, automobiles, ships, and airplanes should not ignite in the presence of flames.

10. ATH releases water when heated.

11. Captax, 2-mercaptobenzothiazole.

12. To prevent dust from being attracted to the bottle during storage.

CHAPTER XVIII

The Future of Giant Molecules

18.1 THE POLYMER AGE

We are living in the polymer age, which began when the first living organism appeared on earth. The polymer industry, which is a net exporter and an employer of over 1 million workers in the United States, with an annual payroll of $25 billion and sales of $75 billion, will double in size in the early years of the twenty-first century.

18.2 GROWTH IN THE POLYMER INDUSTRY

The annual production of plastics in the United States increased from less than 100 million pounds a half century ago to over 60 billion pounds in 1988. As predicted by the senior author in 1965, the volume of synthetic polymers produced in the United States exceeded that of all metals by the late 1970s.

   The growth of general-purpose polymers will continue to outpace that of the gross national product, but the most important growth will be in the field of engineering polymers, which exceeded 2.5 billion pounds in 1988. The utility of many of these high-performance polymers will be enhanced by reinforcement with fiberglass, graphite, and ceramic fibers. The sales volume of these so-called advanced composites will exceed $12 billion by the turn of the century.

18.3   ENERGY CONSERVATION

Because of their low specific gravity, polymers replace material many times their weight when substituted for metals. Polymers are also less energy intensive, but more expensive, than general-purpose metals. This conservation of energy pervades all phases of polymer production, fabrication, and recycling.

18.4   GROWTH POTENTIAL FOR POLYMERS

The production of synthetic elastomers (rubber), fibers, and coatings (paints) will continue to increase, but because of the widespread use of plastics in the automotive, construction, electronics, and aerospace industries, its growth will outpace that of all other synthetic materials.

The use of plastics in the U.S. automotive industry should increase from about 2 billion pounds in 1988 to over 4 billion pounds by the year 2000. The functional advantages and weight reduction provided by the use of more plastics will improve both the gasoline economy and style of automobiles.

In addition to the fiberglass-reinforced polyester and epoxy resin car bodies in the current Corvette and Fiero models, automobiles of the future will be mounted on graphite-reinforced frames. Also, accessories such as door panels, roofs, deck lids, hoods and fenders, drive shafts, and springs will be made of fiber-reinforced epoxy resin. Racing cars with polyamide imide combustion engines are already outperforming cars with metal engines. Obviously, automotive mechanics will have to learn to work with these innovative plastics.

Monsanto built a plastics "House of the Future" in Disneyland in the 1950s. This house withstood the wear and tear of millions of visitors until it was dismantled in the late 1970s. The success of this house did not catalyze the construction of comparable houses, but it did increase the use of plastics in recreational vehicles, mobile homes, and permanent residential structures.

General Electric built a "Living Environment" demonstration house in 1988, and presumably this structure will set the stage for an increase in the use of plastics in construction. The basement of this house consists of precast reinforced concrete covered on both sides by plastic panels. The aboveground tongue-and-groove assembled walls consist of reinforced plastic with molded spaces for ducts and conduits. Thus, the builder of the future will be assembling tongue-and-groove panels without the use of hammers, nails, and saws.

Similar simplification can be observed in circuit boards in which all components are included in this small area. Many of these units will consist of three-dimensional circuits based on high-performance plastics. Today's electronics industry is dependent on plastics and the future electronics industry will be even more dependent.

Since fuel accounts for more than half of aircraft operating costs, and this percentage continues to increase, plastics are an essential component of aircraft. Beechcraft and Avek are producing plastics composite business aircraft, and the around the world flight of the Challenger was accomplished in an all-plastic plane.

Commercial airlines could survive but service would be much more expensive without the use of plastics, whereas aerospace missions, satellites, and outerspace platforms would be impossible without these essential materials. Over two-thirds of the weight of commercial aircraft in the year 2000 will consist of plastic composites, but outerspace vehicles will be constructed entirely of polymeric materials.

18.5   RECYCLING OF POLYMERS

Since the percentage of plastics in municipal solid wastes increased from less than 1% in 1960 to almost 10% in 1988, and only 10% of this waste is being recovered, environmentalists are concerned about the accumulation of these valuable waste products. Some have suggested a "quick fix" by use of biodegradable plastics. However, although starch-filled polyethylene is acceptable for plastic mulch and articles such as disposable diapers, nets, and lobster traps, it is not acceptable for containers for gasoline, bleach, sulfuric acid, and other materials.

Fortunately, recycling technology is becoming available and is slowly being utilized where there is a sizable collectable quantity of plastic containers. The aluminum industry has established the profitable collection of aluminum beer and soft drink containers. Although less than 30% of the discarded aluminum cans are recycled, this rate is superior to that of paper products and glass, whose recovery rate is 20 and 7%, respectively.

It should be recognized that the increase in use of plastic packaging has paralleled the decrease in food poisoning and food waste. An additional decrease in the volume of food waste can be anticipated as a result of future improvements in plastic packaging. Although the separation of various types of waste and the source of this waste are important, the components of commingled waste are more likely to be separated on a large scale where labor is cheap, as in some developing countries.

Solid municipal waste is being shipped from Los Angeles to Asian countries, where it is hand-sorted, recycled, molded, and shipped back to the United States as finished plastic products. Obviously, the recycling and the reuse of this valuable resource in the United States are of concern to both the public and the industry, and solutions will emerge as consumers and producers become more responsible.

18.6   ERASING POLYMER SCIENCE ILLITERACY

In his "Of English Verse," written in the seventeenth century, Edmund Waller wrote: "Poets that lasting marble seek must come in Latin or in Greek." Although science came in German in the early years of the twentieth century, it now comes predominantly in English. More and more, an educated person in a developing country must be fluent in English and thus in the language of science.

The educated American of the seventeenth century was fluent in Latin and Greek, but this counterpart in the twentieth century may be illiterate in science and may hide this ignorance by adopting an antiscience attitude. In spite of an annual expenditure of $300 billion on public education in American schools, too few high school students study chemistry and only 55% are exposed to instruction in algebra. Thus, American students rank twelfth among students from developed countries. Obviously, the rank is even lower in polymer science.

If we are to maintain our position as world leaders, we must erase scientific illiteracy and enhance polymer science literacy. Books like *Giant Molecules* will help, but we must match the more formal polymer education programs available to students in other developed countries. Polymer science is emphasized in universities in Japan, USSR, and West Germany yet, there are now only eight American universities with polymer science departments, and this number must be increased dramatically in the 100 or more American colleges that are now offering some training in polymer science and technology. It is up to all sectors interested in this nation's future to cooperate in erasing polymer science illiteracy.

BIBLIOGRAPHY

Brewer, D. (1987). *50 Years of Progress in Plastics*. Denver: HBJ Plastics Publications.

DuBois, J. H. (1972). *Plastics History U.S.A.* Boston: Cahners Books.

Elliott, E. (1986). *Polymers and People*. Philadelphia: Center for History of Chemistry.

Epel, J. N., Margolis, J. M., Newman, S., and Seymour, R. B. (Eds.). (1988). *Engineering Plastics*. Metals Park, OH: ASM International.

Kauffman, G. B., and Szmant, H. H. (Eds.). (1984). *The Central Science*. Ft. Worth, TX: Texas Christian University Press.

Kaufmann, M. (1983). *Giant Molecules: The Technology of Plastics, Fiber and Rubber*. London: Aldus Books.

Mark, H. (1966). *Giant Molecules*. New York: Time–Life Books.

Mark, H. (1967). *Polymers—Past, Present, Future*. Houston: Proceedings of R. A. Welch Foundation Conference, Vol. X.

McMillan, F. M. (1979). *The Chain Straighteners*. London: Macmillan & Co.

Morris, P. J. T. (1986). *Polymer Pioneers*. Philadelphia: Center for History of Chemistry.

Seymour, R. B. (1975). *Modern Plastics Technology*. Reston, VA: Reston.

Seymour, R. B. (1981). *An Introduction to Polymer Science and Technology*, ACS Audio Course. Washington, DC: American Chemical Society.

Seymour, R. B. (1982). *History of Polymer Science and Technology*. New York: Dekker.

Seymour, R. B. (1988). Chemistry and consumer goods. *Journal of Chemical Education*, **5**(1), 32.

Seymour, R. B. (1988). The world of polyethylene. *The World and I*, **3**(2), 189.

Seymour, R. B. (1989). Needed: More polymer scientists and engineers. *Polymer News*, **14**, 130.

Seymour, R. B., and Carraher, C. E. (1988). *Polymer Chemistry: An Introduction*. New York: Dekker.

Seymour, R. B., and Cheng, T. (Eds.). (1986). *History of Polyolefins*. Dordrecht, The Netherlands: Reidel.

Seymour, R. B., and Kirshenbaum, G. S. (Eds.). (1986). *High Performance Polymers: Their Origin and Development*. New York: Elsevier.

Seymour, R. B., and Mark, H. F. (Eds.). (1988). *Applications of Polymers*. New York: Plenum.

Seymour, R. B., Pauling, L., Mark, H. F., *et al.* (Eds.). (1989). *Polymer Science Pioneers*. Dordrecht, The Netherlands: Reidel.

Stahl, G. A. (Ed.). (1981). *Polymer Science Overview*. Washington, DC: American Chemical Society.

Index